Food Waste Reduction and Valorisation

Piergiuseppe Morone · Franka Papendiek
Valentina Elena Tartiu
Editors

Food Waste Reduction and Valorisation

Sustainability Assessment and Policy Analysis

Springer

Editors
Piergiuseppe Morone
Department of Law and Economics
Unitelma Sapienza
Rome
Italy

Franka Papendiek
Abteilung Forschung | Research Policy Unit
German Council of Science and Humanities
Cologne
Germany

Valentina Elena Tartiu
TIK Centre for Technology, Innovation and Culture
University of Oslo
Oslo
Norway

ISBN 978-3-319-50087-4 ISBN 978-3-319-50088-1 (eBook)
DOI 10.1007/978-3-319-50088-1

Library of Congress Control Number: 2016959520

Printed on acid-free paper

This Springer imprint is published by Springer Nature
The registered company is Springer International Publishing AG
The registered company address is: Gewerbestrasse 11, 6330 Cham, Switzerland

Preface

Global economic and population growth trends are placing pressures upon natural resources threatening future economic and social development. Most notably, the world population, standing on 7.2 billion people in mid-2014, is projected to increase by almost one billion people within the next decade, and further to 9.6 billion in 2050 (United Nations 2013). At the same time, large and fast-growing economies (i.e. the BRICS members) will experience increasing wealth. A major consequence of these two trends is higher consumption and demand for food and other goods, increasing in parallel the rate of waste production and depleting the amount of available resources (e.g. demand for several elements, including helium, phosphorus, indium and gallium is predicted to exceed supply in the near future). Overarching all of these issues is the threat of climate change and the concerns about how mitigation and adaptation measures may affect the food system (Schmidhuber and Tubiello 2007; Godfray et al. 2010).

Scientists, analysts and policy makers are taking stock of these trends, trying to push the society towards more sustainable development patterns. An emerging area of enquiry looks with growing interest at food waste reduction and valorisation as a key area of research to provide answers to these emerging challenges. In fact, the valorisation of food waste has many advantages. It is a rich source of functionalised molecules (i.e. biopolymers, protein, carbohydrates, phytochemicals) and contains valuable extracts for various applications (e.g. resins from cashew nut shell liquid), avoiding the use of virgin land and water resources. In addition, it solves a waste management issue and represents a sustainable renewable resource; making the valorisation of food waste doubly green.

Moving from this, the 15 chapters included in this book address these emerging societal challenges building on the idea that food waste reduction and valorisation is fundamental for promoting environmental, economic and social sustainability, in the framework of the growing interdependence between human societies and the natural environment.

The plurality of perspectives considered gives a truly transdisciplinary angle to the book. Indeed, the proposed book is the outcome of rather fertile networking and research activities conducted over the last years by a broad group of experts, coming from different disciplines, most of whom where partnering in the COST action TD1203 on Food Waste Valorisation for Sustainable Chemicals, Materials and Fuels (EUBis) initiated by Prof. James Clark (head of the Green Chemistry Centre of Excellence at the University of York) back in 2012 and that successfully ended on the 22 November 2016.

Within the COST Action TD1203, the fourth working group, in which the editors of this book took part, dealt specifically with 'Technical & Sustainability Assessment and Policy Analysis', focusing on the economic assessment of alternative innovative technologies, including supply logistics and feasibility evaluation of green processes at the industrial level, whilst also exploring the environmental and social impacts of the valorisation technologies.

The book opens with an introductory chapter by James Clark, presenting the genesis, the purpose, and the scope of the work and setting out a roadmap to guide the readers through the book, underlying the common thread linking the remaining 14 chapters comprised in the book.

We hope this book will contribute in shedding new light on social, techno-economic and policy related issues concerning food waste reduction and valorisation, paving the way to new research in this field of enquiry.

Rome, Italy Piergiuseppe Morone
Cologne, Germany Franka Papendiek
Oslo, Norway Valentina Elena Tartiu

References

Godfray HCJ et al (2010) Food security: the challenge of feeding 9 billion people. Science 327:812–818
Schmidhuber J, Tubiello F (2007) Global food security under climate change. PNAS 104 (50):19703– 19708
United Nations (2013) World Population Prospects: The 2012 Revision, New York

Contents

Part I
Sustainability Assessment

Chapter 1
Introduction

James Clark

Abstract The importance of waste as a future feedstock for the chemical and allied industries is considered. In particular food supply chain waste is identified as a valuable source of useful chemical functions. While waste can contribute to resources globally it is likely to have an especially strong role in the industrial development of the emerging economies.

Keywords Waste · Food waste · Bio-resources · Bio-based chemicals · Recycling

Waste is a major global issue and is becoming more important in developing countries, such as the B(razil)R(ussia)I(ndia)C(hina)S(outh Africa) nations, as well as in Europe. According to the World Bank, world cities generate about 1.3 billion tonnes (1.3 GT) of solid waste per year. This is expected to increase to 2.2 GT by 2025. Waste generation rates will more than double over the next twenty years in lower income countries. Globally, solid waste management costs will increase from today's annual ca. $200 billion to close to $400 billion in 2025. Cost increases will be most severe in low income countries (more than 5-fold increases) and lower-middle income countries (more than 4-fold increases). Global governments need to put in place programmes to reduce, reuse, recycle or recover as much waste as possible before burning it (and recovering the energy) or otherwise disposing of it.

Waste can be categorised into industrial, agricultural, sanitary and solid urban residues, based on its origin. The distribution of the content of these waste streams changes significantly from country to country so that any figures have to be treated with caution. What is clear is that few countries have a positive waste management policy involving significant waste valorisation (although reliable data are not easily available from developing countries other than anecdotal evidence that in some countries such as India many people may make a basic living from collecting and selling waste). To make matters worse, there is a growing concern over the long

J. Clark (✉)
Department of Chemistry, Green Chemistry Centre of Excellence,
University of York, York, UK
e-mail: james.clark@york.ac.uk

P. Morone et al. (eds.), *Food Waste Reduction and Valorisation*,
DOI 10.1007/978-3-319-50088-1_1

term availability of many critical substances. It is not simply a desire to move away from non-renewable fossil resources as sources of carbon, as quickly as possible (as was reaffirmed by all nations in the most recent Paris Climate Change Conference) we also need to review our use of many other elements in the periodic table. The European Union (EU) has identified a critical element list since the demand for several essential elements, including helium, phosphorus, indium and gallium is predicted to exceed supply in the near future. As with fossil sources of carbon, we are not necessarily running out of global reserves of any elements but we have already taken the easiest pickings. The more recent petroleum extractions in areas such as the Gulf of Mexico have begun to demonstrate the high price–both environmentally and economically–that we will have to pay for continuing to depend on these traditional resources. Such increased awareness of resource depletion and the catastrophic influence it could have on economies is beginning to make people realise that the traditional linear economy model of extract-process-consume-dispose is unsustainable and that instead we need to move towards a circular economy whereby we make use of the resources in articles that have reached the end of their useful life or are simply rejected by the owner. We need to keep all substances in use over much longer periods, ultimately aiming for zero waste whereby everything is kept as a resource either via moderate length natural cycles such as for carbon (I would suggest typically less than the average human lifetime) or via man-made processes such as for minerals.

Waste produced by food processing companies is a good example of a pre-consumer type of waste generated on a large scale all over the world but something that also represents a source of valuable substances, notably functional organic molecules that otherwise we have to build up from hydrocarbons (involving chemical processes that consume more resource and generate even more waste). Food supply chain waste can easily account for more than 50% of the total waste produced in many countries. Some 60% of food waste is organic matter which can be a public safety issue if not managed properly. More effort has gone into efficient waste treatment and disposal than waste valorisation. In fact food waste is a rich source of functionalised molecules (i.e. biopolymers, protein, carbohydrates, phytochemicals) and contains valuable extracts ranging from terpenes in citrus waste to phenolics in winery waste and with applications from food additives to solvents and bio-based chemicals. Various waste streams contain valuable compounds, including antioxidants, which could be recovered, concentrated and re-used in applications such as food and lubricants additives. Avoiding the use of virgin land and water resources to produce non-food products is a further major advantage especially as we have become wary of the hidden costs from 1st generation biofuels that committed large areas of arable land to non-food production leading to distortions in the food markets. In addition, food waste valorisation solves a waste management issue and represents a sustainable renewable resource; making the valorisation of food waste doubly green. Despite this there are few large volume commercial examples that show us how to get away from or supplement landfilling or first-generation low-value, recycling practices such as composting, animal feed production and/or re-use of organic matter.

In terms of economic competitiveness it is essential that we seek to promote advanced methods to process food waste residues in order to produce high added value and readily marketable end-products. We also need society to undergo a major change of mentality and perception on waste seeing it more as a resource rather than a pollution problem. Such a paradigm shift needs to be steered by governments and trans-government agencies worldwide as well as through education at all levels. One strong incentive for change is the increase in costs of waste disposal: the EU landfill Directive, for example, has caused landfill gate fees to substantially increase. We need to properly assess the environmental and economic impacts of food waste valorisation and properly assess the sustainability indicators that should be used to monitor the impact of this approach.

This book aims at addressing this very new area of enquiry, building on the idea that food waste reduction and valorisation is a crucial factor to promote sustainability and with significant value in chemical manufacturing and other critical industries. In the first part of the book, we look at sustainability assessment from a number of different perspectives. This includes impacts on climate change, and on the production of high value chemicals and other useful products. We use the 'three pillars' of sustainability–environmental, economic, and social, each with its own specific features. Seven chapters give an overview of available strategies to assess indicators for food waste reduction and valorisation and also of useful tools to facilitate the comparison between the three pillars of sustainability. We look at real examples of where food waste has been converted to valuable fuels, chemicals and materials, going beyond 'first generation' products such as compost and energy from burning or anaerobic digestion (useful though these may be in the overall zero waste bio-refinery, we must try to capture some if not all of that wonderful molecular value in food waste).

The sustainability assessment of food waste reduction and valorisation is complemented with another layer of assessment: policy analysis, aimed at identifying the drivers of change of food waste reduction and valorisation technologies, by looking, for instance, at the regulatory framework and at policy actions undertaken by local and global actors. A further seven chapters cover food waste under the law, policy analysis and how we can lower the barriers that inhibit the implementation of the circular economy.

The development of knowledge-based strategies to unlock the enormous potential of food waste can help satisfy an increasing demand for renewably sourced products, leading to sustainable, bio-derived chemicals, fuels and materials, and probably effecting waste management regulations over the years to come. The valorisation of food supply chain waste is necessary in order to improve the sustainability and cost-effectiveness of food supply. Together with the associated ethical and environmental issues and the drivers for utilising waste, the pressures for such changes are becoming very powerful.

Our future as a flourishing, creative and increasingly affluent society is completely dependent on how we treat the limited resources we have available to us on this one beautiful but finite planet. If we try to apply the same resource exploitation model we have used in the developed world to the whole world then we will quite simply consume ourselves out of existence. Whether the tipping point is one of the many pollution impacts we are becoming aware of or based on an economic meltdown due to 'resource wars', is a moot point. Food waste is an exemplar of the challenge we face; if we both see it as a resource opportunity rather than a waste threat and apply our creative science and technological energies to its valorisation, we show the way to other waste-to-resource opportunities. There has never been a better time to see that 'where there's waste there's wealth'.

Chapter 2
Cutting Through the Challenge of Improving the Consumer Experience of Foods by Enabling the Preparation of Sustainable Meals and the Reduction of Food Waste

Wayne Martindale

Abstract The communication of sustainability and Corporate Social Responsibility measurements by food industry campaigns has identified key areas of activity that dominate sustainable thinking in the food industry. The purpose of this chapter is to show that one of these areas of activity, the intensity of resource use and resulting food waste, can be used as a universal connector of sustainability practice across supply chains and between them. This requires an assessment of food waste production because it is an attribute consumers are familiar with and as such; these connectors are often overshadowed by high-level issues such as global food security, climate change and the loss of biodiversity. While these high-level issues rightly dominate the policy arena they will often take the attention away from issues that practically relate sustainability to us as consumers when we prepare, present, preserve and consume three or four meals a day. This situation presents a major challenge that is tackled here by providing sustainability and security metrics that relate to meals. The Six-Function-Model (*6fm*) is a model developed to assess the sustainability of food and it can be used to overcome the stifling of sustainability thinking by methods that do not enable practical application in retail, kitchen and restaurant situations. The use of the *6fm* by manufacturers, retailers and consumers will stimulate the 'designing-in' of sustainability attributes into meals. The model and a benchmarking analysis of the *6fm* are presented here to account for resource use and food waste associated with meals. The future goal of *6fm* is to stimulate the use of it in the food and beverage industry so that it 'builds-in' sustainable thinking to product design and consumer experience.

Keywords Sustainability · Consumption · Consumers · Responsibility · Consumer goods

W. Martindale (✉)
Sheffield Business School, Sheffield Hallam University,
Sheffield, S1 1WB, UK
e-mail: w.martindale@shu.ac.uk

P. Morone et al. (eds.), *Food Waste Reduction and Valorisation*,
DOI 10.1007/978-3-319-50088-1_2

2.1 Introduction

The development of the Six-Function-Model (*6fm*) for the sustainability and security assessment of meals is derived from ground-breaking work that has combined expertise from academic and industrial sectors. This was initiated by the development of methodologies that have assessed how consumers purchase, prepare and utilise food for each meal (Martindale 2014a). This type of approach not only identified sustainability criteria associated with meals, it also provided specific insight into the 'values attributes' consumers associated with meals. This supply chain focussed methodology has been refined by using a model of household food waste production from different preserved food categories (Martindale 2014b). These studies have demonstrated that food waste can be reduced by half in households if resource use is optimised using freezing as a means of food preservation. In demonstrating this, the requirement to understand how we consumers make meals and associate values of convenience, taste and affordability with them was identified. A model of food waste production from household meals presented by Martindale (2014b), showed the impact of doing this was considerable because frozen preservation in households currently removes 5.5 million tonnes of household food waste across the European Union for seafood, meat, fruit and vegetable food product categories. This begins to establish the goal of achieving zero household food waste when we consider frozen food purchases are lower volume than fresh food purchases and consumer awareness of food preservation can be communicated with greater impact if sustainability outcomes are utilised.

The goal is for the *6fm* to enable manufacturers and retailers to communicate the sustainability credentials of meals and for us consumers to use the *6fm* to action sustainable meal choices. The *6fm* model combines 'values attributes' and 'technical attributes' associated with food choice in a model that is tested in this study with particular focus on manufacturing scenarios that will provide an improved consumer experience. This is because manufacturing is where the outcomes of understanding sustainability in product development are high-impact in the supply chain because they result in stronger manufacturer-retailer relationships and ultimately decide how a large proportion of consumers utilise food products for meals (Jones et al. 2014). Indeed, it is a control point and while the aim is for the *6fm* to be used in all parts of the food supply chain it is important to identify control points in the test of *6fm* presented here. The approach is also to use *6fm* to obtain feedback on the use of foods from consumers so that it can be used to re-design products for meals.

2.2 The Need for a New Approach in Creating Resource Efficient Food Supply Chains

The 'technical attributes' include in the *6fm* are those obtained by using carbon footprinting and Life Cycle Assessment (LCA) methods that are now well established in food industry sustainability practice (Hornibrook et al. 2013). Experience of delivering these carbon footprinting investigations with respect to consumer trends has most notably been obtained in the non-meat protein product marketplace (Martindale et al. 2014). These meat-replacer products dominate the emergent flexitarian markets and provide an important case-study where consumer 'values attributes' and LCA 'technical attributes' must be integrated for application (Green et al. 2010). The need to assess the Greenhouse Gas Emission (GHG) credentials of meat and non-meat foods is an important aspect of meeting global sustainability and security goals. It has provided the opportunity to begin to understand how we as consumers interact with high-level sustainability measures such as eating less meat. While GHG reduction goals are important 'technical attributes' in the meat reduction debate, health issues dominate it and these are the 'values attributes' that consumers associate most favourably with (De Boer et al. 2013). This type of scenario results in a complex decision framework of technical and consumer values associated with choice of meat and non-meat products that has provided the industry with important dietary models (Macdiarmid 2013). The debate on meat free products has stimulated the growth in the flexitarian meal trend in Europe and it has shown how food sustainability criteria can often force us to consider extending scientific evidence to consumer values-driven goals (De Boer et al. 2014).

This work in the meat-free arena has enabled an understanding of trends in marketplaces and the impact of dietary transitions on sustainability outcomes. This has provided insight into more realistic applications of carbon footprinting because models show diets with lower GHG emissions that have lower meat with more fresh fruit and vegetable content can result in greater production of household food waste (Martindale 2014a). Extending such LCA data to populations using consumption models is important because they show us sustainability is not a single issue movement and it is dependent on several values with which scientific evidence must interact. This is precisely what the *6fm* is aiming to achieve by enabling sustainability practice for everyone rather than considering the challenge 'too big' to be solved.

The consumer issues associated with sustainability are often values-driven goals that are focussed on food choices that align with specific environmental issues. These often become a platform for policy development through channels such as the activist programmes of Non-Governmental Organisations (NGOs). This is an important aspect of policy development for the food industry to understand because it does not always result in efficient consumer communications because the industry and consumers are often removed from developing them. Indeed, if science evidence-based and consumer values-based criteria do not align it creates a situation where there is a clash and the policy makers have to take the approach that evidence must be adhered to at all costs, in which case consumer values can seem to be ignored. The retailer to consumer relationship in the food supply chain is another control point challenge that the *6fm* will tackle because it aims to create transparent measurement for sustainability criteria and communication. This is achieved, not by developing feelings of guilt in consumer activities or aligning them to the goals of NGO's but to identify the benefit of using diverse food choices and preparation practices that are sustainable and can be measured.

The approach of *6fm* has evolved because if we do fail to integrate trend-led consumer data into LCA methodologies we will stifle the use of sustainable principles by both industry and consumers alike. It is well known that LCA and footprinting techniques do not integrate well with Fast Moving Consumer Good (FMCG) businesses because market trends and how consumers consume foods are rarely considered by them (O'Rourke 2014). Indeed, this is at the very core of the *6fm* approach, where we determine why consumers like a particular food and then ask, can that product be re-designed in a more sustainable or security conscious way so that preparation and consumption aligns with these values. This approach will communicate the measures of sustainability and security criteria back into the supply chain from consumers so that redesign can be applied. Standard or classical techniques of sustainability assessment such as carbon footprinting or LCA only identify where criticality occurs and act by communicating reduction of consumption alone and I believe that this approach will not work. The *6fm* offers an alternative that can be used by producers, manufacturers, retailers and consumers for an innovative and engaging redesign of practice.

An example of our challenge is provided with livestock products where LCA techniques have shown they embody more energy and result in greater GHG emissions than consuming plant products directly because livestock consume crop products as feeds (Martindale 2014a). In this instance, a weakness of classical LCA is apparent because of two key factors;

- the diversity of livestock production enterprise types means that a single type of LCA will never define the environmental impact of animal production;
- consumers will utilise varying amounts of protein and carbohydrate food groups in meals and these are determined by lifestyle and preference.

In its simplest form these two factors can be described in the fact that comparing feed-lot reared beef to grass-fed grazed beef is not possible; when the production system is normalised, studies might be used to compare meat products but this is rarely done (Ridoutt et al. 2011). Thus, the inflexibility of LCA creates confusion which is compounded by a complete misunderstanding of the functional unit used to communicate LCA results which is usually kilograms of product and this does not easily relate to ingredients of meals.

This inflexibility of LCA in food FMCG supply chains results in organisations reporting sustainability and security goals using biased data for specific causes. For example, when organisations have required the reporting of the more intensive GHG measures for beef production they can use the energy intensive rearing for premium markets such as Wagyu beef or production systems based in areas where grazing is limited by rainfall (Ogino et al. 2007). This does not represent an evidential view of all livestock production enterprises with respect to the diets we consume. While efforts are made to provide normalised perspectives on diets there is still much scope to improve (Garnett 2013). The changes in resource use intensity associated with climate and different production systems are the very basis of optimal grazing and husbandry that are overlooked here. Indeed, the words of

Norman Borlaug ring in our ears with ‘there are no miracles in agricultural production’, an understanding you can’t achieve optimal crop or livestock production in areas where varieties and breeds are not suited-to, seems forgotten by some applications of LCA. However, this happens time and time again for consumer communications creating a confusion and a distrust of data and evidence in the marketplace.

As such, this leads us to observe that attempts to standardising diets, meals and foods using LCA has potentially confused consumers by making comparisons of different food products when they should not be made. This has often only promoted values-driven communications that undermine sustainability and security targets globally. The *6fm* approach breaks through the high-level issues such as meat versus non-meat product use to consider how we as consumers make and consume meals while still maintaining a line of evidence to sustainability and security criteria. The *6fm* structure differs from LCA in that it is designed to firstly to benchmark and consider consumption in marketplaces. This approach will consider meal portion and an outcome will be improved sustainability and reduced food waste in meal preparation and consumption. Finally, the *6fm* develops a measure of sustainability and security that can be used as a dash-board measurement or index within businesses or homes that wish to respond to sustainability and security criteria by redesigning product or meal formats.

2.3 The Six Function Model Approach to Assessing Food Use

The six function model (*6fm*) method presented here tackles two key themes that have vexed food sustainability and security thinking for several years and these are:

- Measuring the potential to provide an adequate supply of nutrients in foods we enjoy as consumers that are accessible, affordable and assured (as the established ‘Triple A’ standard).
- Developing resource efficient ways of guiding us to consume sustainable meals (Martindale 2014a).

The identification of these themes has been made possible by mapping food supply chain sustainability attributes and identifying areas where there are likely to be the most effective resource utilisation gains. While many organisations consider this too complex a task we have obtained data on food supply chains from food manufacturers in semi-structured interviews regarding resource use and waste reduction that have a benchmark that is practical and usable in situ. The need for this has identified six functions in food supply chains that have resulted in developing the *6fm*. It is based on a food supply chain model which is shown in Fig. 2.1, which demonstrates how the *6fm* methodology interacts with supply chain operators.

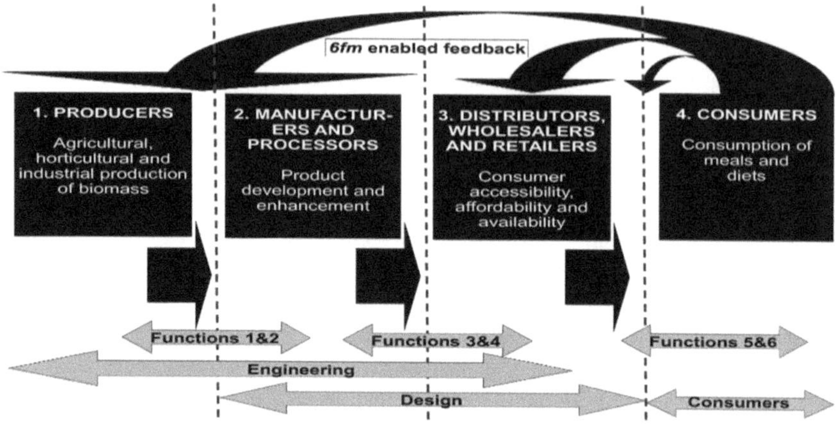

Fig. 2.1 The supply chain approach of the *6fm* is shown here. The supply chain operators of producer, processor and manufacturer, retailer, distributor, wholesaler; and, consumer are shown in a linear function. The *6fm* enables feedback across the whole supply chain as shown in the model developed by Martindale (2014a)

Figure 2.1, shows the four production, manufacture, retailer and consumer operators of the food supply chain and where the six functions of the *6fm* operate in it to provide a measure of sustainability and security. The *6fm* benchmarks meals in terms of energy and resources used in accessing foods, preparing meals and preparing what we eat. The *6fm* functions described here are shown in their typical role of influence in Fig. 2.1, for each supply chain operation (i.e. production, manufacture, retail or consumption) together with the principle skill-sets that dominate in these supply chain operations (i.e. engineering, design or consumers). The six functions have been in part selected through my experience of delivering sustainability research projects that have carbon footprinted individual food products, mapped food supply chains and assessed the impact of consumer use of foods in making meals (Martindale 2014a). They represent a practical set of functions that can be used to measure the sustainability of meals and they were selected using the feedback from semi-structured interviews with food industry professionals. The goal reflected in the supply chain model in Fig. 2.1, is that the *6fm* will be used by the whole food supply chain from producers through to consumers to create a much greater understanding of both security and sustainability associated with meal choices. The six functions of the *6fm* whose operation is shown in Fig. 2.1, are as follows.

1. **Import and export volumes of ingredients (I + E)**. These are based on traded volumes of food ingredients and commodities that are available from FAOSTAT. The volumes determine pressure points and these can be used to project trends when compared to the national crop or livestock yield per unit area data. Tailored ingredient folios would be used for specific manufacturers and they highlight mass-flow sensitivity in delivering food to consumers.

2. **Nitrogenous fertilisers utilised in production of ingredients (Nproduction).** The assimilation of nitrogen into protein is dependent on the production of both organic and mineral fertilisers. The function is based on typical nitrogen requirements for crops and livestock that is quantified using the fertiliser recommendations for crop ingredients and feed conversion factors for livestock ingredients. Nitrogen requirements are well tested and reviewed by Smil (2002) and in this context plant proteins are given a value of 1.0, and livestock products are multiple of this with fish or seafood meat being 1.5, chicken meat being 2.3, pig meat being 5.9, beef meat being 12.7, eggs being 3.8 and milk being 0.7 using feed conversion ratios. Nitrogen use is a convenient function for food because it is converted into protein and accounts for up to a third of the energy balance of the production operator of the food supply chain, it is typically a third of the agricultural energy balance.

The first two functions are concerned with producers and the producer-to-manufacturer and processor relationship within the supply chain model presented in Fig. 2.1. The first and second functions assess the production of ingredients.

3. **Protein content of the food ingredient used in meals (Pcontent).** This is a key determinant of dietary security, it is based on the ability of meals to provide 800 mg of protein per kg of body weight per day and represents the point in the *6fm* where ingredients are constructed into meals (Martindale 2014a). A key quality determinant of meals are their protein content, this is emphasised in the *6fm* because protein supply is a critical component of maintaining food security. Protein values of foods are benchmarked with plant proteins are given a value of 10 and livestock products are multiple of this with fish or seafood meat being 18, chicken meat being 20, pig meat being 14, beef meat being 15, eggs being 13 and milk being 3.5 based on edible protein content (Smil 2002).
4. **The micronutrient content of the food ingredient (Microcontent).** The functional nutritional properties of foods are classically determined by protein, fat and carbohydrate content. Meta-analysis of micronutrient contents have been used to guide the benchmarking process for this function. However, the range of micronutrients is extensive and often misrepresented (Brandt et al. 2013). This means the micronutrient value of meals is not easily communicated to consumers and we require a robust indicator of micronutrient nutrition for this function. We have chosen to use the social or value-attributes in the selection criteria for measuring micronutrients. The use of the Google Trends tool has identified that 'Vitamin C', 'Vitamin D' and 'essential fatty acid' are the terms that have greatest social interest for a range of micronutrients. The *6fm* is reflective of consumer awareness and we have used these trends to measure the benchmark for this function. Thus, Vitamin C, Vitamin D and essential fatty acid content are the micronutrients used to benchmark this function and the micronutrients used can change in respect to trends identified. This function is the one within *6fm* that is subject to greatest change by users because of the

changes in trends and Google Trends manages syntax searches to standardise the approach of selecting the trends.

The third and fourth functions are concerned with manufacturers and retailers relationship within the supply chain model presented in Fig. 2.1. The third and fourth functions measure the use of ingredients in meals to provide quality attributes, in this case protein and micronutrient nutrition. Notably the quality qualifier terms are determined by Google Trends citation volumes.

5. **The energy utilised by the supply chain that delivers foods from farm to consumer (Esupply).** The embodied energy of supply provides and important indicator of food affordability and efficiency. The energy consumed in producing; processing and manufacturing; and, distributing and retailing foods is well characterised with peer review and open-access data-bases (Nielsen et al. 2003). The *6fm* uses this data to assess the amount of energy consumed in the supply chain to produce specific meals. This is reflected in LCA data where the products are produced in manufacturing arenas and the data reported considers energy used to the point of sale by a retailer (Wallén et al. 2004).
6. **The energy utilised in preparing and consuming foods for meals (Eprep).** This is the area of the supply chain that is often disregarded because of the complex decisions that take place during food preparation and consumption. Our approach has been to simplify this decision based system to assess specific 'flagship' meals and in this test of the *6fm* we select six meals shown in Table 2.1, that are selected to represent different culinary management the marketplace. Eprep is a critical point of the *6fm* model because the supply chain prior to preparation of meals is well understood through the application of LCA and carbon footprinting. The *6fm* extends this understanding to meal preparation using meal groups with an assessment of ingredients and food waste. Table 2.1, also considers food waste intensity from different meal types because waste is a critical part of the energy balance for the consumer experience.

The fifth and sixth functions are concerned with consumers and the retailer-to-consumer relationship within the supply chain model presented in Fig. 2.1. The functions measure the use of meals in diet and lifestyle, they are indicative of the food experience of the consumer.

2.4 Testing the Six Function Model

The testing of the *6fm* has been framed using the six specific meal groups that are shown in Table 2.1, they have been selected by reviewing recipe databases from chefs and retailers. The meal groups are constantly under consideration and are representative of the UK trends in meal popularity (Defra 2014). The selection of meal groups is strengthened by this study undertaking semi-structured interviews of nine practitioners in the food industry working with food trends and product

Table 2.1 The description of the meal groups used to test the *6fm* in this study considers the Esupply and Eprep functions together with the energy embodied in potential domestic food wastes

Meal group	Principle ingredients	Proportion of typical ingredient groups (% of meal weight in brackets)[a]	Energy consumed supply chain	Energy consumed domestic preparation	Energy embodied in domestic wastes	Notes
Fish and seafood	Fish		Intermediate	Low	Low	Fillet and high quality meat may be used, there is variability
Pasta	Cereals, Eggs	Cereal 70, egg 5, vegetables 20, meat 20	Low	Low	Intermediate	Optional processed meat can be used, potentially reduce protein if vegetables are used
Curries, mince and stews	Meat, Pulses, Spices	Vegetables 40–60, rice 20, meat 40–60	High	Intermediate	Low	Optional processed meat can be used, potentially reduce protein if vegetables are used
Salads and fruit	Leaf, Pulses, Fruit, Eggs	Vegetables 50–90, egg 20, oil 10	Low	Low	High	Optional processed meat can be used, potentially reduce protein if vegetables are used
Roasts	Meat, Veg	Meat 20–30, vegetables 60–70	High	High	Intermediate	Fillet and high quality meat may be used, there is variability
Breakfasts and bakery	Cereal, Dairy, Sugar, Eggs	Cereal 40–60, sugar 25, egg 5, fat 20, dairy 5–20, fruit 5–20	Intermediate	Low	High	Flexible preparation is built into this meal design

[a]Proportions have been derived from recipes reported in Martindale et al. (2008)

development. The practitioners included three development chefs, three food manufacturers and three retailers. The development chefs were independent specialists who regularly develop meals for restaurants and develop food solutions for food manufacturers that are from the Small Medium Enterprise (SME less than 250 employees) to large company (international and pan-European) scales. The food manufacturers were SME's with 50–100 employees and included a bakery, a ready meal provider; and, a confection and snack supplier. The retailers were regional or local retailers who had at least 20 employees in the specific retail-outlet identified and included two general retail grocery outlets and a farm shop specialising in local food. The focus of this sample was regional because the goal of the *6fm* is SME accessibility to sustainability assessment and it is well established that large multinational companies have robust sustainability reporting measures in place for internal and external affairs.

The Google Trends web-crawler tool has added rigor to the six functions used by the *6fm* because it has provided a search system that can rank specific consumer-used terms associated with each of the functions. The use of the web-crawler enables benchmarking and testing of the responses from the nine practitioners. The most popular terms selected are shown in Table 2.2, they are those terms that have greatest citation volume from Google searches and they are used to guide the selection of the six functions and guide the benchmarking for each meal identified in Table 2.1.

The benchmarking analysis of *6fm* is carried out by asking end-users of *6fm* to relate their understanding of the six functions to a scale ranging from 0 to 9 points (a rank of 1–10). In taking the test a participant is asked to mark on a scale of 0–9 the score for each function. This is guided by the 'technical' or LCA derived attributes and the 'values' or Google Trends attributes described in this methodology.

Using the meal groups identified in Table 2.1, and descriptors of each of the six functions in Table 2.2, we can begin to develop an index that effectively sums the functions by scoring them. The scoring is influenced by product type and the security and sustainability criteria associated with each function to infer ethical outcomes by users and the choice of meal. The testing of the *6fm* here used six meal groups obtained by interviewing nine practitioners involved with developing meal design strategies. The sample meal groups selected are as follows.

1. **Pasta dishes**. The pasta trend is still relevant in providing the basis for the Mediterranean diet (Kearney 2010).
2. **Salad dishes**. The salad meal has changed dramatically in the United Kingdom and it still provides an important health driven trends in diets (Leslie et al. 2014).
3. **Mince, stews and curry dishes**. These represent those that are convenient and flexible in that the meals can be both vegetable and meat focussed (Scarborough et al. 2014).

Table 2.2 The selection of keywords used to support the assessment of the functions in the *6fm*, these keywords are used for the micronutrient function in this study

Function	Key words identified by Google Trends
Import and export volumes	Food trade, coffee trade are used as keywords These were filtered from a group that also included the keywords imported fruit, imported vegetables, imported meat, exported fruit, exported vegetables, exported meat; fruit trade, vegetable trade, meat trade, soy trade, corn trade, cereal trade, sugar trade, banana trade
Nitrogen used in production	Intensive agriculture, organic food are used as keywords These were filtered from a group that also included the keywords nutritional quality, meat and environment, meat and greenhouse gas, low greenhouse gas food, sustainable agriculture, fertiliser and food, (low impact food) was changed to organic food. Greenhouse gas references with food provided no trend volume
Protein content of foods	Protein and disease, protein and muscle; protein for aging; protein and the elderly; meat alternatives, meat consumption are used as keywords These were filtered from a group that also included the keywords sustainable protein, synthesised protein, waste protein, protein and research, protein and capacity, protein and government decisions, protein and global warming/climate change, protein availability, protein and exercise, protein and deficiency, protein and child development, protein and carbon footprint, protein and eating demographics
Micronutrient content of foods	Vitamin C, Vitamin D, essential fatty acid are used as keywords These were filtered from a group that also included the keywords iron nutrition, essential amino acid, folic acid, carotene, biotin, phosphorus nutrition, calcium nutrition, copper nutrition, zinc nutrition
Energy used in food supply chain	Transport cost and manufacturing cost are used as keywords These were filtered from a group that also included the keywords fuel cost (too general), refrigeration cost, resource efficiency, frozen food, chilled food
Energy used in preparation of food	Cost of preparation and healthy food are used as keywords These were filtered from a group that also included the keywords environmental food, cost of recipes, convenient food (preparation), family food, food for ageing (no trend), food for young

4. **Roasts**. The roast meal is the celebration meal in the United Kingdom, the meal will typically have a meat ingredient as the centre-of-the-plate and the preparation and cooking processes provide an important test for the *6fm*.
5. **Fish and seafood**. These meal trends are also driven by health issues and are included here because they do present specific sustainability issues.
6. **Breakfast and bakery**. The breakfast and snacking trend is dominant in recent years and it is an important test group for the *6fm* because of this (Hoyland et al. 2012).

2.5 Benchmarking Meals with the Six Function Model

A summary of the results for the testing of *6fm* by the assessment of nine food industry specialists is shown in Fig. 2.2, and it can be seen that an index of sustainability can be developed using the sum of rank for the six functions. Figure 2.2, shows the highest *6fm* scores are for meat containing meal groups and variation between meal groups can be relatively small. This means that meal choices over longer periods of time are likely to only show large differences and the *6fm* will convey this to consumers using it. That is, *6fm* needs to be built into lifestyle and used constantly for the impacts of sustainable meals to be realised by consumers. The actual format and design of the *6fm* will be decided in future with a range of clients in manufacturing and retail arenas. Herein lies the challenge to us currently, it is well established that meat products have increased GHG emissions, it is likely that the differences between different meat and non-meat products will continue to be identified and discussed but there is a current opportunity to relate these analyses to meal making and diet planning with the *6fm*. Figure 2.2, shows the *6fm* can be used to inform meal planning within diets across supply chain operations using consumer experience as a focus of application.

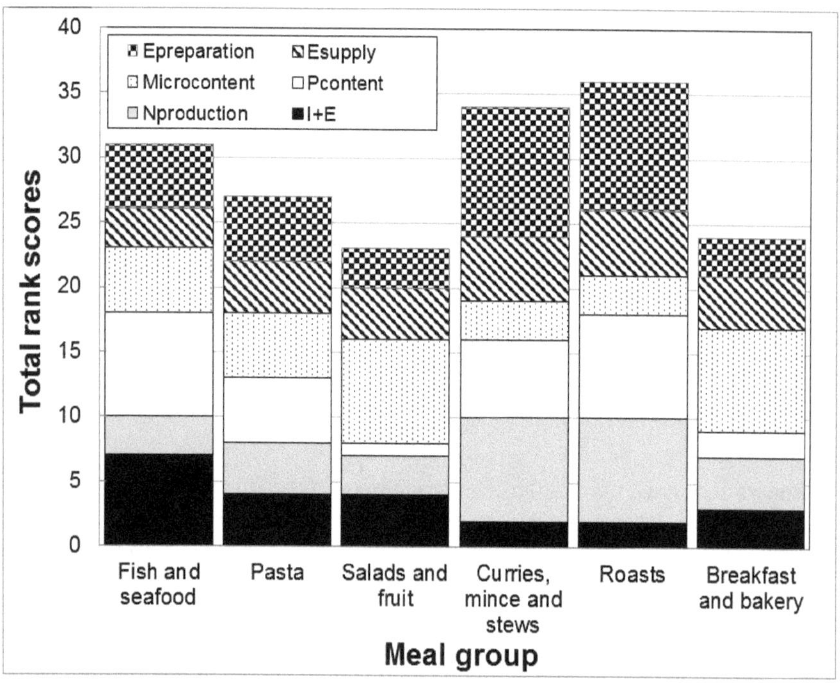

Fig. 2.2 The *6fm* test summary for the meal groups considered in Table 2.1. The scoring was obtained by analysing the responses of nine specialists that tested the *6fm* approach

A sustainable outcome for every meal is the ultimate goal of the *6fm*, with this being coupled to the 'Triple A' standard of affordability, assurance and assurance already being built-into the design of most food ingredients and products it is possible to achieve the most sustainable outcome. A sustainable meal will be defined as one that enables consumers to live within limits of available resources so that they are not destroyed. Assessing and defining this as a practice is what the *6fm* guides end-users to do by enabling access to good culinary preparation.

The *6fm* can help benchmark high-level sustainability and security issues with the consumption of meals. An important aim for *6fm* is to create change in diets where there is sufficient or over supply of protein so that the consumption is balanced. As such, we considering transitions in protein use or changing protein preference in order to reach a sustainable goal rather than limiting protein itself. Typical summaries of the functions include the following.

1. Meal groups containing fish and seafood had the greatest import and export rank with pasta, and, salad and fruit, meal groups having notably greater ranking than other meal groups.
2. Meal groups for mince, stews and roasts that typically include livestock ingredients have increased ranking for nitrogenous fertiliser use.
3. Meal groups for seafood, pasta and breakfasts show increased micronutrient content. The protein content rank of meal groups containing seafood and livestock products is greater than the other meal groups.
4. The meal groups show relatively constant energy of supply whereas the energy for preparing stews and roasts is increased compared to other meal groups.

2.6 The FMCG Supply Chain Challenge

While supply chain operations can be simplified, analysis of how supply chains interact with consumer choices are complex because of the scale of behaviours in populations where there are millions of consumers who exert choice editing, preference and demand pressures. The *6fm* approach provides an opportunity for end users to cut through this complexity with individual meal choices they make. This is achieved by identifying universal connectors to assess resource utilisation at all control points of the supply chain and packages these connectors into the six functions of the *6fm*. This enables the user of the *6fm* to ask structured questions about the ingredient and meal choices made for specific meals and relate this understanding to values of security, sustainability and ethics of diets. The six functions we identify to assess the meal groups are well characterised in literature and we use this evidence to support our reasoning for including them in the *6fm*.

Consumer use of the *6fm* will also enable feedback into the whole supply chain and Fig. 2.1, shows this feedback being most effective when transferred back to the producer operators so that it impacts across the whole supply chain. Currently, feedback is more advanced way for the consumer to retailer and manufacturer operators and this is reflected in Fig. 2.1 (Hutchinson et al. 2015). There are notable successes of feedback where specialist science has complemented cultural interest in preparing food and communicated dietary changes effectively. Such an example is the Total Well Being (TWB) Diet from CSIRO in Australia that has used dietary trials and recipe listing to promote health and sustainability in domestic food preparation (Noakes and Clifton 2013).

The use of preservation and preparation practices methods will be critical to ensuring nutritional quality and this is a well-established practice for consumers (Hounsome et al. 2008). We must now integrate these understandings with the excellent body of GHG emission, water-use and energy use studies that have established data initiated by Nielsen et al. (2003), Wallén et al. (2004) and strengthened by the meta-analysis published by Tilman and Clark (2014). The energy used to supply meals in the test of the *6fm* here is relatively constant because it relates to supply chain resource efficiency where sites of processing and manufacture can optimise resource utilisation. The energy used in food preparation by consumers is highly variable because energy used in preparing meals is rarely measured and managed in the same way as in manufacturing or service environments. This emphasises the requirement to stimulate the use of sustainability assessment in situ, when meals are actually being prepared whether the users are producers, manufacturers, retailers or consumers.

While food waste is not one of the six functions in the *6fm*, it is universally connected to each of them through resource use and an important consideration here is the waste of food by consumers. Consumers' waste food because

- we have too much,
- we do not like,
- we have forgotten about it while it has been stored.

The *6fm* was developed to capture this information so that meals can be designed efficiently for preservation or portion control. A food waste model that utilises preparation data has been developed and the data supporting this model is reported by Martindale (2014b). It considers the use of frozen and fresh foods and shows the use of frozen foods results in less domestic food waste. This study highlights the importance of understanding how meals are prepared because much of the food waste debate has tended to overlook the importance of preparation and preservation in tackling food waste and focus on the problem of food waste rather than the solutions to it. Our future efforts on applying the *6fm* will be in the consumer use environment because ultimately this is where the *6fm* will change food preparation and storage practices that in turn will have positive impacts on security and sustainability.

An important consideration in applying the *6fm* is the relationship it has with other methods of assessing sustainability practice for preparing meals. The dominant methods in this arena are LCA's and these are not necessarily the most practical for informal use because they require in-depth analysis of data quality and processes. However, sustainability criteria associated with foods and meals often requires us to consider attributes that are not measured by LCA methods, these include food waste, transportation, purchasing frequency, preservation method and so on. The *6fm* provides flexibility to include these attributes and provide a broad measure of sustainability through the sum of the six functions as an index. The sustainability assess arena has established standards such as the Carbon Label and certifications for many different products that develop communication and transparency for the sustainability of the supply chain. The *6fm* is primarily concerned with the use of food by consumers even though the data obtained can be utilised by manufacturers and retailers; and this differentiates the *6fm* from LCA which is designed to be specific. It is important for us to consider how we might link successful brand communication techniques to sustainability communications and the household management of food groups that we consume. The development of such systems that utilise meal groups is already well established with nutritional profiling tools that have been successful at linking food product development with consumer driven nutritional outcomes and meal design (Vlassopoulos et al. 2016). The *6fm* aims to utilise these types of techniques used by nutrient profiling to group meals, but to extend this thinking to a sustainability theme. In maintaining the flexibility and reducing the requirement for detailed investigation in the *6fm* it also exposes the weakness of the *6fm*; that is, the data output will be variable and dependent of the intrinsic knowledge the end-user holds. This caveat in the use of *6fm* also provides a need to introduce learning and self-assessment feedback techniques which are the subject of future development of the model.

The *6fm* approach does relate more closely to the broader established certifications of the food industry in that it has to remain flexible and informal so that it can be used by many different end-users. This approach has been used in general FMCG arenas with the Lifestyle of Health and Sustainability (LOHAS) principles that are used internationally (Lifefstyle of Health and Sustainability 2016). They provide a focus on the health of consumers and food consumption using attributes of freshness and dietary standards, as well as sustainability actions that include reaching lower or zero-waste goals. The LOHAS movement has been focussed on the USA marketplace where some 13% of citizens are considered LOHAS consumers, if we consider the European consumer, some 15% fit into the aiming for LOHAS. Indeed, the scorecard approaches used by LOHAS are incorporated into the *6fm* because the scorecard method provides flexibility of use and reduces the requirement for in-depth investigation (Crawford and Scaletta 2006). The LOHAS platform partners with trade organisations and Non-Governmental Organisations (NGOs) so that the actions delivered achieve purchasing awareness across supply chains. The *6fm* differentiates its approach here because it is focussed on how foods are utilised when they are delivered to consumers from the supply chain, this is a particularly important difference because it is consumer focussed and it is not associated with the goal of certification.

The *6fm* moves us towards an important point in the development of sustainability of food FMCG which is to enable consumers to demonstrate what a sustainable meal is using the *6fm* methodology. The *6fm* can be used universally across food and meal groups by all supply chain operators and there is a focus on consumers when they are considering how to make meals. In conclusion, our goal is to stimulate a transition from 'LCA-thinking' to one of 'consumer experience-thinking' by 9 billion consumers if we are to develop successful dietary scenarios for meal design and food consumption. A critical future consideration for the *6fm* approach is to understand how it might be delivered and promoted across food supply chains. This requires future investigation into how consumers will utilise the *6fm* methodology and how the data obtained from consumers will be most effectively fed-back into the supply chain in order to stimulate the re-design of sustainable products.

References

Brandt K, Średnicka-Tober D, Barański M, Sanderson R, Leifert C, Seal C (2013) Methods for comparing data across differently designed agronomic studies: examples of different meta-analysis methods used to compare relative composition of plant foods grown using organic or conventional production methods and a protocol for a systematic review. J Agri Food Chem 61(30):7173–7180

Crawford D, Scaletta T (2006) The balanced scorecard and corporate social responsibility: aligning values for profit. Financ Manage Inst J 17(3):39–42

De Boer J, Schösler H, Boersema JJ (2013) Climate change and meat eating: an inconvenient couple? J Environ Psychol 33:1–8

De Boer J, Schösler H, Aiking H (2014) "Meatless days" or "less but better"? Exploring strategies to adapt Western meat consumption to health and sustainability challenges. Appetite 76:120–128

Defra (2014) Food statistics pocketbook 2014. https://www.gov.uk/government/collections/food-statistics-pocketbook. Last accessed 26 Oct 2015

Garnett T (2013) Food sustainability: problems, perspectives and solutions. Proc Nutr Soc 72(1): 29–39

Green L, Costello L, Dare J (2010) Veganism, health expectancy, and the communication of sustainability. Aust J Commun 37(3):51

Hornibrook S, May C, Fearne A (2013) Sustainable development and the consumer: exploring the role of carbon labelling in retail supply chains. Bus Strat Environ

Hounsome N, Hounsome B, Tomos D, Edwards-Jones G (2008) Plant metabolites and nutritional quality of vegetables. J Food Sci 73(4):48–65

Hoyland A, McWilliams KA, Duff RJ, Walton JL (2012) Breakfast consumption in UK schoolchildren and provision of school breakfast clubs. Nutr Bull 37(3):232–240

Hutchinson K, Donnell LV, Gilmore A, Reid A (2015) Loyalty card adoption in SME retailers: the impact upon marketing management. Eur J Mark 49(3/4):467–490

Jones P, Hillier D, Comfort D (2014) Assurance of the leading UK food retailers' corporate social responsibility/sustainability reports. Corp Gov 14(1):130–138

Kearney J (2010) Food consumption trends and drivers. Philos Trans R Soc B Biol Sci 365(1554): 2793–2807

Leslie WS, Eunson J, Murray L, Lean ME, Hankey CR (2014) What, not just salad and veg? Consumer testing of the eatwell week. Publ Health Nutr 17(07):1640–1646

Lifefstyle of Health and Sustainability (2016) See http://www.lohas.com/about. Accessed 25 Feb 2016

Macdiarmid JI (2013) Is a healthy diet an environmentally sustainable diet? Proc Nutr Soc 72(1): 13–20

Martindale W (2014a) Global food security and supply. John Wiley & Sons, Oxford, UK, p 220

Martindale W (2014b) Using consumer surveys to determine food sustainability. Brit Food J 116(7):1194–1204
Martindale W, McGloin R, Jones M et al (2008) The carbon dioxide emission footprint of food products and their application in the food system. Aspects Appl Biol 86:55–60
Martindale W, Finnigan T, Needham L (2014) Current concepts and applied research in sustainable food processing. In: Tiwari BK, Norton T, Holden NM (eds) Sustainable food processing. John Wiley & Sons, Oxford, UK, pp 9–38
Nielsen PH, Nielsen AM, Weidema BP, Dalgaard R, Halberg N (2003) LCA food data base. Representative inventory for resource use and environmental impact of Danish farms and food products. Available on-line at http://www.lcafood.dk
Noakes M, Clifton P (2013) The CSIRO total wellbeing diet. Penguin, UK
O'Rourke D (2014) The science of sustainable supply chains. Science 344(6188):1124–1127
Ogino A, Orito H, Shimada K, Hirooka H (2007) Evaluating environmental impacts of the Japanese beef cow–calf system by the life cycle assessment method. Anim Sci J 78(4):424–432
Ridoutt BG, Sanguansri P, Harper GS (2011) Comparing carbon and water footprints for beef cattle production in southern Australia. Sustainability 3(12):2443–2455
Scarborough P, Appleby PN, Mizdrak A, Briggs AD, Travis RC, Bradbury KE, Key TJ (2014) Dietary greenhouse gas emissions of meat-eaters, fish-eaters, vegetarians and vegans in the UK. Clim Change 125(2):179–192
Smil V (2002) Nitrogen and food production: Proteins for human diets. Ambio 31:126–131 (Royal Swedish Academy of Sciences 2002)
Tilman D, Clark M (2014) Global diets link environmental sustainability and human health. Nature 515(7528):518–522
Vlassopoulos A, Masset G, Charles VR et al (2016) A nutrient profiling system for the (re)formulation of a global food and beverage portfolio. To describe the Nestlé Nutritional Profiling System (NNPS) developed to guide the reformulation of Nestlé products, and the results of its application in the USA and France. Eur J Nutr 1–18. Published on-line doi: 10.1007/s00394-016-1161-9
Wallén A, Brandt N, Wennersten R (2004) Does the Swedish consumer's choice of food influence greenhouse gas emissions? Environ Sci Policy 7(6):525–535

Chapter 3
The Economic Case for the Circular Economy: From Food Waste to Resource

Corrado Topi and Magdalena Bilinska

Abstract One of the major challenges decision and policy makers face when trying to introduce sustainable food waste management strategies is to isolate high value waste material. In our paper we assess whether it is logistically, economically and socially feasible to isolate exhaust coffee grounds from the catering industry in one British district and use them as raw material for a novel process to produce alternative high added value products in a near-perfect circular economy cycle making use of reverse logistics and generating near-zero waste. We chose coffee as the product because it is the most traded food commodity in the world, and the second most traded commodity in general, which makes the impact of the outcomes particularly significant. Due to resource and time constraints we had to limit the range of high added value products and to constrain the geographic area, hence we focused on the production of high quality compost for the amateur and professional growers market and on the geographic catchment area of the York municipal waste collection service. To do so, we developed a series of theoretical scenarios corresponding to the different possible logistic and process options that stakeholders could identify and we evaluated the economic indicators. We conclude that the process is technically feasible with available technology within current infrastructure and modest investments and the economic case is very attractive to investors. The outcomes of our research can be used as a model for similar developments in other geographical areas.

Keywords Circular economy · Economics · Reverse logistics · Waste management · Food waste · Ecosystem services · CICES · ALCHEMY · Scenarios · Coffee

---

C. Topi (✉) · M. Bilinska
Green Economics Research Group (GECO), Stockholm Environment Institute, University of York, York, UK
e-mail: corrado.topi@york.ac.uk

C. Topi
York Centre for Complex System Analysis (YCCSA), University of York, York, UK

P. Morone et al. (eds.), *Food Waste Reduction and Valorisation*,
DOI 10.1007/978-3-319-50088-1_3

3.1 Introduction

Notwithstanding some substantial market oscillations coffee remains the largest traded food commodity in the world and the second most traded after oil, making the impact on economies at the global and local scale of the activities and processes related to the production, use and disposal of the coffee commodity very significant. Equally significant are the impacts on the environment, in particular on those ecosystems on which coffee related processes rely, and on society, in particular those communities that live around and rely on the different stages of the value chain (Elkington 1997; Norman and Mcdonald 2004).

In the crop year 2016, the total production was estimated to be 143,371,000 standard 60 kg bags, equivalent to 8,602,260 tons of coffee, 100,217,000 of which were destined to the global market (International Coffee Organisation 2016b). In Great Britain, the market we considered in our case study, imports have been slowly but steadily growing to 4,206,000 bags/year (International Coffee Organisation 2016a), with an approximate consumption of around 511 million cups of coffee a week (Walsh 2011). Whilst no significant change in coffee consumption patterns and trends has been observed in the mid-term (e.g. Ng et al. 2012) expansion in the market has been detected in the short-term and coffee shops have been the most rapidly growing niche market in the B2C catering sector over the last few years.

Whilst there is no precise estimate of the amount of exhaust coffee ground that remains as food waste, or a clear industrial framework allowing us to produce estimates that are accepted by all stakeholders and boundary partners (Parfitt et al. 2010), it is possible to produce a crude estimate of the magnitude of the problem. Quantities of coffee grounds necessary to brew one cup of coffee range from 7 to 20 g depending on the method used. Since weight lost during the extraction process is negligible, a conservative estimate is that between 186 kilotons and 531 kilotons of exhaust coffee ground produced in a year, which is in line with the consumption estimates of the International Coffee Organisation (International Coffee Council 2014) if we keep into account the amount of water trapped in the exhaust coffee ground through the brewing process. Currently, the exhaust coffee grounds are normally disposed of in landfill or through generic food waste reprocessing. Yet, academic research has shown that spent coffee grounds can be used as raw materials for the production of high value added intermediate or final consumer products. Among other things, spent coffee grounds may be utilised to produce clay bricks (Eliche-Quesada et al. 2011); chemicals, oils, materials and biofuels (e.g. Pfaltzgraff et al. 2013; Lin et al. 2013); high value specialist compost because of the substantial amount of precious nutrients that promote plant growth and improve soil structure (e.g. Liu and Price 2011). There is also a long history of traditional use in the informal economy and in the households (e.g. DIY 2012).

In our paper we focus on the development of real world circular economy scenarios which could be implemented in the short term at local economy level. The Circular Economy (CE) concept and principles started gaining popularity in modern Europe in the late 1990s (e.g. Allen 2007; Cooper 1999). The publication in 2002 of the book *Cradle to cradle: Remaking the way we make things* by Braungart (2000) and McDonough and Braungart (2010) and the creation of the Cradle To Cradle Product Certification[1] kick-started the process of formalising the integration of the CE principles into business practice. In China, CE was first embedded into national policy with the approval of the 11th 5-year plan in 2006 (e.g. Zhijun and Nailing 2007). In the English speaking world, interest in CE has been growing since the early 2010 thanks to high commodity prices and exposure of the concept by the Ellen MacArthur Foundation (e.g. EMF 2012, 2013, 2014). In December 2015, the European Commission published the Circular Economy Action Plan, putting CE close to the core of the new mainstream sustainable development policy for Europe (European Commission 2015).

In practice, Circular Economy aims at minimizing waste and excessive resource use by turning goods at the end of their lifespan and waste generated during the manufacturing and use of goods into resources for the manufacturing of other products (Stahel 2016; EMF 2012; Lehmann et al. 2014; Taranic et al. 2016).

Within this context, we decided to use the value chain of coffee as a case study to investigate whether Circular Economy scenarios are implementable today using commercially available technologies and commercial market ready solutions. In particular, we decided to look at closing the loop of the value chain of coffee by using the exhaust coffee grounds for the production of high quality compost in a large case study catchment area, and on co-developing feasible scenarios in collaboration with market actors and stakeholders using a participatory collaborative approach. The aim was to produce an economic investigation of the different scenarios to evaluate their feasibility and attractiveness to investors and some of their key environmental and social impacts.[2] Last but not the least, in the building of our circular economy scenarios and in our quest to minimise environmental impacts we have eschewed traditional distribution strategies to make use of reverse logistics strategies. The close-loop value chains facilitate end-of-life recovery in order to recapture the value in used products (Kumar and Putnam 2008; Pokharel and Mutha 2009) whilst reducing environmental and social impacts (Nikolaou et al. 2013; Sarkis et al. 2010). As far as we are aware, this is the first case of reverse logistic strategies to develop green transition scenarios on cash crop waste products. Research is currently ongoing to include and compare alternative scenarios which include the production of other high value added products.

Beyond the researcher team, five other groups of stakeholders were involved: coffee shops and their staff; the waste collection and management company who had the responsibility for the York catchment area and their staff; the officers in the local council in charge of waste collection; academic experts from Green Chemistry and from Department of Biology; representatives of two companies which

[1]Information available at http://mbdc.com/certification-overview.

[2]In a qualitative way.

manufacture and distribute compost to the amateur and professional growers markets, a larger global enterprise with international reach and a limited concern with national reach. Two categories of coffee shops in the catchment area of the York waste collection service were involved: coffee shops belonging to national or global chains, e.g. Costa, and local coffee shops. Coffee shops that had multiple retail points that were all local where assimilated into the second group, whilst other categories of enterprises which were providing coffee as an additional service were excluded at this stage, e.g. restaurants, hotels, patisseries and tea rooms. Our investigation covered the planning, preparation, data collection and analysis stages, but it excluded activities to transform research into a viable option.

3.2 Methodology

Our analysis starts from the point where the shop staff extract the exhaust coffee ground from the portafilter in the espresso machines. We first evaluate the scope for action by calculating the average total amount of exhaust coffee grounds generated in the catchment area over the year by the catering industry, which we use to develop a series of 4 scenarios with the stakeholders and boundary partners. To do so we use ALCHEMY, a novel participatory collaborative approach to mapping the interactions of the value chain with the environment and society, developed at the Stockholm Environment Institute over the last 5 years details of which can be found in Topi (2015). In essence, the approach consists in the development of a series of conceptual process based maps which represent the processes and interactions of the stage of the value chain under examination, the environment, represented through the Ecosystem Services Framework (Daily 1997; De Groot et al. 2002; Millennium Ecosystems Assessment 2005a, b, c, d; Haines-Young and Potschin 2010, 2011; UK National Ecosystem Assessment 2011; Haines-Young and Potschin 2013a, b), and the society that are affected. That is done through a participatory collaborative process which is framed by specific rules including the use of an internationally standardised graphic modelling language, BPMN. Over those conceptual maps, we overlay economic information, as well as quantitative and qualitative information on environmental and social impacts.[3]

In the case of the present paper by using the approach we develop the maps for 5 scenarios. The first scenario, or reference or business as usual (BAU) scenario, corresponds to the existing process, where the exhaust coffee is routed through waste collection to reprocessing. We then developed a series of 4 theoretical circular economy scenarios (described in detail in the results section) which progressively take us from a less customised and dedicated process, with no modifications to existing infrastructure, equipment and personnel, to progressively

[3]Additional material on the approach is available on the dedicated website at http://www.alchemy.info.

more customised, full circular economy processes[4] with additional capital investment and staff employment.

For each scenario map we then calculate the value of key economic indicators that were identified in collaboration with the stakeholders as significant to support the process of deciding whether the green transition option proposed was feasible and which of the options was more convenient in the specific situation. They were:

- Upfront Capital Investments
- Running Costs
- Revenues
- Payback Period

Revenues were split in three components: revenues from collection, revenues from gate fees and revenues from the sale of the high quality compost.

Concerning environmental and social indicators we evaluated them qualitatively and we used a marginal impacts and benefits approach, where we highlighted only the differences, i.e. gains and losses, when compared to the other scenarios rather than evaluating them in absolute terms. In Table 3.1 we have summarised the main characteristics of the 4 Circular Economy (CE) scenarios, whilst in Table 3.2 we have summarised the unitary marginal costs and revenues by item and by type, including an estimate of the uncertainty.

In particular, the average running costs for the composting process including operations and maintenance are of £14.5 (£13 to £16) per tonne of waste. Revenues come through three channels: a fixed collection fee of £10 per collection per bin is applied to the consumer generating the waste; an average gate fee of £20.5 (£17 to £24) per tonne of waste when waste enters the site; the sale of the compost ranging, with prices ranging from £0, i.e. for compost giveaways to raise awareness of environmental issues, to £5 per tonne for low quality compost, £9 per 40 l bags for high quality compost for the specialist market.

It is important to highlight the fact that the alternative Circular Economy scenarios are constructed on a theoretical basis. They could be easily replicated and applied to other catchment areas by recalculating the values of the indicators. Recalculating the economic indicators may change the final decision on what scenario is most convenient in different situations, and outcomes of the analysis would be highly dependent on local and global market conditions, e.g. consumption patterns or financial instability. For these reasons, an estimate of the uncertainty for all values seems to be beyond the scope of the present paper, because effects due to external drivers would be order of magnitudes larger than the intrinsic uncertainty, and because of the difficulty of estimating the true uncertainty in a reliable way, for example due to the fact that only certain equipment can be used in specific

[4]It must be noted that whilst the City of York, which is the catchment area in our case study, has organised a green composting reprocessing plant where organic waste is transformed into generic low grade and low quality compost, many other catchment areas do not have such facilities.

Table 3.1 Description of the 4 alternative circular economy (CE) scenarios for the exploitation of exhaust coffee grounds (ECG)

Scenario	Equipment	Land	Process
CES 1	No changes	No changes	Route ECG through existing green waste process Same site Separate collection Composting together with green waste Separate distribution (by reverse logistics) Compost sold at low quality compost prices
CES 2	No changes	No changes	Same site. Separate collection No composting Separate distribution (by reverse logistics) Compost free as additive
CES 3	No changes	On site changes	Same site. Separate collection Separate composting Separate distribution (by reverse logistics) Compost for sale at high quality compost prices
CES 4	New equipment	New land purchased	Separate site. Separate collection Separate composting Separate distribution (by reverse logistics) Compost for sale at high quality compost prices

Table 3.2 Table summarising the marginal costs (and revenues) by type as used in the scenarios

Item	Type	Average value	Uncertainty
Running costs	Cost	£14.5 per tonne	£13 to £16
High quality compost	Revenue	£9 per 40 l bag	Fixed
Collection fee	Revenue	£10 per collection per bin	Fixed
Low quality compost	Revenue	£5 per tonne	Fixed
Compost giveaway	Revenue	£0 per tonne	Fixed
Gate fee	Revenue	£20.5 per tonne	£17 to £24

situations, with one manufacturer and one costs, or that the costs and availability of land is specific to certain areas, with one supplier that may be in monopoly conditions.

3.3 Results

To produce a realistic estimate of the average total amount of exhaust coffee grounds generated in the catchment area over the year by the catering industry we worked closely with the stakeholders: the coffee shops management and staff involved provided the necessary data on exhaust coffee grounds generation in their outlets, which were either used for direct calculations or to produce projections for outlets which we were not able to involve.

In some cases, the shop had already collected data on their output, but information was reported in different units of measure and time scales: some reported daily, weekly, monthly or yearly production. To homogenise the results, the group decided to use average production in tonnes on a yearly basis.

Finally, we estimated the total number of coffee shops in the area in two categories: coffee shop chains and local coffee shops. The average total amount of exhaust coffee ground produced in the catchment area was estimated to be ~165 tonnes a year, 99 tonnes of which (~60%) were produced by coffee chains and 66 tonnes (~40%) by local coffee outlets.

In the following sections we discuss the reference scenario and we compare it in detail the four circular economy scenarios.

3.3.1 The Reference Scenario

In the reference scenario we investigate the existing process and we assess existing technologies and behavioural options. The exhaust coffee is currently collected through the general waste collection service together with generic waste and channelled to landfill.

Yet, mapping this process alone would be misleading, because the same waste management company manages also wood and other green waste from private households through a separate process on the same site, which makes use of equipment that is currently not being used to its full potential and which has a trained workforce. We decided to present the full picture by mapping the available processes for green waste collection and management as part of the reference scenario (Fig. 3.1).

Wood and green wastes are collected from household recycling centres, kerbside collections and commercial collections by a transport company and delivered to the waste management site (collection in the map). At the arrival on the site the waste is visually inspected and contaminants are removed by hand and routed to other waste disposal processes (inspection in the map). Part of the input materials which the stakeholders estimate at approximately 50% present themselves is in too large a format and requires shredding by one of the many shredding machines operated by the company on site (shredding in the map). The material output from shredding and the material which was already of the correct size and form are mixed and routed to composting. The site operates an open windrow composting system hence the material needs to be prepared and then formed into a windrow using a wheeled loader. The same wheel loader is used to turn the windrows approximately once a week to maintain aerobic condition and ensure even composting, and the cycle is repeated for 8–12 weeks depending on external conditions (windrow composting in the map, represented as a subprocess collapsed to outline the presence of several subprocess tasks). When the composting cycle is deemed to be finished, the output is channelled to a double drum rotary compost screening trommel (40 and 10 mm) removing contaminants and oversized materials to bring the compost to specifications (screening in the map) and compliance is checked by an independent third party, i.e. a PAS 100 accredited laboratory (quality control in the map). The 10 mm compost can be sold in bulk to the professional growers' market and to the public in bags. The 40 mm compost can be supplied to the farming market through agricultural trade channels.

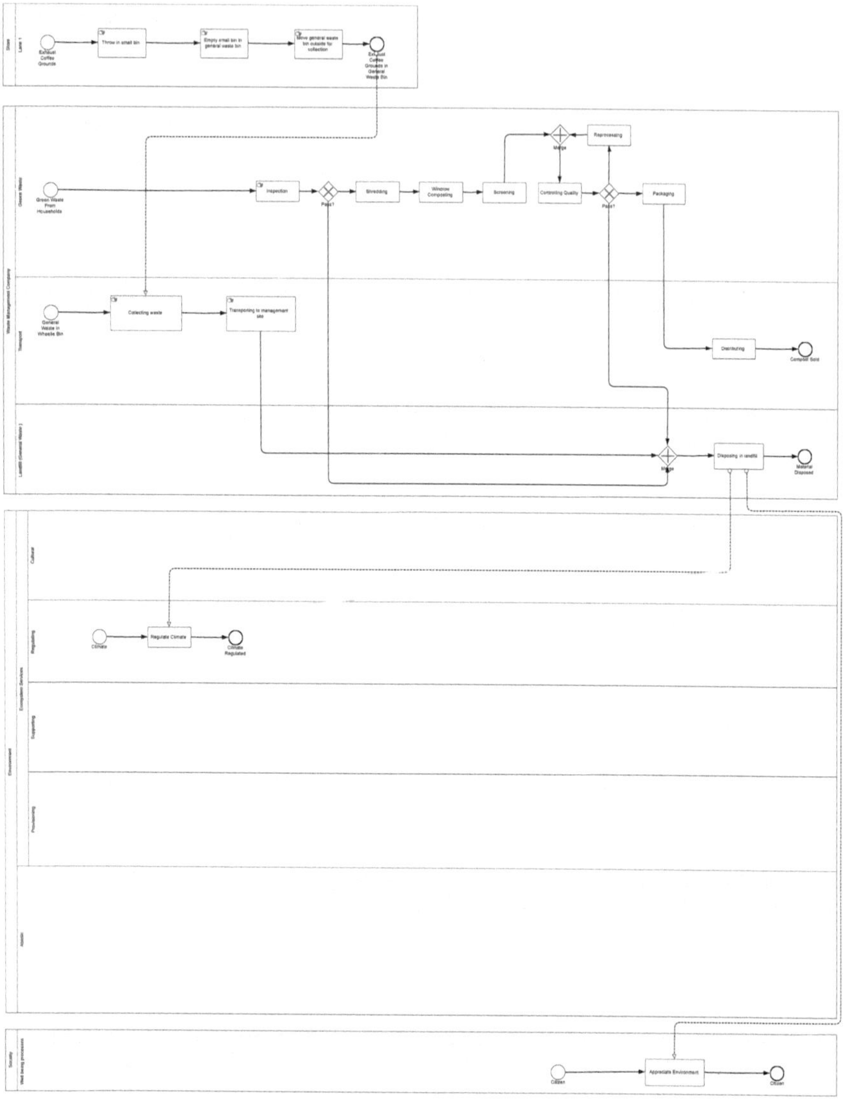

Fig. 3.1 Reference scenario

3.3.2 *First Circular Economy Scenario (CES 1)*

In the first circular economy scenario (see Table 3.3) we route the ground coffee exhaust through the green waste lane, we apply the reverse logistic approach and resell the compost to the customers of the waste management service, minimising the environmental impact of distribution. The low grade compost is routed back to the garden owner through the existing infrastructure to be resold, increasing the

Table 3.3 Summary of economic indicators for circular economy scenario 1

Upfront capital investment	Running costs	Revenues	Payback period
£20,000	£2393/year	£39,358/year	<1 year

amount of compost that is reused locally, reducing final waste and generating revenues. The scenario ends when the compost is resold to the final user.

In this case there are no changes to operational processes on site, but there would be changes to the processes at the point of collection and to transport segments of the chain.

At the point of collection, a separate waste collection for exhaust coffee waste would need to be set up. A 240-l dedicated orange wheelie bin and two under the counter kitchen caddies in which staff can dispose of the coffee grounds directly from the coffee machine before transferring them to the orange bins will be distributed to participating stores.

The collection truck available at the company can carry 4–5 t of waste. Considering that an average of 0.45 t of exhaust coffee grounds are produced every day in the catchment area, that coffee grounds preserve well for over two weeks and that the truck can visit 120 sites in a day, we planned a one-day collection every 10 days.

From an economic point of view, the additional upfront capital investments are minor, the only new assets being the orange wheelie bins and the caddies, with an estimated cost of £8000, whilst additional hidden and missing costs, i.e. store information and recruitment campaigns and staff training sessions would be larger, with estimated costs at £12,000. These costs have been integrated into an estimated cumulative upfront capital investment of £20,000.

The exhaust coffee grounds would be treated as other compostable waste. Treating the additional 165 t of material would add to the running costs of the waste management plant an average of £2393/year using the running cost per tonne from Table 3.2 (£14.5).

From the point of revenues the additional collection fee £10 per collection from Table 3.2 will generate £35,150/year, considering that we will collect once every 10 days, i.e. 37 collection rounds a year over 95 outlets, for a total of 3150 collections. The gate fee would bring an average of £3383/year using the £20.5 per tonne gate fee from Table 3.2.

The average marginal revenues derived from the symbolic sale of the additional tons of low quality compost would be £825/year.

The payback period would be less than one year, making the investment profitable and attractive to the investors.

Table 3.4 Summary of economic indicators for circular economy scenario 2

Upfront capital investment	Running costs	Revenues	Payback period
£20,000	£2393/year	£38,533/year	<1 year (longer than CES 1)

3.3.3 *Second Circular Economy Scenario (CES 2)*

In the second scenario (see Table 3.4), we send the exhaust coffee grounds through a separate collection process but we resell it through a reverse logistic system without composting it. Exhaust coffee grounds have been traditionally used as fertiliser providing nitrogen to plants and improving moisture and water retention. Still, because they have been not composted, they usability and requirement would be lower, and they would only be used as additives. In this scenario, coffee grounds would be stored at the waste management site for 3 consecutive collection cycles to achieve a critical mass and then redistributed through the reverse logistic chain. Because the process does not add value to the exhaust coffee ground they would be simply redistributed either free of charge or at a symbolic price.

From the economic point of view changes to both the marginal operational costs and the revenues for the composting process would be insignificant when compared to scenario CES1 (see Sect. 3.3.2), with the exception of the revenue from compost sale, which would be zero in this scenario, since non composted coffee grounds would be distributed as 'giveaways' to raise awareness. The remaining indicators would remain the same as in the previous scenario, and the case remains attractive to investors.

3.3.4 *Third Circular Economy Scenario (CES 3)*

We use exhaust coffee ground to produce high quality compost through a separate collection and composting process in the same facility and using the existing equipment (see Table 3.5).

To do so we need to determine a generic composting recipe with the appropriate balance of macro- and micronutrients, of granularity and of moisture content. Exhaust coffee grounds have a C:N ratio of $\sim$20:1 which need to be balanced to between 25:1 and 30:1 to reach a good composting ratio and avoid excessive release of odour. It can be done by adding appropriate amendments rich in carbon, e.g. cardboard chips (350:1 ratio) or wood chips (400:1) (Liu and Price 2011; Rynk et al. 1992). These are dry amendments which will also correct the high moisture content which would have led to anaerobic conditions. Cardboard chips and wood

Table 3.5 Summary of economic indicators for circular economy scenario 3

Upfront capital investment	Running costs	Revenues	Payback period
£20,000	£7503/year	£96,025/year	<6 months

chips are already available at the collection site. Systematic quality testing of all batches is required by the market to keep variability under control, but frequency could be reduced to statistical control if quality turns out to be stable (Woodbury and Breslin 1996).

The exhaust coffee grounds are mixed with the amendments for separate windrow composting, which requires a dedicated windrow area which must be sizes optimal for the composting process to be efficient (Rynk et al. 1992). We calculated the optimal size of the area to be <400 m^2, holding two dedicated windrows and the areas for manoeuvring the equipment. In this scenario we assume that the waste management company has enough space at the waste management site to implement the separate composting process without buying any additional land. With this set up, we estimated that approximately 255,500 l of compost can be produced each year by the two windrows with a rotation period of 50 days and can be bagged into 6388 standard 40 l bags using on site existing machinery.

From an economic point of view, there are no additional capital investments related to assets compared to scenarios CES 1 and CES 2 since the land is owned and existing equipment could be used, whilst operational costs can be estimated at the same level as for the other scenarios. The process of bagging the compost into standard 40 l bags costs £0.8 per bag, adding £5110 to the running costs of the waste management plant.

Revenues from fees are similar to the previous scenarios, whilst revenues from sales change substantially: assuming that a standard 40 l bag of compost could be sold for £9 to the specialist market (see Table 3.2), we estimate the revenues from sales to be £57,492/year, which in addition to the collection fees revenues of £35,150/year and gate fees revenues of £3383/year brings the total revenues to £96,025/year. With this level of revenues and a payback period well under a year the case is extremely attractive to the local investors.

3.3.5 Fourth Circular Economy Scenario (CES 4)

The fourth is a theoretical scenario where a separate composting facility is set up to produce the high quality spent coffee grounds-based compost. This entails acquisition of land, new equipment and a waste management license to operate the composting process, whilst the labour force would be shared between facilities.

From an economic point of view, the major voices for capital investments are land and equipment acquisition. For the former, we need to add space for the storage of equipment to the estimate we produced previously which brings the total amount of land required to ~500 m^2. The land should be situated at a certain distance from communities and be easily isolated from ground water. With an average estimated cost of £444,000 per hectare[5] for industrial grounds in the

[5]At the time of writing, see Valuation Office Agency (2011).

Table 3.6 Summary of economic indicators for circular economy scenario 4

Upfront capital investment	Running costs	Revenues	Payback period
£95,450	£7750/year	£95,447/year	~1 year

Yorkshire and Humberside we can estimate the cost of land acquisition to be £22,200.

To purchase the equipment, we have estimated costs of £53,250 where we have used average market prices to produce the estimate (McCloskey International Ltd. 2015; Machine Solutions 2015; Organic Recycling Group 2009) (Fig 3.2).

Figure 3.2 shows an itemised breakdown of those costs overlaid on the map where the hotspots of capital investment are highlighted using the Semaphore Artifact.

These figures bring the total amount of capital which needs to be invested upfront to £95,450 where we have also included the costs of the wheelie bins, the caddies and the marketing campaign as previously calculated. The equipment we purchase would not be used at full capacity and would allow us to cope with an amount of waste at least one order of magnitude larger.

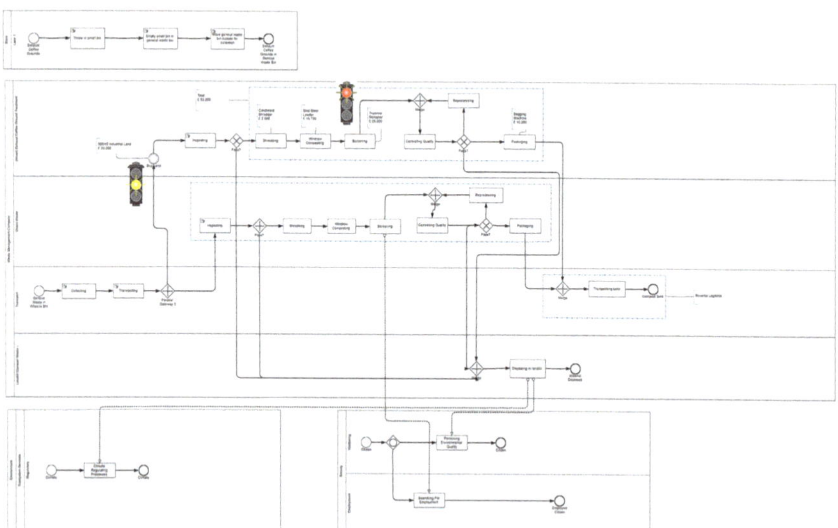

Fig. 3.2 Extract of the map with the itemised breakdown of equipment costs for scenario 4. The Semaphore Artifact has been used to visualise the hotspots of capital investment, in *red* the top hotspot and in *yellow* the second hotspot. It is important to note that the *yellow* hotspot, albeit being the second highest investment in absolute terms does relate to a long term asset, i.e. land purchase

Concerning running costs, to develop a conservative estimate and in view of the fact that we would be using a new plant, we decided to use the higher end running costs per tonne at regime previously estimated (£16 per tonne instead of the average £14.5 per tonne), which brings the costs to £2640 if we only collect from the 95 initial outlets. When considered in addition to the £5110 cost for the bagging process, the total running costs would be £7750. This is of course a conservative approach where we assume the highest costs and the lowest source of revenues. The operational costs include the manpower which is shared with the general waste composting facility.

From the point of view of revenues and considering again the very conservative case of no new customers, collection fees will bring £35,150, and the sale of compost £57,492 as in CES 3. To make the case conservative also for gate fees we use the lowest, and hence more conservative, value of £17 per tonne which brings the estimated revenues to a total of £2805/year: setting the gate fee at the lowest end would also make separate composting much more attractive to stores.

Even with these very conservative assumptions the payback period is approximately a year (Table 3.6).

Whilst this may seem a less attractive proposition compared to the previous scenarios it is still very attractive and it opens the gates to developing a transferable and scalable model. For example, increasing by four times the number of collection points would be easily feasible within the waste management plant set up of CES 4 but not with the setup of CES 1, 2, 3 and would bring the payback period to approximately 6 months with annual revenues at regime to ~£500,000.

3.4 Conclusion

Our research suggests that the case for applying circular economy principles to the chain of coffee is strong and attractive to investors under all of the scenarios analysed.

In particular, we have investigated the case of using exhaust coffee ground waste from the catering industry in the catchment area of the City of York's waste collection and management service, in the North East of England, as raw material to produce high quality, high added value compost for the horticultural professional and amateur industry using existing technology and reverse logistics. To do so we investigated the costs and benefits of four alternative scenarios based on a set of conservative assumptions on waste production, technology deployment and costs and we compared them to a reference scenario where waste coffee grounds follow the existing waste management route to identify the most feasible and attractive scenario. The four scenarios are summarised in Table 3.7.

All scenarios carry the same environmental benefits which are due to a reduction of the quantify of organic waste that goes to landfill, the corresponding reduction of gas emissions and permanent land use. We have not quantified these indicators because they are assumed to be the same in all scenarios and hence cannot be used

Table 3.7 Comparative summary of economic, environmental and social indicators for the four circular economy scenarios (CES), where environmental and social indicators are evaluated only qualitatively

Scenario	Capital investment	Running costs	Revenues	Payback period	Environmental	Social
CES 1	£20,000	£2393	£39,358	<1 year	+	
CES 2	£20,000	£2393	£38,533	<1 year	+	
CES 3	£20,000	£7503	£96,025	<6 months	++	
CES 4	£95,450	£7750	£95,447	~1 year	++	+

as discriminants. CES 3 and CES 4 carry the additional benefit of a reduction in use of peat in compost.

The social case was evaluated through qualitative considerations: only the last scenario carries direct social benefits in terms of additional jobs (the only indicator we considered), hence making it favourable compared to the others from a social point of view.

Whilst CES 1 and CES 2 require low upfront capital investment the opportunity for additional revenues is missed, and so are the environmental, cultural and social benefits.

From an economic point of view CES 3 is the most attractive within the current setup with marginal initial investments and maximum revenues from sales of the highest quality compost. Scenario 3 has intrinsic limitations, in particular it has limited scalability and it cannot be replicated well. The problem with scalability affects the possibility to extend the catchment area.

CES 4 whilst having initially the worst payback period and requiring much larger upfront capital investments shows the greatest potential from the point of view of scalability and replication and carries the highest social benefits.

The limitations to our approach are linked to the specificity of the context and set up and should not represent major obstacles to replication in different situations. First, in all scenarios we have assumed varying degree of sharing of land, facilities, equipment and labour force between the existing waste management processes and the proposed exhaust coffee grounds management processes. That would require the processes to be run in geographically adjacent or not to distant locations. Whilst this is possible in the case under consideration, it may be difficult to implement in other specific cases. The only scenario which could be modified easily to keep into account locations which geographically very distant would be Scenario 4, to which would be necessary to add separate labour force related costs. Secondly, our calculations are based on the assumption that the market demand for compost is constant throughout the year, whilst we know it to be seasonal. That being the case, additional costs to store composted material before distribution should be included in a specific case, whilst from an academic point of view, it would be worth extending the evaluation of the environmental and social cases to include a full analysis both of direct and second order effects, and our research group is currently trying to address this issue.

Finally, alternative scenarios for using the exhaust coffee grounds to produce alternative high added value products should be considered and developed using the participatory mapping approach and their economic, environmental and social cases compared with compost production scenarios. Alternative options may indeed be even more attractive. Research is currently underway to address the limitations and extend the scope.

References

Allen J (2007) Cradle-to-Cradle. Achieving the vision through remanufacturing. Available at http://www1.calrecycle.ca.gov/Archive/IWMBMtgDocs/mtgdocs/2008/05/00023461.pdf. Accessed 29 May 2016

Braungart M (2000) Cradle to Cradle®. Mc GrawHill

Cooper T (1999) Creating an economic infrastructure for sustainable product design. J Sustain Prod Des 7–17

Daily GC (1997) Nature's services, societal dependence on natural ecosystems. Island Press, California

De Groot RS, Wilson MA, Boumans RM (2002) A typology for the classification, description and valuation of ecosystem functions, goods and services. Ecol Econ 41:393–408

DIY (2012) Notcutts promotes benefits of coffee as garden fertiliser. Available at http://www.diyweek.net/news/news.asp?id=15480&title=Notcutts+promotes+benefits+of+coffee+as+garden+fertiliser. Accessed 1 Aug 2013

Eliche-Quesada D, Pérez-Villarejo L, Iglesias-Godino FJ, Martínez-García C, Corpas-Iglesias FA (2011) Incorporation of coffee grounds into clay brick production. Adv Appl Ceram 110:225–232

Elkington J (1997) Cannibals with forks. The triple bottom line of 21st century business. New Society Publishers. Oxford

EMF (2012) Towards the circular economy vol. 1: an economic and business rationale for an accelerated transition. Available online at https://www.ellenmacarthurfoundation.org/publications/towards-the-circular-economy-vol-1-an-economic-and-business-rationale-for-an-accelerated-transition

EMF (2013) Towards the circular economy vol. 2: opportunities for the consumer goods sector. Available online at https://www.ellenmacarthurfoundation.org/publications/towards-the-circular-economy-vol-2-opportunities-for-the-consumer-goods-sector

EMF (2014) Towards the circular economy vol. 3: accelerating the scale-up across global supply chains. Available online at https://www.ellenmacarthurfoundation.org/publications/towards-the-circular-economy-vol-3-accelerating-the-scale-up-across-global-supply-chains

European Commission (2015) Communication from the commission to the European parliament, the council, the European economic and social committee and the committee of the regions closing the loop—An EU action plan for the Circular Economy

Haines-Young R, Potschin M (2010) Proposal for a common international classification of ecosystem goods and services (CICES) for integrated environmental and economic accounting

Haines-Young R, Potschin M (2011) Common international classification of ecosystem services (CICES): 2011 Update

Haines-Young R, Potschin M (2013a) CICES V4. 3–Revised report prepared following consultation on CICES Version 4, August-December 2012. EEA Framework Contract No EEA. IEA/09/003

Haines-Young R, Potschin M (2013b) Common international classification of ecosystem services (CICES): consultation on version 4, August-December 2012

International Coffee Council (2014) World coffee trade (1963–2013): a review of the markets, challenges and opportunities facing the sector. International Coffee Organisation, London, England, Great Britain

International Coffee Organisation (2016a) Exports by crop year. International Coffee Organisation, London, England, Great Britain. http://www.ico.org/historical/1990%20onwards/PDF/1e-exports.pdf. Accessed on 1 Apr 2016

International Coffee Organisation (2016b) Total production by crop year. International Coffee Organisation, London, England, Great Britain. http://www.ico.org/historical/1990%20onwards/PDF/1a-total-production.pdf. Accessed on 1 Apr 2016

Kumar S, Putnam V (2008) Cradle to cradle: Reverse logistics strategies and opportunities across three industry sectors. Int J Prod Econ Inst Perspect Supply Chain Manag 115:305–315. doi:10.1016/j.ijpe.2007.11.015

Lehmann M, de Leeuw B, Fehr E, Wong A (2014) Circular economy. Improving the management of natural resources: Swiss academies of arts and sciences. Available online at http://www.satw.ch/publikationen/schriften/kreislaufwirtschaft/a__circulareconomy_with_references_EN.pdf

Lin CSK, Pfaltzgraff LA, Herrero-Davila L, Mubofu EB, Abderrahim S, Clark JH, Koutinas AA, Kopsahelis N, Stamatelatou K, Dickson F et al (2013) Food waste as a valuable resource for the production of chemicals, materials and fuels. Current situation and global perspective. Energy Environ Sci 6:426–464

Liu K, Price GW (2011) Evaluation of three composting systems for the management of spent coffee grounds. Bioresour Technol 102:7966–7974

Machine Solutions (2015) Cardboard shredders, corrugated shredders, cardboard box shredders, carton shredder [WWW Document]. http://www.machine-solution.com/Cushion-Pack-PacMasters-Cardboard-Box-Shredders. Accessed 20 Feb 2015

McCloskey International Ltd. (2015) Mobile crushers, vibratory screeners, trommel screens, and stacking conveyors for the recycling and aggregate industries—512 A/R Trommel Screener [WWW Document]. http://www.mccloskeyinternational.com/trommel-screeners-512-ar.php. Accessed 20 Feb 2015

McDonough W, Braungart M (2010) Cradle to cradle: remaking the way we make things. MacMillan

Millennium Ecosystem Assessment (2005a) Ecosystems and human well-being: synthesis

Millennium Ecosystem Assessment (2005b) Ecosystems and human well-being: biodiversity synthesis

Millennium Ecosystem Assessment (2005c) Ecosystems and human well-being: opportunities and challenges for business and industry

Millennium Ecosystem Assessment (2005d) Ecosystems and human well-being. The World Resources Institute, Washington DC

Ng SW, Ni Mhurchu C, Jebb SA, Popkin BM (2012) Patterns and trends of beverage consumption among children and adults in Great Britain, 1986–2009. Br J Nutr 108:536–551

Nikolaou IE, Evangelinos KI, Allan S (2013) A reverse logistics social responsibility evaluation framework based on the triple bottom line approach. J Clean Prod Sustain Manag Beyond Corp Bound 56:173–184. doi:10.1016/j.jclepro.2011.12.009

Norman W, Macdonald C (2004) Getting to the bottom of triple bottom line. Bus Ethics Q 14:46–54

Organics Recycling Group (2009) Buyers' guide: screening and bagging equipment. Available at http://www.organics-recycling.org.uk/uploads/article1762/Screening%20and%20Bagging%20guide.pdf. Accessed 19 Aug 2013

Parfitt J, Barthel M, Macnaughton S (2010) Food waste within food supply chains: quantification and potential for change to 2050. Philos Trans R Soc Lond B Biol Sci 365:3065–3081. doi:10.1098/rstb.2010.0126

Pfaltzgraff LA, Cooper EC, Budarin V, Clark JH et al (2013) Food waste biomass: a resource for high-value chemicals. Green Chem 15:307–314

Pokharel S, Mutha A (2009) Perspectives in reverse logistics: a review. Resour Conserv Recycl 53:175–182

Rynk R, Kamp M, Willson G, Singley M, Richard T, Kolega J, Gouin F, Laliberty L, Kay D, Murphy S, Hoitink H, Brinton W (1992) On-Farm composting handbook. Coperative Extension, New York

Sarkis J, Helms MM, Hervani AA (2010) Reverse logistics and social sustainability. Corp Soc Responsib Environ Manag 17:337–354. doi:10.1002/csr.220

Stahel WR (2016) The circular economy. Nature 531:435

Taranic I, Behrens A, Rizos V, Topi C (2016) (forthcoming) Understanding circular economy: from resource efficiency to sharing platforms. CEPS Policy Brief. CEPS. Brussels

Topi C (2015) Towards a blueprint for the development of integrated green transition strategies at the local level. University of York, York, UK

Topi C, Vega Barbero J, Ghosh A, Bilinska M (2016) Embedding ecosystem services in private enterprises: a novel participatory collaborative approach. In submission

UK National Ecosystem Assessment (2011) The UK national ecosystem assessment technical report. UNEP-WCMC, Cambridge, UK

Valuation Office Agency (2011) Property market report 2011: the annual guide to the property market across England, Wales and Scotland. Available at http://www.voa.gov.uk/dvs/_downloads/pmr_2011.pdf

Walsh J (2011) Britain's caffeine boom: Why can't we wake up without smelling the coffee? The independent. Available at http://www.independent.co.uk/life-style/food-and-drink/features/britains-caffeine-boom-why-cant-we-wake-up-without-smelling-the-coffee-2342104.html. Accessed 3 July 2013

Woodbury P, Breslin V (1996) Municipal solid waste composting: key aspects of compost quality assurance. Cornell Waste Management Institute. Available at http://compost.css.cornell.edu/MSWFactSheets/msw.fs7.html. 27 Aug 2013

Zhijun F, Nailing Y (2007) Putting a circular economy into practice in China. Sustain Sci 2:95–101. doi:10.1007/s11625-006-0018-1

Chapter 4
Decreasing Greenhouse Gas Emissions of Meat Products Through Food Waste Reduction. A Framework for a Sustainability Assessment Approach

Thomas Winkler and Ralf Aschemann

Abstract The global food production industry is responsible for producing high levels of greenhouse gas (GHG) emissions. Along the entire food supply chain (FSC), potential for mitigation exists because approximately one-third of all food globally produced is wasted, equivalent to 1.3 billion tons per year. On a global scale, emissions from livestock production are about 4600–7100 Mt CO_2-eq/year when considered over the whole life cycle. These numbers represent roughly 9.4–14.5% of the total global GHG emissions. In Austria, the livestock sector was responsible for producing about 11.6% of the total GHG emissions in 2012 as a result of the production of about 909,000 t of meat. A high potential for mitigation of GHG emissions from livestock production exists, especially during the farming and production phases. A reduction in meat waste would, in the long-term, directly reduce GHG emissions stemming from livestock production. Two scenarios were considered to assess the GHG mitigation potential of waste from meat production: a business-as-usual (BAU) scenario and a reduction (RED) scenario (assuming a one-third reduction in waste from meat production in Austria). Because food waste is influenced by several phenomena along the FSC, taking an approach such as the life cycle assessment (LCA) offers only a partial solution. By using a Sustainability Impact Assessment (SIA) approach, researchers can consider social, economic and ecological impacts. It is possible to analyze and compare food waste reduction potentials through the use of such a tool, which can support GHG mitigation efforts in terms of their social, environmental and economic contribution to the livestock and meat processing sector. This approach allowed the identification of indicators that contribute to all sustainability dimensions and support the conclusion that preventing waste from meat processing would save at least 4.8 Mt CO_2-eq emissions per year in Austria, which represented 6% of Austria's total CO_2-eq emissions in 2012.

T. Winkler · R. Aschemann (✉)
Institute of Systems Sciences, Innovation and Sustainability Research,
University of Graz, Merangasse 18, 8010 Graz, Austria
e-mail: ralf.aschemann@uni-graz.at

P. Morone et al. (eds.), *Food Waste Reduction and Valorisation*,
DOI 10.1007/978-3-319-50088-1_4

Keywords Food supply chain · Food waste · Sustainability impact assessment · Greenhouse gas emissions · Austria

4.1 Introduction

The '*Loi Macron*' is a French Act[1] (Senat 2015), which, inter alia, is banning supermarkets from wasting food, brought the discussion on food waste to the center of attention in the European Union in 2015. Italy soon picked up on the discussion and introduced the '*Spreco zero*' (zero waste) campaign to its parliament (Partito Democratico 2015). Even though food waste is a topic of with a high level of importance globally, few studies have been conducted, and its media coverage has not been high in recent years. Considering how our society wastes food, particularly during the consumer phase, it is necessary to initiate programs that will prevent food waste entirely. About one-third of the total food produced in our world is being thrown away, representing about 1.3 billion tons yearly. Food waste and food losses are similarly high in both developed and developing countries (assessed by comparing the amount of food produced in a country or global region), but this waste and loss occurs at different stages of the food supply chain (FSC). Many different drivers result in food being wasted: the economic system, legislation, cultural issues, resource limitations and lack of infrastructure, to name a few (Parfitt et al. 2010). In developing countries, food loss often occurs during the first phases of the FSC-agriculture and food processing–due to lack of management skill and technical expertise in food production. In developed countries, in contrast, food is thrown away by members of the wholesale and retail sector as well as end-consumers (households, food services and restaurants). These processes coincide with increasing fragmentation of the global population and multiplication of issues about how food is grown and produced (FAO 2011; Parfitt et al. 2010).

Food waste is gaining increasing amounts of attention as a crucial waste management study area and is recognized more and more as a global problem. Using resources (along the FSC) in a more sustainable and efficient way can effectively decrease greenhouse gas emissions (GHG) and impact global climate change, as well as influence other economic and social factors (Papargyropoulou et al. 2014). In the European Union, it has been estimated that the food sector alone causes about 22% of all GHG emission. Thus, it ranks very high among life-cycle-wide impacts on resources and has a high environmental impact potential (European Commission 2006; Papargyropoulou et al. 2014). Food waste also has many social implications, which tend to involve ethical and moral issues (Salhofer et al. 2008). Interventions in the first stage of the FSC offer the best opportunities for mitigation; this specifically means influencing agricultural practices and preventing food waste at the consumer stage (Papargyropoulou et al. 2014).

[1]The Act will enter into force by July 2016 (Moveforhunger 2016).

This contribution examines the effects of food waste, specifically waste from (national) meat production, on our (global) climate. When considering the entire FSC, food production is a main contributor to the total GHG emission level, erosion, water depletion and deforestation. The effects of food waste in Austria are in the focus of this study due to the rather unique absence of large industrial food production sites and importation of feed from other countries (USDA 2012), as well as the good availability of data. Food production and waste in Austria was quantified and connected to the primary emission sources. One of the largest contributors to emissions is meat production and consumption. The meat production industry in Austria produced about 909,000 t meat in 2013 (Statistik Austria 2013) and resulted in the slaughter of around 83 million animals (92% poultry) (Statistik Austria 2014b). A sustainability impact assessment (SIA) was conducted to evaluate the climate effectiveness of waste from meat production in Austria (in terms of CO_2-eq savings) and identify possible GHG reduction options. By using an SIA, it was possible not only to focus on the ecological implications of food and meat waste in Austria, but also examine social and economic factors.

First, an overview of food waste is given, providing definitions of and statistics for food waste in meat production and meat waste. Next, the SIA procedure, including two scenarios and system boundaries and six indicators, is described. Then, the results of an assessment of these indicators are given. Finally, a discussion is presented, and conclusions are drawn.

4.2 Scope and Statistical Overview of Food Waste

4.2.1 Definitions

Different studies use different definitions of food and food waste, and, equally importantly, set different system boundaries and/or consider different background information (e.g., inclusion of food processing sector, food service sector and restaurants). It is necessary to provide a clear definition of food and food waste in order to be able to reasonably interpret the results of the analysis.

The European Union (European Parliament 2002) defines food (or foodstuff) as ‘any substance or product, whether processed, partially processed or unprocessed […] expected to be ingested by humans’. This broad definition also includes drinks, chewing gum and all additional materials that are intentionally included in food processing (including water).

In one of the deliverables of the EU-funded FP7-FUSIONS project (‘Food Use for Social Innovation by Optimising Waste Prevention Strategies’, 2012–16), food waste is defined as ‘… any food, and inedible parts of food, removed from the food supply chain to be recovered or disposed (including composted, crops ploughed in/not harvested, anaerobic digestion, bio-energy production, co-generation, incineration, disposal to sewer, landfill or discarded to sea)’, cf. Fusions (2014).

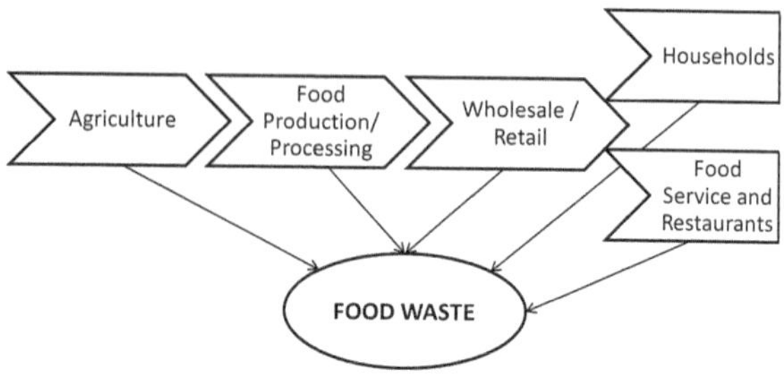

Fig. 4.1 Food supply chain (FSC) and food waste (based on Kranert et al. 2012)

According to Kranert et al. (2012), food waste is (broadly) defined as waste from[2]:

- agricultural production
- food production and processing
- wholesale and retail markets
- food service and restaurants
- households
- (raw and processed food that is potentially edible).

Furthermore, food waste can be divided into avoidable, partly avoidable and unavoidable waste. Avoidable food waste includes edible food that is thrown away at some point along the FSC (please see Fig. 4.1), and food that would have been edible if it has been eaten before it spoiled. Unavoidable food waste is basically waste from food production that occurs somewhere along the FSC and is considered inedible under 'normal' circumstances (e.g., banana skins, bones, intestines). Defining partly avoidable food waste is rather difficult as this is quite often a subjective topic. In general, it can be argued that partly avoidable food waste includes food that is eaten by some and treated as waste by others (e.g., apple cores, bread crusts, potato skins); leftovers are also included in this category (Kranert et al. 2012; Monier et al. 2010; Quested and Johnson 2009). Due to this subjective classification of partly avoidable food waste, this study only categorized the food as avoidable or unavoidable food waste (e.g., leftovers were included in avoidable food waste). By-products in food processing are not defined as waste as long they are used for a different purpose later in the FSC (Kranert et al. 2012; European Parliament 2008).

Food waste and food loss are often treated synonymously. In most studies, the term 'food loss' refers to (food) waste that occurs:

- at the beginning of the FSC;
- during the agricultural stage;
- during food production
- during processing.

---

[2]Please see Fig. 2.1 for a graphical depiction.

The term *food waste* in the literature is associated with behavioral patterns observed during the retail and consumer (household) stages (Parfitt et al. 2010). This study addressed both food losses and food waste, but did not distinguish between these two terms. The term *food waste*, therefore, refers to both food losses and food waste.

4.2.2 Meat Production and Meat Waste

Globally, food production is responsible for about 9.4–14.5% of total GHG emissions (IPCC 2014; Steinfeld et al. 2006). About 52 billion animals were slaughtered in 2004 (not including marine animals) worldwide. Chicken represented about 90% of these animals, which were slaughtered for meat production (based on FAO statistics, Humanresearch 2015). In the European Union Member States, 15.6 million tons of animals were slaughtered in 2013, which represents–more or less–the livestock population for meat production in the EU (slightly higher numbers of imports of living animals were reported than exports) (Eurostat 2015). The maintenance of these animals (including their slaughter, but excluding all stages after the slaughterhouse) results in GHG emissions of 616–852 Mt CO_2-eq/year. In addition to these high levels of GHG emissions, livestock rearing contributes to erosion, eutrophication of water bodies and has a high water footprint.

In Austria, 200,000 cattle and calves, almost 500,000 pigs, 7700 goats and sheep and 125,000 poultry were slaughtered for national use in 2012 (Statistik Austria 2014b). These data do not include meat that is exported after slaughter out of Austria. Emissions from livestock production in Austria are approximately 9.3 Mt CO_2-eq/year, which represent roughly 11.6% of Austria's total GHG emissions in 2012.

In the EU, 35 million tons of animal and vegetal waste was produced by the food, beverage and tobacco processing sectors in 2008 (7.5 million tons of which were animal waste). Household waste amounts of animal and vegetal waste in the EU was estimated[3] to be 23.8 million tons in 2008, which represents about 48 kg per capita and 10.8% of all household waste (European Union 2011).

It has been estimated that between 89 and 178.3 million tons of food waste accumulates each year in the EU, which will generate roughly 70–170 Mt CO_2-eq/year of emissions (Monier et al. 2010). Until 2020, it has been assumed that emissions from food waste will represent up to 240 Mt CO_2-eq per year (Monier et al. 2010). A large part of these emissions (21%) stems from animal and meat waste, although meat waste accounts for less than 5%, and vegetables for almost 25%, of total food waste (FAO 2013). This study assessed GHGs from food waste, and respectively waste produces as a result of food production, along the FSC.

[3]It is believed that these levels are underestimated (European Union 2011).

4.2.3 Food Waste in Austria

In terms of food waste per capita/year, Austria is not among the top 20 EU member states (Croatia not included). Eurostat data indicate a range from 4.51 kg/capita/year in Greece to 56.03 kg/capita/year of food waste in the Netherlands. The average for all EU-27 countries is 17.7 kg/capita/year of food waste (the official Eurostat data was used and complemented with data from various national sources that were provided by the EU member states, cf. Monier et al. 2010). Even though these data may not be trusted completely, it is obvious that Austria, at least in terms of food waste mitigation, is not among the leading countries (please see Table 4.1).

In Austria, slightly less than 1.5 million tons of animal and vegetal waste was produced by the processing industry, and 300,000 t came from food preparation. The amount of animal and vegetal waste as compared to total household waste in Austria is, at 18.7%, above the EU average (this value also includes food packaging material). In total, Austrians wasted 1,185,800–1,956,240 t of food in 2008, which is equal to about 21.7–22.8 kg of food wasted per person and year. 34–66% of the total food waste in Austria is produced by households, and another 30–48%, by food manufacturers (agriculture, food production, food processing). The remainder can be allocated to retail, wholesale and large-scale consumers such as restaurants and hospitals (European Union 2011; Monier et al. 2010; Selzer 2010). Unfortunately, no complete data set is available on avoidable food waste in Austria. Several studies have been conducted, each of which has focused on specific areas and/or stages of the FSC and/or waste categories (Bernhofer 2009; Obersteiner and Schneider 2006; Schneider and Lebersorger 2009; Selzer 2010).

4.3 Sustainability Impact Assessment (SIA) Procedure

No clear guidelines for how to assess food waste on a global and/or regional level have yet been created. Several international standards have been set and, frequently, life-cycle analysis (LCA) is used as a method to supplement household diaries, 'waste-bin research' and surveys. However, as Katajajuuri et al. (2014) mentioned, 'no commonly approved standard or communication method for evaluating a foodstuff's climate impacts are available'. Moreover, LCAs do not consider the social and economic impacts of food waste on the implications for climate change.

Therefore, different forms of Sustainability Impact Assessment (SIA) methodologies have been developed which address all three dimensions of sustainability, see OECD (2010) for example. Singh et al. (2012) reviewed many sustainability assessment methodologies, as "sustainability indicators simplify, quantify, analyze and communicate otherwise complex and complicated information". Singh and his colleagues collected 61 different indices and ratings to assess the sustainability of various subjects such as development, products, cities, policies, industries and/or nations.

Table 4.1 Emissions stemming from all stages of the FSC–from the agricultural to the slaughterhouse stage–for livestock globally, in the EU and Austria (based on Winkler and Winiwarter 2015)

	Global	European union	Austria			
	Mt CO_2-eq/year	Share of total GHG emissions	Mt CO_2-eq/year	Share of total GHG emissions	Mt CO_2-eq/year	Share of total GHG emissions
GHG emissions	4600–7100	≙9.4–14.5%	616–852[b]	≙13.5–18.75%	9.3	≙11.6%
Sources	FAOSTAT (2015a, b, c, d) and Steinfeld et al. (2006)	European Environment Agency (2014) (data for 2012), Lesschen et al. (2011) and Weiss and Leip (2012)	Leip et al. (2010) and Umweltbundesamt (2014a) (data for 2012)			

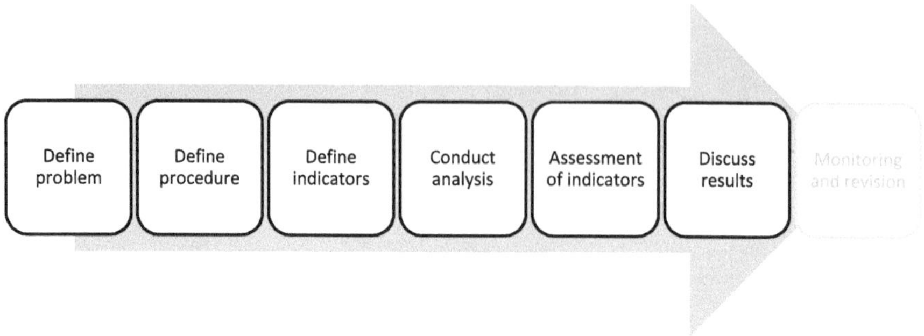

Fig. 4.2 Sustainability Impact Assessment (SIA) procedure followed during this study (based on ARE 2004)

Regarding the use indicators for SIAs, the authors referred to Ness et al. (2007), who distinguished between non-integrated ('meaning they do not integrate nature-society parameters') and integrated indicators ('meaning the tools aggregate the different dimensions'). Referring to the latter, Pope et al. (2004) identified two options for the development of assessment for sustainability criteria, namely:

- a simultaneous achievement of a series of environmental, social, and economic goals or objectives;
- the development of assessment for sustainability criteria using a 'top-down' generation of criteria.

We began our study by focusing on overall sustainability goals in the context of food waste reduction and derived indicators from these. By transferring the integrated/non-integrated classification of Ness et al. (2007) to indicators for a food waste SIA, we were able to choose between developing and applying non-integrated indicators, integrated indicators or a combination of both. Both approaches resulted in advantages and disadvantages: when examining particular sustainability dimensions, we could compare their individual economic, social and environmental contributions and contrast these with each other, but not examine the contributions holistically. The approach of integrated indicators was eventually chosen in order to maintain that holistic dimension and avoid the disadvantages that resulted from the inclusion of non-integrated indicators (Fig. 4.2).

4.3.1 Scenarios

Two scenarios related to meat waste in Austria were created to illustrate how changes in behavior can lead to a decrease of waste and connected GHG emissions. One business-as-usual scenario (BAU) is compared to a reduction scenario (RED). In the BAU scenario, it was assumed that no behavioral changes occurred and that members of society produced meat waste as currently observed. In the RED scenario, in contrast, the total meat waste was reduced by one-third, which is the amount of estimated avoidable meat waste in Germany (Kranert et al. 2012) and also seems to be realistic for Austria. It was assumed that less meat was wasted and, therefore, fewer animals were slaughtered for meat-consumption. Therefore, all parameters concerning meat production were decreased by one third. The reduction focused on aspects of meat production and did not consider the economic or environmental implications of by-product production. It is necessary to understand that these assumptions are purely a theoretical experiment. Even though, the potential mitigation of meat waste is one third it is not realistic (at the moment) that no meat is wasted at all. As no data was available about the type of meat that is generally wasted, an equal distribution over all types of meat was assumed. In terms of waste reduction, the RED scenario is considered to be the best case scenario.

4.3.2 System Boundaries

The food sector is a highly complex industrial branch and focusing on more details, such as of meat waste production, introduces still more complexity. It was not possible, therefore, to include all factors and elements of meat waste along the FSC in this study. Due to the interconnectivity of factors, multiplier effects were observed when considering the different ways to reduce food and meat waste. For example, if reducing meat waste results in fewer animals slaughtered, we assume that fewer livestock will be reared in total, and this will lead to a reduction in the production of animal by-products. As a potential side effect, this could mean that certain products would need to be fabricated using different materials, which could possibly have a higher global warming potential (which was not considered in this study).

The SIA presented has clear boundaries, and three input factors that were necessary for the whole FSC to function were considered: energy, water and feedstuff. The outputs of the FSC are the products themselves, including meat; certain emissions into the air, soil and water; by-products such as fat (i.e., products from livestock production that can be used afterwards); and food waste (see Fig. 4.3).

We applied the SIA to the output *meat* and *food waste* and showed how to achieve a potential reduction in CO_2-eq emissions through meat waste mitigation. This method can easily be extended and adapted to address more aspects and include other food products.

4.3.3 Indicators

The following indicators (Fig. 4.4) were chosen to represent inputs and outputs of meat production. However, it was not possible to consider every factor and every parameter, such as certain drug residues in the meat or the effect of less slaughtering on the production of by-products such as carcass meal. All indicators were chosen

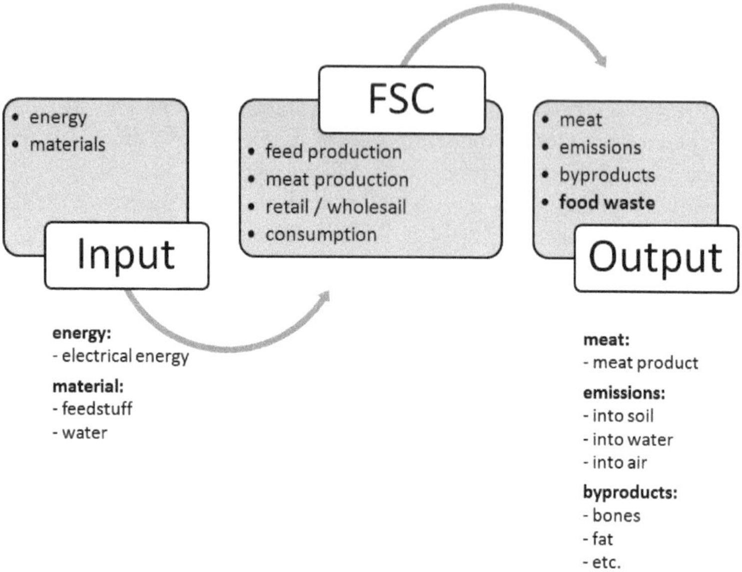

Fig. 4.3 System boundaries

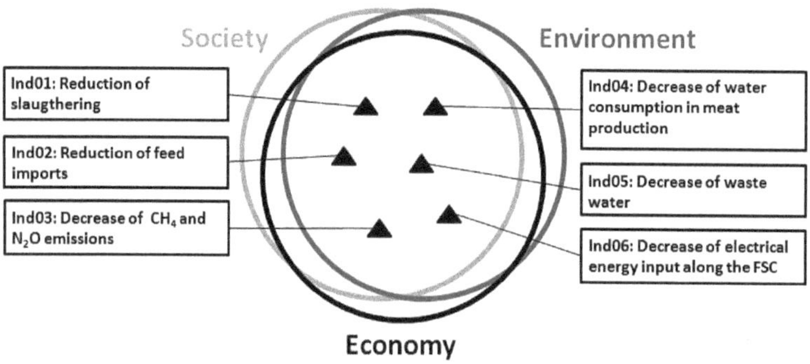

Fig. 4.4 Indicators of integrated SIA for meat waste in Austria

Table 4.2 National net consumption of meat in Austria (Statistik Austria 2013, 2014a)

	BAU	RED
	National net consumption (t)	National net consumption (t)
Cattle	201,452	134,302
Calves	6453	4302
Pigs	491,313	327,542
Sheep and goats	7454	4969
Poultry	121,515	81,010
Total	828,187	552,125

to represent meat production and its effects in Austria over one year (the locations of the triangles within the circles are random and do not indicate a connection with one of the three sustainability circles).

4.3.3.1 Reduction of Slaughtering (Ind01)

Food waste, and specifically meat waste, has many social, economic and ethical implications (Salhofer et al. 2008). In 2014, more than 83 million animals were slaughtered in Austria, the majority of which were poultry (77 million animals). From 2009 to 2013, the number of slaughtered cattle, which have the highest impact on the climate, decreased by ~1% (7414 animals) according to Statistik Austria (2014b). This 1% represents animals that have an individual average weight of 753 kg and carcass weight of about 452 kg (60%). Assuming an amount of 14.2 kg of CO_2-eq/kg of Austrian beef, this 1% reduction was equivalent to 48,000 t CO_2-eq/year (FAO n.d.; Leip et al. 2010). The net meat production in 2014 was roughly 909,000 t, and the national consumption approximately 828,200 t (see Table 4.2), respectively representing 97.7 kg per person and year (Statistik Austria 2013, 2014a). Almost 60% of the total meat consumption in Austria is pork, 25% is beef, 15% is poultry and the rest is goat and sheep (cf. Table 4.2). A reduction in meat waste would, consequently, lead to a reduction in slaughter numbers.

4.3.3.2 Reduction in Feed Importation (Ind02)

75% of the imported feed in Austrian comes from other European Union Member States, and about 10% are imported from MERCOSUR countries-mostly Southern American countries (Statistik Austria 2012). About 570,000 t of soy meal and 100,000 t of soybeans are fed to Austrian livestock. In 2011, about 104,000 t of soybeans were harvested in Austria and, thereof, 32,000 t (27%) were used for feed. Most of this soy meal and these soybeans come from South America. In total, it can be assumed that at least 500,000 t of soy meal is imported every year (Global 2000 n.d.), cf. Table 4.3. An LCA study of pork (Hinterberger et al. 2011) showed that 80% of the climate impact is caused by deforestation of rainforest areas. Changing feedstuff to local (soy) meal could reduce GHG emissions stemming from pig meat production by 50%. When examining the RED scenario, soy import was cut by one-third and the demand was met by national and European soy production.

Table 4.3 Austrian imports of soybeans and soy meal (Castanheira and Freire 2013; Global 2000 n.d.)

BAU			RED		
Soy import (kg)	CO_2-eq min (kg)	CO_2-eq max (kg)	Soy import (kg)	CO_2-eq min (kg)	CO_2-eq max (kg)
1	0.3	17.8	1	0.3	17.8
500,000,000	150,000,000	8,900,000,000	333,333,333	100,000,000	5,933,333,333

It is difficult to get a clear picture of the environmental impact of Austrian soy imports from MERCOSUR countries. On one hand, no data on how much soybeans and soy meal is imported is available, and on the other hand, GHG emissions from soy production in South America mainly depend on emissions from land-use change and vary greatly depending on where the soy is planted. Castanheira and Freire (2013), in an LCA study on soy-bean production in Brazil and Argentina, showed that GHG emission per kg of product varied between 0.3 and 17.8 kg CO_2-eq (including emissions from cultivation, land-use change and transport). Due to missing data for Austria, the emissions from soy production from Argentina and Brazil were used to calculate the impact on the environment by imports.

4.3.3.3 Reductions of CH_4- and N_2O-Emissions (Ind03)

The main sources of methane (CH_4) and nitrous oxides (N_2O) are agricultural processes and emissions from waste systems. Manure from livestock emits CH_4 and N_2O, but the manure characteristics vary according to the animal species and feedstuff. These two climate gases have a global warming potential that is several times higher than CO_2: CH_4 is 34 times higher and N_2O is 298 times higher (IPPC 2014).

In Austria, emissions from enteric fermentation, manure management, animal manure applied to agricultural soil, and pasture, range and paddock manures are responsible for producing 4.5 Mt CO_2-eq/year,[4] which represents 5.6% of the total GHG emissions for Austria. Enteric fermentation from ruminants is responsible for the majority of the emissions in this sector.

According to the official emission inventory data for Austria (Umweltbundesamt 2014a, data from 2012), cattle farming is responsible for the majority of CH_4 and N_2O emissions produced in this sector, cf. Table 4.4. In addition, emissions from animal manure applied to agricultural soil (inventory subsector 4D1.2) and pasture, range and paddock manures (inventory subsector 4D2)[5] need to be considered. The N_2O emissions from these areas are 2150 and 300 t per year, respectively, which represent an amount of 730,100 t CO_2-eq/year.

Landfills are one of the largest sources of methane emissions (Nguyen 2012), and these include biodegradable waster (i.e., biowaste). Although landfilling is probably the worst waste management strategy to use when dealing with biowaste, it is still the method most frequently used (30–40%) in the EU (European Commission 2008). Biowaste is, in general, defined as any waste that can be anaerobically or aerobically digested such as vegetal material, kitchen waste and paper (European Parliament 1999).

---

[4]Using the global warming (GWP) potential as calculated in the IPCC Second Assessment Report achieves consistency with the Austrian inventory report; however, using GWP values from AR5 increases national livestock emissions by approx. 2 Mt CO_2-eq/year.

[5]Subsectors defined as in UNFCCC (2006).

Table 4.4 CH_4 and N_2O emissions from livestock in Austria (Umweltbundesamt 2014a)

	BAU			RED		
	CH_4 ent. ferm. (t)	CH_4 man. mgmt. (t)	N_2O man. mgmt. (t)	CH_4 ent. ferm. (t)	CH_4 man. mgmt. (t)	N_2O man. mgmt. (t)
Cattle	79,890	5880	1290	53,260	3920	860
Pigs	4470	3500	180	2980	2333	120
Sheep and goats	2920	70	80	1947	47	53
Poultry	280	1050	230	187	700	153

In total, approximately 76.5–102 million tons of green and food waste is produced annually in the EU, while another 37 million tons of waste produced by the food and drink industry are categorized as biowaste (European Commission 2008). It has been estimated that up to 29,000,000 t CO_2-eq emissions could be saved by preventing the production of bio-waste (European Union 2011).

In Austria, separate bio-waste collection is supported by a waste management system initiated by the government. In 2014, almost 80,000 t of biowaste (including food) were collected in the city of Vienna. About 21% of the food waste in Austria's capital is treated through anaerobic digestion (biogas production) and about 77% is sent to a biological treatment plant to be transformed to compost (MA 48 2014). Unfortunately, little data on food waste in the Austrian biowaste collection system exists. Therefore, emissions from this sector were not included.

4.3.3.4 Reductions of Water Consumption in Meat Production (Ind04)

In general, agriculture accounts for about 92% of the total global water footprint and about one-third of this is related to livestock production (Gerbens-Leenes et al. 2013). In 2005, the average global water footprint of meat production was 2422 Gm^3/year, whereas the majority of this water was needed for feedstuff production (Mekonnen and Hoekstra 2012). The values for the average water footprint of a live animal measured at the end of its lifetime, and the average annual water footprint of one animal are presented in Table 4.5.

Mekonnen and Hoekstra (2012) assessed the water footprints of several food products in their study. These data included water from feed production, drinking water and service water (e.g., for cleaning), but not water from processing (processing water is included in Ind05). Beef had the highest water footprint measured: about 15.4 million l per ton of product.

Ridoutt et al. (2011) argued that Mekonnen and Hoekstra's estimations are rather high because they included water produced as a result of evapotranspiration from crops and pasture grasses, which enhanced the footprint. Thus, only the grey and blue water footprint has been considered for calculating the water footprint of Austrian meat. By combining those footprint data with the Austrian net

Table 4.5 Average annual water footprint of on animal from 1996–2005 (Mekonnen and Hoekstra 2012)

Animal category	Average water footprint of a live animal at the end of its lifetime (m^3/ton)	Average annual water footprint of one animal (m^3/year/animal)	Annual water footprint of animal category (Gm^3/year)
Cattle	7477	630	798
Pigs	3831	520	458
Sheep	4519	68	71
Goats	3079	32	24
Broiler, layer chickens	3364	59	422
Total (excluding water footprint from horses and dairy cattle)			1773

Table 4.6 Water footprint of Austrian meat products (Mekonnen and Hoekstra 2012; Statistik Austria 2013, 2014a)

Meat product	BAU		RED
	Water footprint per ton (m^3/ton)	National water footprint (m^3) of meat	National water footprint (m^3) of meat
Beef	15,415	208,112,905	138,742,604
Pig meat	5988	531,109,353	354,072,902
Sheep/goat meat	8763	3,801,540	2,534,190
Chicken meat	4325	94,781,700	63,187,800
Total		837,805,498	558,537,496

consumption of meat, the national water consumption caused by meat production equals 838 million m^3 (or 838 billion liter), please see Table 4.6.

However, when considering the RED scenario, approximately 279 billion liter of water could be saved (Mekonnen and Hoekstra 2012; Statistik Austria 2013, 2014a). Due to the lack of data for Austria, the footprint data Mekonnen and Hoekstra (2012) had to be used in the analysis.

4.3.3.5 Reduction of Waste Water from Slaughterhouses (Ind05)

Food production generally has a large influence on bodies of water. Water run-off from farming and rearing livestock leads to eutrophication and leaches fertilizers into the environment, causing an increase in nutrient levels and algal blooms in larger bodies of water. As a direct consequence, water quality can be jeopardized and hypoxia of the lifeforms in the water bodies might occur (Chislock et al. 2013).

Waste water is primarily produced during the slaughterhouse stage of the production chain. A European Commission (2005) report assumed that the production of one chicken as delivered to the supermarket results in the production of 70–130 L

Table 4.7 Waste water from slaughterhouses (European Commission 2005; Statistik Austria 2014a)

Animal	BAU		RED
	Waste water (l per t of carcass)	Waste water (l per total animal prod.)	Waste water (l per total animal prod.)
Cattle	1623–9000	658,944,872–3,654,038,106	439,296,581–2,436,025,404
Pig	1600–6000	1,047,294,301–3,927,353,628	698,196,201–2,618,235,752
Sheep	5556–8333	78,513,225–117,755,706	52,342,150–78,503,804
Poultry	5070–67,400	858,344,916–11,410,739,120	572,229,944–7,607,159,413
Total		2,643,097,230–19,109,886,560	1,762,064,876–12,739,924,373

of waste water by the slaughterhouse. The waste water from poultry production carries a high microbial load and increases the risks of microbial infections (e.g., Salmonella). The RED scenario data in Table 4.7 show that the amount of waste water created by the meat industry in Austria can be reduced by one-third. The total potential reduction of waste water from slaughterhouses in Austria, as a result of reducing meat production and, thus, meat waste, ranges from 881 million to 6.37 billion liters (European Commission 2005; Statistik Austria 2014a).

4.3.3.6 Reduction of Energy Input Along the FSC (Ind06)

In addition to water and various materials, energy represents another input factor in food production. De Vries and de Boer (2010) compared several LCA studies of animal products and the energy intensity of pork, beef and chicken meat. 18–34 MJ of energy are needed to produce 1 kg of pork; 34–52 MJ, for 1 kg of beef; and 15–29 MJ, for 1 kg of chicken meat. These high values are derived from wide system boundaries. De Vries and de Boer (2010) also included the energetic input from feed production, for example. Winkler et al. (2016) examined Austrian pork production more narrowly, considering only the energy input on-farm, and calculated an energetic input of 1.75 kWh/kg of pork ($\sim$6.3 MJ) for electricity, heat and mechanical energy.

This study focused on the entire life cycle of meat, and the cradle to gate-data from de Vries and de Boer (2010) was used to calculate the energy input of the Austrian meat processing industry, see Table 4.8. The RED scenario data show that a reduction of about 8.13 million MJ (or 2.26 GWh) per year (calculated for all meat products, but excluding sheep and goat) can be achieved (de Vries and de Boer 2010; Statistik Austria 2014a). Due to a lack of data, the energy input of sheep was not included, but is considered to be negligible as sheep represent only 0.8% of the total Austrian meat industry.

Table 4.8 Energy input for meat production (de Vries and de Boer 2010; Statistik Austria 2014a)

	BAU		RED
	MJ per t (av. value)	MJ of total meat prod.	MJ of total meat prod.
Beef	43,000	8,939,915,000	5,959,929,000
Pork	26,000	12,774,138,000	8,516,092,000
Chicken meat	22,000	2,673,330,000	1782.220,000

4.4 Assessment of Indicators

Data gathered from the evaluation of the indicators was converted into CO_2-eq emission, and the two scenarios were compared and evaluated (for detailed results, please see Annex). In the RED scenario, Ind01 shows a decrease in animals slaughtered of more than 26 million animals per year (most are chicken) and of 2.13 million t CO_2-eq/year (Leip et al. 2010; Statistik Austria 2014b).

Data from Castanheira and Freire (2013) showed that emissions from feed imports (Ind02) had an extremely high range and that their variability was mainly due to the effects of land-use change. In the RED scenario, it was assumed that one-third less soy (beans and meal) would need to be imported and that this amount could be substituted by soy and other high-energy crops sourced from Europe (a substitution was not calculated in this study). In total, a decrease in feed imports from South America was estimated to reduce GHG emissions by 50,000–2,966,667 t CO_2-eq/year (Castanheira and Freire 2013; WWF 2014). Our analysis did not consider emissions that occurred as a result of possible substitutions, but solely took the consequences of decreases in imports into account.

Ind03 shows that emissions from enteric fermentation and manure management were one of the biggest contributors to GHG emissions from food production. Data for this indicator included CH_4 emissions from enteric fermentation and manure management as well as N_2O emissions from manure management, animal manure applied to agricultural soil and pasture, range and paddock manure. In total, the reductions in the CH_4 and N_2O emissions in the RED scenario led to total emission reductions of 1,531,527 t CO_2-eq/year (Umweltbundesamt 2014a). Due to missing data, emissions from sheep and goats were not included.

Ind04 shows a potential water reduction in the RED scenario of 279 million m^3 (Mekonnen and Hoekstra 2012; Statistik Austria 2013, 2014a). On the basis of data gathered in Germany, a calculation of the average emissions of water production with 0.82 g CO_2/l of fresh water (Stadtwerke Karlsruhe 2014) indicated a possible reduction of approximately 229,000 t CO_2-eq/year in the RED scenario.

Ind05 describes the waste water produced as a result of meat production in the slaughterhouse phase. Waste water per ton of carcass was combined with the total slaughter numbers in Austria and emissions from waste water production. The amount of waste water needed during slaughter depends on the animal and ranges from 1600 to 67,400 l per t of carcass. The highest amount of water is needed during poultry production because of the risk of microbial infection (e.g., Salmonella)

Table 4.9 Savings of RED scenario

Indicator	Savings in RED scenario (t CO_2-eq)		% of Austrian CO_2-eq emissions 2012
	Min	Max	Min max
Ind01—reduction of slaughtering	2,129,355		2.7
Ind02—reduction of feed imports	50,000	2,966,667	0.06 3.7
Ind03—decrease of CH_4 and N_2O emissions	1,531,527		1.9
Ind04—decrease of water consumption	228,998		0.3
Ind05—decrease of waste water	2973		0.004
Ind06—decrease of energy input along the FSC	826,463		1.03
Total savings (Mt CO_2-eq)	4.8	7.7	6.0 9.6

(European Commission 2005). In total, the reduction in waste water by decreasing meat waste was calculated to lead to GHG savings of 2973 t CO_2-eq/year (Stadtwerke Karlsruhe 2014).

Ind06 describes the potential reduction in energy input as a result of reducing meat waste. Using data from de Vries and de Boer (2010), who estimated the necessary energy input of meat production to be 0.366 kg CO_2-eq/kWh (Umweltbundesamt 2014b), a possible reduction of approximately 0.83 Mt CO_2-eq per year could be achieved in the RED scenario. Due to missing data, Ind06 did not consider sheep and goats.

Table 4.9 shows the summary of all indicators and the total savings achieved through the application of the RED scenario, ranging from 4.8 to 7.7 Mt CO_2-eq/year, as compared to the BAU scenario, which is equivalent to minimal 6.0% or maximal 9.6% of Austria's total CO_2-eq emissions in 2012.

4.5 Discussion

The assessment of indicators conducted in this study highlights the potential impacts of food waste reduction on social, economic and environmental factors. The environmental impact on the climate is discussed in more detail than economic and social impacts. It was difficult to clearly assess the social and ethical impacts, because this would have required us to assess or place a price on life itself, which has considerable moral and ethical implications. Even though the animals (and their products) are subject to continual "pricing", it is on conviction that an appropriate price can never be determined. Our study indicates that the social and ethical impacts of a reduction in food waste, and particularly meat waste, can result in a reduction in the numbers of slaughtered animals. When considering the system boundaries of this study, the decrease in Austrian feed imports could indirectly help indigenous people living in or near rainforest areas in South America, who might be affected by enlargements in agricultural areas.

The economic impact of decreasing food waste is obvious because the production of less meat waste, as estimated in this scenario, would directly lead to less meat production and, consequently, lower income levels in the agriculture (e.g., for farmers), meat processing and retail sectors. Assessing the overall cost of food waste is an extremely complex task. The FAO (2013) attempted to price each impact from food waste (including social aspects) on a global scale and arrived at total costs of 2.625 billion US-$. The highest costs arise from production of food (which is subsequently wasted), social factors such as loss of livelihoods and the increasing risk of conflicts and from GHG emissions. Many of these values can be easily contested and, therefore, this assessment focused on the environmental impact of meat waste and, specifically, the impact on our climate. Nevertheless, the important components of the meat supply chain, namely energy, feedstuff and water, were fully taken into account and treated within the system boundaries of the SIA.

However, some restrictions of this study should be considered. Not all indicators cover all three sustainability dimensions, although none of these is an indicator that focuses only on environmental impact. One of the indicators selected did not cover the whole supply chain, because Ind05 only considered waste water produced during the slaughterhouse stage. For this indicator, it was not possible to calculate CO_2-eq emissions and, instead, an average value extracted from the literature was used. Due to missing data, the emissions from landfills in Austria could not be considered, but it is verifiable that a certain amount of food waste is landfilled. For Ind02, a high degree of uncertainty had to be accepted, because only LCA data on soybean production in Brazil and Argentina could be used, which ranged from 0.3–17.8 kg CO_2-eq per kg feedstuff. Data from dairy cows were included because it was not possible to exclude them. The average soy use for beef in the OECD represents less than 1% of the total use of soy as feed due to high numbers of pigs and chicken (WWF 2014), but because only approximately one-fourth of all cattle in Austria are dairy cows, this minor error was not considered to have an impact on the overall results. Ind04 was based on water footprint data for livestock reported in Mekonnen and Hoekstra (2012). The critique of Ridoutt et al. (2011) was considered, as they have argued that Mekonnen and Hoekstra's estimations are rather high because they included water produced as a result of evapotranspiration from crops and pasture grasses, which enhanced the footprint. Therefore, only the blue and grey fraction of the water footprint was taken into account, resulting in a lower mitigation potential of the BAU scenario regarding water consumption, when compared to the entire water footprint.

Despite these restrictions, uncertainties and missing data, the SIA presented can provide researchers with a rough picture, revealing the huge potential for GHG emissions savings that would exist if the total meat waste in Austria were reduced by one-third. The largest factors that could contribute to this potential are reducing feed imports (e.g., particularly soybeans and meal), reducing the number of animals slaughtered and decreasing CH_4 and N_2O emissions. Our study demonstrated that by decreasing the energy input and the water consumption, GHGs emissions could be reduced further, whereas waste water decreases were negligible.

Future research could focus on modifying the SIA to address the restrictions listed above, to gain a more precise and accurate forecast for GHG emissions resulting from food waste (as well as the impact of food waste reductions). In this context, it might be of interest to examine the situation in other countries and/or for other food products and gather more data. Moreover, including a greater variety of indicators would allow researchers to test the reliability of these results. For this purpose, including approaches used in other fields such as technology assessment could be supportive [cf. e.g., the study to determine the requirements for a sustainability product label, which was developed by the Office of Technology Assessment in the German Parliament (TAB 2015)].

4.6 Conclusion

This paper addresses a topic with high societal relevance. The reduction in food waste is a sub-goal of one of the United Nations 'Sustainable Development Goals' (SDGs), which have been adopted by the United Nations Sustainable Development Summit (25–27 September 2015). According to SDG 12 ('Ensure sustainable consumption and production patterns'), EU member nations are required to 'By 2030, halve per capita global food waste at the retail and consumer levels and reduce food losses along production and supply chains, including post-harvest losses' (United Nations 2015).

In many European countries, food waste is a topic of active discussion. One of the goals of this study was to gain a clear picture, how food waste in Austria—and in particular meat waste—is affecting the climate. Our data indicated that, by making 'simple' changes in behavior, Austria could potentially save at least 4.8 Mt CO_2-eq emissions per year (i.e., 6% of Austria's total CO_2-eq emissions in 2012), without considering making any improvements in meat production or taking into account a growing number of Austrians that switch to vegetarianism or veganism. By combining all these factors, a significant reduction in Austrian (and global) GHG emissions could be achieved, and a step towards the goal to keep the rise in global average temperature below 2 centigrades compared to pre-industrial times by 2100 could be taken.

Annex: Overview of SIA Indicators and Their Values for Austria for BAU and RED Scenarios

		BAU scenario	RED scenario						
Ind01[a]		Beef/cattle	Pork/pig	Sheep (& goat)	Poultry	Beef/cattle	Pork/pig	Sheep (& goat)	Poultry
	Net national consumption (t)	207,905	491,313	7454	121,515	138,603	327,542	4969	81,010
	kg CO_2-eq/kg meat	14.2	6.0	8.4	3.5	14.2	6.0	8.4	3.5
	kg CO_2-eq of net consumption	2,952,251,000	2,947,878,000	62,613,600	425,302,500	1,968,162,600	1,965,252,000	41,739,600	283,535,000
	Sum (t CO_2-eq)				6,388,045				4,258,690
Ind02[b]		Soy import (kg)	CO_2-eq min (kg/kg of soy product)	CO_2-eq max (kg/kg of soy product)		Soy import (kg)	CO_2-eq min (kg/kg of soy product)	CO_2-eq max (kg/kg of soy product)	
	Total soy (meal) feed = 530,000 t (30,000 t from Austria)	500,000,000	0.3	17.8		333,333,333	0.3	17.8	
			150,000,000	8,900,000,000			100,000,000	5.933.333.333	
	Sum (t CO2-eq)		150,000	8,900,000			100,000	5.933.333	
Ind03[c]		Beef/cattle	Pork/pig	Sheep (& goat)	Poultry	Beef/cattle	Pork/pig	Sheep (& goat)	Poultry
	CH_4 emissions from enteric fermentation (t)	79,890	4470	2920	280	53,260	2980	1947	187
	CH_4 emissions from manure management (t)	5880	3500	70	1050	3920	2333	47	700

(continued)

(continued)

		BAU scenario	RED scenario						
	N_2O emissions from manure management (t)	1290	180	80	230	860	120	53	153
	CO_2-eq (t)	3,300,600	324,620	125,500	113,760	2,200,400	216,413	83,667	75,840
	Addit. N_2O em. (t) to agric. soil	2150				1.433			
	Addit. N_2O em. (t) to pasture, manures	300				200			
	CO_2-eq (t)	730,100				486,733			
	Sum (t CO_2-eq)				4,594,580				3,063,053
Ind04[d]		Beef/cattle	Pork/pig	Sheep (& goat)	Poultry	Beef/cattle	Pork/pig	Sheep (& goat)	Poultry
	Estimated Austrian water footprint (m^3)	208,112,905	531,109,353	3,801,540	94,781,700	138,742,604	354,072,902	2,534,190	63,187,800
	CO_2-eq (t)	170,652	435,509	3117	77,721	113,769	290,340	2078	51,814
	Sum (t CO_2-eq)				686,999				458,001
Ind05[e, g]		Beef/cattle	Pork/pig	Sheep (& goat)	Poultry	Beef/cattle	Pork/pig	Sheep (& goat)	Poultry
	Average total waste water (l)	2,156,491,489	2,487,323,964	98,134,466	6,134,542,018	1,437,660,993	1,658,215,976	65,422,977	4,089,694,679
	CO_2-eq (t)	1768	2040	80	5030	1179	1360	54	3354
	Sum (t CO_2-eq)				8919				5946

(continued)

(continued)

		BAU scenario	RED scenario						
Ind06[f]		Beef/cattle	Pork/pig	Sheep (& goat)	Poultry	Beef/cattle	Pork/pig	Sheep (& goat)	Poultry
	MJ of total meat production (average value)	8,939,915,000	12,774,138,000	No data	2,673,330,000	5,959,929,000	8,516,092,000	No data	1782.220,0000
	kWh	2,483,309,722	3,548,371,667	No data	742,591,667	1,655,535,833	2,365,581,111	No data	495,061,111
	CO_2-eq (t)	908,891	1,298,704	No data	271,789	605,926	865,803	No data	181,192
	Sum (t CO_2-eq)				2,479,384				1,652,921.79

[a]Leip et al. (2010) and Statistik Austria (2013, 2014a)
[b]Castanheira and Freire (2013), Global 2000 (n.d.) and WWF (2014)
[c]Umweltbundesamt (2014a)
[d]Mekonnen and Hoekstra (2012), Stadtwerke Karlsruhe (2014) and Statistik Austria (2013, 2014a)
[e]European Commission (2005), Stadtwerke Karlsruhe (2014) and Statistik Austria (2014a)
[f]de Vries and de Boer (2010), Statistik Austria (2014a) and Umweltbundesamt (2014b)
[g]Eutrophication potential included in Ind01

References

ARE (2004) Sustainability assessment: conceptual framework and basic methodology [cited 9 June 2015]. Available from: http://www.are.admin.ch/themen/nachhaltig/00270/index.html?lang=en

Bernhofer V (2009) Monetäre Bewertung von Lebensmittelabfällen im Restmüll aus Konsumentensicht im Untersuchungsgebiet Salzburg. Master's thesis, Wien

Castanheira EG, Freire F (2013) Greenhouse gas assessment of soybean production: implications of land use change and different cultivation systems. J Clean Prod 54:49–60

Chislock MF, Doster E, Zitomer RA, Wilson AE (2013) Eutrophication: causes, consequences, and controls in aquatic ecosystems. Nat Educ Knowl 4(4):10

De Vries M, De Boer IJM (2010) Comparing environmental impacts for livestock products: a review of life cycle assessments. Liv Sci 128:1–11

European Commission (2005) Integrated pollution prevention and control—reference document on best available techniques in the slaughterhouses and animal by-products industries

European Commission (2006) Environmental impact of products (EIPRO). Spain

European Commission (2008) Green paper on the management of bio-waste in the European Union, Brussels

European Environment Agency (2014) Approximated EU GHG inventory: proxy GHG estimates for 2013. EEA Technical report, No 16/2014, European Environment Agency, Copenhagen

European Parliament (1999) Council directive 1999/31/EC of 26 April 1999 on the landfill of waste EU Parliament, Brussels

European Parliament (2002) Regulation (EC) No 178/2002 of the parliament and of the council of 28 January 2002 laying down the general principles and requirements of food law, establishing the European food safety authority and laying down procedures in matters of food safety. EU Parliament, Brussels

European Parliament (2008) Directive 2008/98/EC of the European parliament and of the council of 19 November 2008 on waste and repealing certain directives. EU Parliament, Brussels

European Union (2011) Food: from farm to fork. European Union, Luxembourg

Eurostat (2015) Slaughtering in slaughterhouses—annual data [cited 31 May 2016]. Available from: http://ec.europa.eu/eurostat/web/products-datasets/product?code=apro_mt_pann

FAO (n.d.) Cattle body weights [cited 10 Jun 2015]. Available from: http://www.fao.org/wairdocs/ilri/x5522e/x5522e0b.htm

FAO (2011) Global food losses and food waste—extent, causes and prevention. FAO, Rome

FAO (2013) Food wastage footprint—impact on natural resources. Technical report. FAO. Rome

FAOSTAT (2015a) Emissions agriculture/enteric fermentation [cited 5 Jan 2015]. Available from: http://faostat3.fao.org/browse/G1/GE/E

FAOSTAT (2015b) Emissions agriculture/manure management [cited 5 Jan 2015]. Available from: http://faostat3.fao.org/browse/G1/GM/E

FAOSTAT (2015c) Emissions agriculture/manure applied to soils [cited 5 Jan 2015]. Available from: http://faostat3.fao.org/browse/G1/GU/E

FAOSTAT (2015d) Emissions agriculture/manure left on pasture [cited 5 Jan 2015]. Available from: http://faostat3.fao.org/browse/G1/GMP/E

Fusions (2014) FUSIONS definitional framework for food waste. Full report [cited 3 Sept 2015]. Available from: http://www.eu-fusions.org/index.php/publications?download=5:fusions-definitional-framework-for-food-waste

Gerbens-Leenes PW, Mekonnen MM, Hoekstra AY (2013) The water footprint of poultry, pork and beef: a comparative study in different countries and production systems. Water Resour Ind 1–2:25–36

Global 2000 (n.d) Fleischatlas Österreich—Zurück zum Sonntagsbraten [cited 21 Aug 2015]. Available from: https://www.global2000.at/sites/global/files/import/content/fleisch/Sonntagsbraten_Hintergrundpapier4.pdf_me/Sonntagsbraten_Hintergrundpapier4.pdf

Hinterberger F, Burger E, Sellner G (2011) Schweinfleischproduktion in Österreich—Klimaauswirkung und Ressourceneffizienz. SERI Nachhaltigkeitsforschung- und -kommunikation, Wien

Humanresearch (2015) Worldwide animal slaughter statistics [cited 10 Jun 2015]. Available from: http://www.humanresearch.org/content/worldwide-animal-slaughter-statistics

IPCC (2014) Summary for policymakers. climate change 2014: Mitigation of climate change. In: Edenhofer O, Pichs-Madruga R, Sokona Y, Farahani E, Kadner S, Seyboth K et al (eds) Contribution of working group III to the fifth assessment report of the intergovernmental panel on climate change. Cambridge University Press, Cambridge, New York

Katajajuuri J-M, Silvennoinen K, Hartikainen H, Heikkilä L, Reinikainen A (2014) Food waste in the Finnish food chain. J Clean Prod 73: 322–329. doi:10.1016/j.jclepro.2013.12.057

Kranert M, Schneider F, Hafner G, Lebersorger S, Barabosz J, Scherhaufer S et al (2012) Ermittlung der weggeworfenen Lebensmittelmengen und Vorschläge zur Minderung der Wegwerfrate bei Lebensmitteln in Deutschland [Identification of food waste amounts and suggestions for minimizing food waste in Germany]. Bundesministerium für Ernährung, Landwirtschaft und Verbraucherschutz, Stuttgart

Leip A, Weiss F, Wassenaar T, Perez I, Fellmann T, Loudjani P et al (2010) Evaluation of the livestock sector's contribution to the EU greenhouse gas emissions (GGELS), Administrative Arrangements AGRI-2008–0245 and AGRI-2009-0296. European Commission, Joint Research Center, Italy

Lesschen JP, van den Berg M, Westhoek HJ, Witzke HP, Oenema O (2011) Greenhouse gas emission profiles of European livestock sectors. Anim Feed Sci Technol 166–167:16–28

MA 48 (2014) Leistungsbericht 2014. MA 48—Abfallwirtschaft, Straßenreinigung und Fuhrpark, Wien

Mekonnen MM, Hoekstra AY (2012) A global assessment of the water footprint of farm animal products. Ecosystems 15(3):401–415

Monier V, Mudgal S, Escalon V, Reisinger H, Dolley P, Ogilvie S, et al (2010) Preparatory study on food waste across EU 27. Technical report—2010-054, European Commission, Brussels

Moveforhunger (2016) France cracks down on food waste [cited 30 May 2016]. Available from https://moveforhunger.org/france-cracks-down-on-food-waste/

Ness B, Urbel-Piirsalu E, Anderberg S, Olsson L (2007) Ecol Econ 60:498–508

Nguyen TP (2012) Greenhouse gas emissions from composting and anaerobic digestion plants. Dissertation. Rheinische Friedrich-Wilhelms-Universität, Bonn

Obersteiner G, Schneider F (2006) NÖ Restmüllanalysen 2005/06. Studie im Auftrag des NÖ Abfallwirtschaftsverein. Wien

OECD (2010) Guidance on sustainability impact assessments. OECD, Paris

Papargyropoulou E, Lozano R, Steinberger JK, Wright N, Ujang ZB (2014) The food waste hierarchy as a framework for the management of food surplus and food waste. J Clean Prod 76:106–115

Parfitt J, Barthel M, Macnaughton S (2010) Food waste within food supply chains: quantification and potential for change to 2050. Phil Trans R Soc B 365:3065–3081

Partito Democratico (2015) Spreco zero, in commissione la legge Pd [cited 27 Aug 2015]. Available from: http://www.mariachiaragadda.it/spreco-zero-in-commissione-la-legge-pd/

Pope J, Annandale D, Morrison-Saunders A (2004) Environ Impact Asses 24:595–616

Quested T, Johnson H (2009) Household food and drink waste in the UK. WRAP, Oxon

Ridoutt BG, Sanguansri P, Harper GS (2011) Comparing carbon and water footprints for beef cattle production in Southern Australia. Sustainability 3(12):2443–2455

Salhofer S, Obersteiner G, Schneider F, Lebersorger S (2008) Potentials for the prevention of municipal solid waste. Waste Manag 28(2):245–259

Schneider F, Lebersorger S (2009) Untersuchung der Lebensmittel im Restmüll in einer oberösterreichischen Region. Amt der OÖ Landesregierung, Direktion Umwelt und Wasserwirtschaft, Linz

Selzer MM (2010) Die Entsorgung von Lebensmittel in Haushalten: Ursachen—Flüsse—Zusammenhänge. Diploma thesis. Wien

Senat (2015) Séance du 26 mai 2015 (compte rendu intégral des débats) [cited 8 Jun 2015]. Available from: http://www.senat.fr/seances/s201505/s20150526/s20150526_mono.html#par_116
Singh RK, Murty HR, Gupta SK, Dikshit AK (2012) An overview of sustainability assessment methodologies. Ecol Indic 9:189–212
Stadtwerke Karlsruhe (2014) Karlsruher Trinkwasser—nahezu klimaneutral [cited 21 Aug 2015]. Available from: www.stadtwerke-karlsruhe.de/swka-de/.../trinkwasser-klimaneutral.pdf
Statistik Austria (2012) Der Außenhandel Österreichs. Statistik Austria, Wien
Statistik Austria (2013) Versorgungsbilanzen für tierische Produkte. Statistik Austria, Wien
Statistik Austria (2014a) Lebend- & Schlachtgewichte—Jahresergebnis 2014. Statistik Austria, Wien
Statistik Austria (2014b) Statistik der Landwirtschaft. Statistik Austria, Wien
Steinfeld H, Gerber P, Wassenaar T, Castel V, Rosales M, de Haan C (2006) Livestock's long shadow: environmental issues and options. Food and Agriculture Organization of the United Nations (FAO), Rome
TAB (2015) Büro für Technikfolgenabschätzung beim Deutschen Bundestag. TAB-Fokus No. 6 regarding Report No. 163. Opportunities and criteria for a sustainability level [cited 3 Sept 2015]. Available at: http://www.tab-beim-bundestag.de/en/pdf/publications/tab-fokus/TAB-Fokus-006.pdf
Umweltbundesamt (2014a) Austria's National Inventory Report 2014. REP-0475. BMLFUW, Wien, Vienna
Umweltbundesamt (2014b) Berechnung von Treibhausgas (THG)-Emissionen verschiedener Energieträger [cited 21 Aug 2015]. Available from: http://www5.umweltbundesamt.at/emas/co2mon/co2mon.htm#Berechnungen
UNFCCC (2006) Updated UNFCCC reporting guidelines on annual inventories following incorporation of the provisions of decision 14/CP.11. United Nations Framework Convention on Climate Change, Note by the secretariat, FCCC/SBSTA/2006/9
United Nations (2015) United Nations sustainable development summit 2015 [cited 3 Sept 2015]. Available from: http://www.un.org/sustainabledevelopment/sustainable-development-goals/
USDA (2012) Food processing ingredients—an overview on the austrian food processing sector. USDA Foreign Agriculture Service, Washington
Weiss F, Leip A (2012) Greenhouse gas emissions from the EU livestock sector: a life cycle assessment carried out with the CAPRI model. Agric Ecosyst Environ 149:124–134
Winkler T, Schopf K, Aschemann R, Winiwarter W (2016) From farm to fork—a life cycle assessment of fresh Austrian pork. J Clean Prod 116:80–89. doi:10.1016/j.jclepro.2016.01.005
Winkler T, Winiwarter W (2015) Greenhouse gas emission scenarios of livestock in Austria. J Int Env Sci 12:107–119. doi:10.1080/1943815X.2015.1110186
WWF (2014) Soy report card—assessing the use of responsible soy for animal feed in Europe. WWF, Gland, Switzerland

Chapter 5
Fodder Legumes for Green Biorefineries: A Perspective for Sustainable Agricultural Production Systems

Franka Papendiek

Abstract Peak oil is forcing our society to shift from fossil to renewable resources. However, such renewable resources are also scarce, and they too must be used in the most efficient and sustainable way possible. Biorefining is a concept that represents both resource efficiency (waste reduction) and sustainability. This approach initiates a cascade use, which means food and feed production before material use, and an energy-related use at the end of the value-added chain. However, sustainability must already start in the fields, on the agricultural side, where the industrially-used biomass is produced. The highest premise of the study was to develop an agricultural production system that is more sustainable than existing ones. Fodder legumes, produced in expanded crop rotations are cultivated. They have a very positive environmental impact in agricultural production systems. They are used as bio-industrial feedstock and fodder in the Green Biorefinery approach. Following evidence that both intermediate products are suitable in the biorefining process, a cost-benefit analysis, comparing different production scenarios on a farm, showed that for large farm sizes in particular, the potential profits are high. Therefore, all three pillars of sustainability in agricultural production systems can be improved.

Keywords Lactic acid · Lucerne · Field trials · Cost-Benefit-Analysis

5.1 Introduction

If we want to stop the anthropogenic impact on climate change, as the last climate conference in Paris 2015 concluded, our society faces the necessity to shift from fossil to renewable resources. Biorefinery is a concept that can help to implement this shift. Picking up the idea of oil refineries, the aim is to maximise outputs in the

F. Papendiek (✉)
Head Office of the German Council of Science and Humanities, Research Infrastructure (Forschungsinfrastrukturen), Brohler Straße 11, 50968 Cologne, Germany
e-mail: papendiek@wissenschaftsrat.de

P. Morone et al. (eds.), *Food Waste Reduction and Valorisation*,
DOI 10.1007/978-3-319-50088-1_5

processing of feedstocks, in this case biomass and residuals (Lin et al. 2013). This approach reduces waste to a minimum and fosters a cascade use. This means building blocks, existing in the plants, are used with the highest added value for food and feed production before material use and an energy-related use at the end of the value-added chain. The necessity of sustainability already starts at the raw material provision, on agriculture fields where the industrially-used biomass is produced. The point is not to increase pressure on agricultural land by raising demand for biomass, but instead should support sustainable production systems.

The aim of the study presented in this chapter was to develop a sustainable value chain for the Green Biorefinery approach (Fig. 5.1). Put another way, the approach may be thought of as a concept to connect biorefineries with a sustainable supply of feedstock.

Fodder legumes from temporary and permanent grasslands in the federal state of Brandenburg (Germany) were used as feedstock. They have a positive environmental impact in agricultural production systems and therewith support sustainability.

The generated products of the value chain were also chosen for their sustainability potential. Indigenous feed for ruminants is needed. Only 3% of arable land in the EU is occupied by protein plants, producing only 30% of protein crops consumed as animal feed (Committee on Agriculture and Rural Development 2011). The generated lactic acid can be used as substitute for fossil based products, such as plastics (Papendiek and Venus 2014).

To ensure the sustainability of the value chain, the following research objectives had to be answered:

- What is the biomass potential from grasslands in Brandenburg (Germany) for utilisation within Green Biorefineries?
- Are the examined fodder legumes, namely alfalfa and clover/grass a worthwhile feedstock for Green Biorefineries?
- Following evidence that they are a worthwhile feedstock for Green Biorefineries,

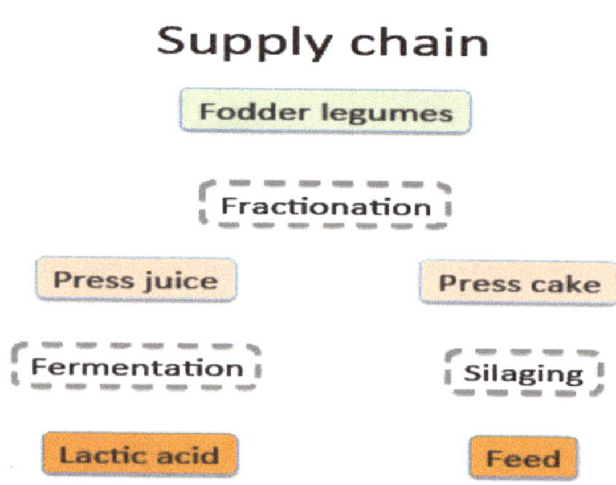

Fig. 5.1 Sustainable supply chain for a green biorefinery

- Can they be produced in an economically sound manner?

The remainder of the chapter is structured as follows: Sect. 5.2 gives an overview on 3rd generation Biorefinery and the associated circular economy concept; Sect. 5.3 describes the methods; in Sects. 5.4 and 5.5 the sustainability of the value chain is analysed; results on the field trials and the economic analysis of the value chain are illustrated in Sect. 5.6; before a conclusion finalizes this chapter.

5.2 Circular Economy in 3rd Generation Biorefineries

As already addressed in the Introduction Section, biorefinery is an approach that reduces waste to a minimum and embodies the circular economy. The maximum value of the resource is extracted by using it for products with a high added value and afterwards using spent articles as resource for other manufacturing (WRAP 2015).

In a biorefinery any kind of biomass, including waste is converted into a spectrum of marketable products (food, feed, chemical, biomaterials) and energy (de Jong et al. 2009). 3rd generation biorefineries are the most developed/advanced type of biorefineries, using various types of feedstocks and processing technologies to generate a diverse range of products (Clark and Deswarte 2015).

Biorefinery systems are classified by quoting the involved platforms, products, feedstocks or processes (de Jong et al. 2009). Currently, five of those 3rd generation biorefinery systems are examined in research and development, namely lignocellulosic feedstock Biorefinery, whole crop Biorefinery, two-platform Biorefinery, marine Biorefinery and Green Biorefinery (Clark and Deswarte 2015).

This chapter deals with the Green Biorefinery. Here, feedstocks formerly used (literally) as feed are processed (Fig. 5.1). The fresh green biomass is fractionated with a screw press. The arising press cake can be used to produce, for example feed (Bryant et al. 1983; Lu et al. 1979). The press juice, on the other hand, can be used as substitute for synthetic compounds in existing biotechnological processes like the lactic acid production (Venus 2006).

5.3 Methods

5.3.1 Biomass Potential from Grasslands in the Federal State of Brandenburg (Germany)

The biomass potential for grasslands in the region of Brandenburg (Germany) was determined, using numbers from the Office for statistics Berlin-Brandenburg between 1990 and 2012 (Papendiek et al. 2012). To get a more detailed view on how the grasslands are used today, we carried out a survey of farmers in

Brandenburg, having more than 200 ha of agricultural area managed. We asked them how much permanent and temporary grassland they have and how they use it.

5.3.2 Fodder Legumes as Feedstock for the Green Biorefinery

To analyse if alfalfa and clover/grass are valuable feedstocks to produce feed and lactic acid from it (Fig. 5.1), field trials were carried out at different study sites in the federal state of Brandenburg (Germany) (Papendiek and Venus 2014). Alfalfa (*Medicago sativa*) was cultivated on arable land at field stations of the Leibniz-Centre for Agricultural Landscape Research (ZALF) in Muencheberg (coordinates: 52.516045, 14.124929) and Paulinenaue (coordinates 52.683381, 12.685897). In addition to planting alfalfa on arable land, a clover/grass mixture (*Lolio-Cynosuretum*) was cultivated on permanent grassland at Paulinenaue. Therewith, we obtained information on biomass quality and quantity depending on the crop, study site and harvest time. Afterwards, the biomass was fractionated, varying the pressing process in an effort to analyse differences in the composition of both resulting compounds (juice and cake). The juice was utilised as fermentation medium in lactic acid production and the feed potential of the press cake was determined (Papendiek and Venus 2014).

5.3.3 The Economic Viability of Fodder Legume Cultivation for Green Biorefineries

After providing evidence that both lactic acid and feed can be produced from the biomass observed, the development of a cost-benefit model allowed us to analyse the economic profitability of the approach (Papendiek et al. 2015). The cost-benefit analysis compared different production scenarios on a farm. Two standard crop rotations for Brandenburg, producing either only market crops or market crops and fodder legumes for ruminant feed production were compared to a system that uses the cultivated fodder legumes for the Green Biorefinery value chain instead of only feed production. Two farm sizes (210 and 420 ha), common for many European regions, were chosen to examine the influence of scale. The cost structure of the farms was analysed in detail to assess which farm characteristics make the production of press juices for biochemical industries viable.

The analyses of all research objectives thus build on one another. The core value of this study is the resulting comprehensive perspective on the entire value chain of fodder legumes in the Green Biorefinery approach.

5.4 Sustainability of Feedstock in the Value Chain

5.4.1 Sustainable Biomass Supply from Grasslands

More than 30% of the agricultural land in Germany is grasslands which are often inefficiently used agricultural production systems (DAFA 2015; Koschuh et al. 2003). In many European countries, grasslands are endangered due to abandonment or conversion into arable land (Gerowitt et al. 2013). However, since the proportion of natural conservation areas for grasslands is high compared to all agriculturally used land, these areas are very important for the conservation of nature (Becker et al. 2014). Therefore, the challenge for the use of grasslands is to combine provisioning services (i.e. feed) and non-provisioning services (i.e. biodiversity) to make the management of grasslands more attractive for farmers.

Statistics for the federal state of Brandenburg show that the number of cattle using grasslands declined by 50% between 1990 and 2010 while the numbers for permanent grasslands remained stable (Papendiek et al. 2012). The evaluation of the survey showed that over 80% of farmers having permanent grassland sites are using them as extensive pasture. A sustainable intensification seems possible. As in Green Biorefineries, the maintenance of biodiversity is coupled with the more intensive use of these sites.

Next to permanent grasslands there are temporary sites, using agricultural fields for the production of fresh green biomass. Temporary grasslands deliver feed for ruminants. However, cultivation numbers decreased dramatically since 1990 (Amt für Statistik Berlin-Brandenburg 2010b). Milk yield more than doubled between 1990 and 2010 (Amt für Statistik Berlin-Brandenburg, 2010a), increasing the concentrate feed ratio in cattle feed, while reducing the demand for structurized feed as grass. Using the fresh green biomass for the Green Biorefinery approach could increase cultivation numbers for temporary grasslands. Temporary grasslands, especially cultivated with fodder legumes, are an important part of sustainable crop rotations and therefore needed in agricultural production systems.

5.4.2 Sustainability of Fodder Legumes

Legumes convert and use atmospheric nitrogen by means of nodule bacteria so that, in general, mineral nitrogen fertilisers are not necessary (National Research Council 2002). Legumes previously were an essential element of crop rotations before mineral fertilisers became available at reasonable prices. However, their impact on the agricultural production system is more diverse than just delivering nitrogen. The perennial cultivation of fodder legumes on arable land promotes the accumulation of carbon in soils (Jensen et al. 2012) and impedes the spread of pests and diseases in cereal cultivars (Malézieux et al. 2009). The well-branched root system of the perennial plants increases the water infiltration capacity, reducing erosion risk in heavy rain events (Freyer 2003). In addition, nutrient leaching will only rarely appear, because the root system takes up nutrients before they are transferred to the groundwater or into other ecosystems (Robertson et al. 2011). Moreover, the root system takes up nutrients, i.e. phosphorus, from the deep soil layers (Kahnt 2008). These nutrients can be used by the plant or are stored for following crops, subsequently reducing the demand for mineral fertilisers throughout the whole crop rotation (Parajuli et al. 2015). Along with these benefits, soil fertility is increased; as a result, grain crop yields and grain quality for the succeeding crops are improved (Gooding et al. 2007; Grzebisz et al. 2001; Hejcman et al. 2012).

Unfortunately cultivation figures do not yet reflect these benefits of legume cultivation.

5.5 Sustainability of Products in the Value Chain

Products generated in the analysed value chain are feed for ruminants and lactic acid (Fig. 5.1). Lactic acid (2-hydroxypropionic acid) is a promising platform chemical that can be produced from a carbon source (i.e. cereals) by using press juices as fermentation medium. Lactic acid is for example applied in the production of polylactic acid (PLA), which is a bioplastic that has the potential to substitute ample amounts of petroleum-based plastics in the future (Jim Jem et al. 2010; Madhavan Nampoothiri et al. 2010). There are moves afoot within the European Union to drastically reduce plastic bag utilization (Council of the European Union 2014) and bioplastic is an alternative especially for lightweight plastic carrier bags that are endangering the environment. Already today, bioplastics play an important role in the field of packaging, agriculture, gastronomy and automotive (European Bioplastics 2012). In 2013, the demand for lactic acid was estimated at 714,000 t and it is expected to further increase at an annual rate of 15.5% between 2014 and 2020, mainly as a result of the growing demand for bioplastics (Abdel-Rahman et al. 2013; SpecialChem 2014). Hence, there is a market for lactic acid with positive future prospects.

The production of feed for ruminants is the traditional usage of fodder legumes. To study the potential of re-establishing this use option, it was integrated in the study. Today, typical indigenous fodder legumes, like alfalfa and clover, have been replaced in animal nutrition by soy meal from Latin and South America, and are therefore no longer cultivated in Germany. The increased production of ruminant feed in Germany and Europe would also increase sustainability since transport distances are reduced and land use change in the tropics is reduced. However, economically viable production in conventional farming does not seem to be assured. Politicians have recognised the problem, and strategies for legume support are already in existence or are under development (BMELV 2012; Committee on Agriculture and Rural Development 2011; Schreuder and De Visser 2014). However, these strategies will only make an impact when use options and markets for these crops exist. Therefore, new utilisation concepts generating products with a higher added value are needed.

5.6 Fodder Legumes as Feedstock for Green Biorefineries

5.6.1 *The Feasibility Aspect*

Investigations performed in this thesis proved the suitability of alfalfa and clover/grass for the Green Biorefinery approach. All press juices were a proper feedstock for lactic acid production (Papendiek 2016). Results show that the harvest time is not of high importance for the quality of press juices as a fermentation medium. However, biomass quantity is the limiting factor for the expanse of the harvest window. If plants are still too small or the dry matter content is already too high, juice quantities are negatively affected. Thus, harvest time has to be adapted to weather conditions and other external influences to generate sufficient quantities of biomass.

Regarding crop choice, results show that alfalfa performed better in direct comparison to clover/grass with regard to biomass quantity and quality. However, the results obtained for clover/grass also show a high potential for permanent grassland sites. Biomass quantities are still attractive for the demand in Green Biorefineries, since the quality as fermentation medium is only marginally lower and the press cake is still an appropriate feed for dairy cows during their dry period.

5.6.2 *The Economic Viability*

A key task of this study was to find out if fodder legume production can become profitable for farmers again when a new purchaser, in the form of biochemical industries, appears. Therefore, we carried out a cost-benefit analysis, using data from field trials. This study analyses the potential benefits–and risks–of a new market for legumes. Such information is essential for farmers to assess whether they want to revive the production of fodder legumes on their farms.

Results show that for large farm sizes in particular, the potential profits are high (Table 5.1).

The cost-benefit analysis was carried out for alfalfa on arable land only. The calculated IRRs are not comparable with reality because of the exclusion of baseline

Table 5.1 Internal rate of return for specific scenarios and farm sizes

Scenario	Farm size 210 ha (%)	Farm size 420 ha (%)
State-of-the-art scenario without fodder production	26	26
State-of-the-art scenario with fodder production	41	41
Green biorefinery scenario[a]	15	49

[a]For the most likely juice price of 1300 € t^{-1}

costs. However, the IRR comparability between the three scenarios is still sound, because baseline costs in all scenarios would be the same.

Farmers can therefore integrate both crops into their harvest schedule without overlapping with cereal harvest times. Another benefit of the broad harvest window that we discovered is that not all fodder legume fields need to be harvested at the same time, reducing the demand for industrial presses and respective labour force. As analysed in the cost-benefit analysis (Papendiek et al. 2015), these factors are highly relevant for the viability of press juice production on the farm.

5.7 Conclusions

The overall aim of this chapter is to encourage sustainability in the provision of available feedstock for biochemical industries and to increase resource efficiency. To meet these requirements, a specific value chain, attuned to sustainability issues was analysed (Fig. 5.1).

Perennially produced fodder legumes support non-commodity outputs within the perspective of environmentally sound, sustainable agricultural production systems. First, processing takes place on the farm to generate a high-value juice and to retain the residual press cake as feed for ruminants on the farm itself or a farm nearby.

The accruing products can improve environmental sustainability. The derived feed can partly substitute for imported soy meal, and the lactic acid could be used for bioplastic production, an eco-friendly packaging alternative. The more sophisticated value chain helps a relevant proportion of the added value to stay in the rural area, which improves social sustainability. The economic viability of the value chain, as a crucial part of sustainability, was also explored in this study and the economic potential has been proved.

References

Abdel-Rahman MA, Tashiro Y, Sonomoto K (2013) Recent advances in lactic acid production by microbial fermentation processes. Biotechnol Adv 31:877–902. doi:10.1016/j.biotechadv.2013.04.002

Amt für Statistik Berlin-Brandenburg (2010a) Time series on Viehbestand und tierische Erzeugung im Land Brandenburg (1990–2009). Dataset provided on request

Amt für Statistik Berlin-Brandenburg (2010b) Time series on Bodennutzung der landwirtschafltichen Betriebe im Land Brandenburg (1990–2009). Dataset provided on request

Becker N, Emde F, Jessel B, Kärcher A, Schuster B, Seifert C, Ammermann K, Balzer S, Benzler A, Dietrich K, Dünnfelder H, Ellwanger G, Finck P, Heym A, Job-Hoben B, Krug A, Mues A, Natho S, Schweppe-Kraft B, SSymank A, Strauß C, Vischer-Leopold M, Zueghart W (2014) Grünland-report - Alles im Grünen Bereich? Bundesamt für Naturschutz, Bonn

BMELV (2012) Eiweißpflanzenstrategie des BMELV. Bundesministerium für Ernährung, Landwirtschaft u. Verbraucherschutz (BMELV), Berlin, Germany, p 13

Bryant AM, Carruthers VR, Trigg TE (1983) Nutritive value of pressed herbage residues for lactating dairy cows. New Zealand J Agric Res 26:79–84. doi:10.1080/00288233.1983.10420954

Clark JH, Deswarte FEI (2015) The biorefinery concept: an integrated approach. In: Clark JH, Deswarte FEI (eds) Introduction to chemicals from biomass. Wiley, New York, p 344

Committee on Agriculture and Rural Development (2011) The EU protein deficit: what solution for a long-standing problem? (2010/2111(INI)). European Parliament, Straßburg, France 17

Council of the European Union (2014) Use of plastic bags: agreement on phasing down. P. Release, Brussels

DAFA (2015) The grassland expert forum—draft, Deutsche Agrarforschungsallianz (DAFA) c/o Johann-Heinrich-von-Thünen-Inst., Braunschweig, Germany

de Jong E, Langeveld H, van Ree R (eds) (2009) IEA bioenergy task 42 biorefinery. pp 1–26. Available online at: http://www.biorefinery.nl/fileadmin/biorefinery/docs/Brochure_Totaal_definitief_HR_opt.pdf, last access 15 May 2010

European Bioplastics (2012) What are bioplastics? Fact sheet. http://en.european-bioplastics.org/wp-content/uploads/2011/04/fs/Bioplastics_eng.pdf, last access: 13 Aug 2012

Freyer B (2003) Fruchtfolgen – Konventionell, Integriert. Biologisch Eugen Ulmer GmbH & Co., Stuttgart, Germany

Gerowitt B, Schröder S, Dempfle L, Engels E-M, Engels J, Feindt PH, Graner A, Hamm U, Heißenhuber A, Schulte-Coerne H, Wolters V, Scientific Advisory Board for Biodiversity and Genetic Resources at the Federal Ministry of Consumer Protection F.a.A.B. (2013) Biodiversity in grasslands—indispensable for agriculture and society. Federal Ministry of Consumer Protection, Food and Agriculture (BMELV), p 20

Gooding MJ, Kasyanova E, Ruske R, Hauggaard-Nielsen H, Jensen ES, Dahlmann C, Von Fragstein P, Dibet A, Corre-Hellou G, Crozat Y, Pristerf A, Romeo M, Monti M, Launay M (2007) Intercropping with pulses to concentrate nitrogen and sulphur in wheat. J Agric Sci 145:469–479. doi:10.1017/s0021859607007241

Grzebisz W, Kryszak J, Szczepaniak W, Gaj R (2001) Cultivation of legumes and grass/legume mixtures on arable land as a sustainable management. In: Horst WJ et al (eds) Plant nutrition: food security and sustainability of agro-ecosystems through basic and applied research. Kluwer Academic Publishers, Dordrecht, pp 1006–1007

Hejcman M, Kunzova E, Srek P (2012) Sustainability of winter wheat production over 50 years of crop rotation and N, P and K fertilizer application on illimerized luvisol in the Czech Republic. Field Crops Res 139:30–38. doi:10.1016/j.fcr.2012.10.005

Jensen ES, Peoples MB, Boddey RM, Gresshoff PM, Hauggaard-Nielsen H, Alves BJR, Morrison MJ (2012) Legumes for mitigation of climate change and the provision of feedstock for biofuels and biorefineries. A review. Agron Sustainable Dev 32:329–364. doi:10.1007/s13593-011-0056-7

Jim Jem K, van der Pol JF, de Vos S (2010) Microbial lactic acid, its polymer poly(lactic acid) and their industrial applications. Plastics from Bacteria Microbiology Monographs 323–346

Kahnt G (2008) Leguminosen im konventionellen und ökologischen Landbau DLG Verlags-GmbH, Frankfurt am Main, Germany

Koschuh W, Kromus S, Krotscheck C (2003) Grüne Bioraffinerie - Gewinnung von Proteinen aus Grassäften, Berichte aus Energie- und Umweltforschung, Kornberg Institut. pp 1–256

Lin CSK, Pfaltzgraff LA, Herrero-Davila L, Mubofu EB, Abderrahim S, Clark JH, Koutinas AA, Kopsahelis N, Stamatelatou K, Dickson F, Thankappan S, Mohamed Z, Brocklesby R, Luque R (2013) Food waste as a valuable resource for the production of chemicals, materials and fuels. Current situation and global perspective. Energy Environ Sci 6:426–464. doi:10.1039/c2ee23440h

Lu CD, Jorgensen NA, Barrington GP (1979) Wet fractionation process: preservation and utilization of pressed alfalfa forage. J Dairy Sci 62:1399–1407

Madhavan NK, Nair NR, John RP (2010) An overview of the recent developments in polylactide (PLA) research. Bioresour Technol 101:8493–8501. doi:10.1016/j.biortech.2010.05.092

Malézieux E, Crozat Y, Dupraz C, Laurans M, Makowski D, Ozier-Lafontaine H, Rapidel B, de Tourdonnet S, Valantin-Morison M (2009) Mixing plant species in cropping systems: concepts, tools and models. A review. Agron Sustain Dev 29:43–62. doi:10.1051/agro:2007057

National Research Council (2002) Tropical legumes: resources for the future books for business. New York, Hong Kong

Papendiek F (2016) Fodder legumes for green biorefineries: a perspective for sustainable agricultural production systems. University Potsdam, Germany

Papendiek F, Venus J (2014) Cultivation and fractionation of leguminous biomass for lactic acid production. Chem Biochem Eng Q 28:375–382. doi:10.15255/CABEQ.2013.1854

Papendiek F, Ende HP, Steinhardt U, Wiggering H (2012) Biorefineries: relocating biomass refineries to the rural areas. Landscape Online 27:9

Papendiek F, Tartiu VE, Morone P, Venus J, Hönig A (2015) Assessing the economic profitability of fodder legume production for green biorefineries—a cost-benefit analysis to evaluate farmers profitability. J Clean Prod. doi:10.1016/j.jclepro.2015.07.108

Parajuli R, Dalgaard T, Jorgensen U, Adamsen APS, Knudsen MT, Birkved M, Gylling M, Schjorring JK (2015) Biorefining in the prevailing energy and materials crisis: a review of sustainable pathways for biorefinery value chains and sustainability assessment methodologies. Renew Sustain Energy Rev 43:244–263. doi:10.1016/j.rser.2014.11.041

Robertson GP, Hamilton SK, Del Grosso SJ, Parton WJ (2011) The biogeochemistry of bioenergy landscapes: carbon, nitrogen, and water considerations. Ecol Appl 21(4):1055–1067. doi:10.1890/09-0456.1

Schreuder R, De Visser C (2014) Report EIP-Agri focus group protein crops: final report. European Commission, Brussels, Belgium 49

SpecialChem (2014) Global lactic acid market to grow at a CAGR of 15.5% from 2014–20: Grand view research 2014. http://www.specialchem4bio.com/news/2014/05/23/global-lactic-acid-market-to-grow-at-a-cagr-of-15-5-from-2014-20-grand-view-research, last access: 30.03.2015

Venus J (2006) Utilization of renewables for lactic acid fermentation. Biotechnol J 1:1428–1432. doi:10.1002/biot.200600180

WRAP (2015) What is a circular economy? http://www.wrap.org.uk/about-us/about/wrap-and-circular-economy

Chapter 6
Municipal Waste Treatment, Technological Scale up and Commercial Exploitation: The Case of Bio-waste Lignin to Soluble Lignin-like Polymers

Enzo Montoneri

Abstract The present chapter addresses municipal bio-waste, as worldwide easily available concentrated source of organic matter to convert to and recycle as valuable products for further use. Municipal bio-waste contains polysaccharides and lignin as major components. On the other hand, these are major components of biomass, generally. This implies that technology used for treating municipal bio-waste is likely applicable to other bio wastes, as well. Current biomass treatment technology addresses mainly the production of biofuel by fermentation of the polysaccharide fraction. Lignin is an insoluble recalcitrant material withstanding biochemical and chemical treatment. It inhibits fermentation microorganisms. Thus, the separation of lignin from the fermentable organic fractions is necessary. In addition, the separated lignin is regarded as secondary process waste, which needs disposal. A number of technologies are currently available for this purpose. These include lignin combustion, pyrolysis, hydrocracking, or aerobic fermentation. Yet, the bio-waste lignin fraction has further potential that can be exploited by low energy consumption chemical technology. The valorisation of lignin in this fashion would contribute important economic and environmental improvements to current waste treatment practices. Taking an Italian municipal bio-waste treatment plant as empirical case study, the present chapter reviews work performed in the last decade for the valorisation of lignin originating from the organic humid fraction and gardening residues obtained from the separate source collection of municipal bio-wastes. The work covers also agriculture residues, although in a relatively very limited extent. The chapter reports processes and applications related to new speciality chemicals stemming for research developed at EU technology readiness level 5. The results prospect sustainable processes and products, and the possibility to realize a business model with reduced entrepreneurial risk for the conversion of a municipal bio-waste treatment plant to biorefinery producing fuel and bio-based chemicals. However, the chapter does not provide the reader with a strong methodology for evaluating the potential sustainability. In addition, the proposed

E. Montoneri (✉)
Bio-Waste Processing, Via XXIV Maggio 25, 37126 Verona, Italy
e-mail: enzo.montoneri@gmail.com

P. Morone et al. (eds.), *Food Waste Reduction and Valorisation*,
DOI 10.1007/978-3-319-50088-1_6

business model with reduced entrepreneurial risk is at a very early stage. It relies mostly on assumptions that need validation. The results related to agriculture residues demonstrate that, although mainly focused on municipal bio-waste, the developed technology is applicable as well to other bio-waste types.

Keywords Muncipal bio-waste · Bio-waste valorisation · Bio-based specility chemicals · Biosurfactants · Biopolymers · Biostimulants

6.1 Introduction

Solid bio-waste originates from industrial, agriculture and urban activities. Industrial bio-waste includes wastes from the food processing and pharmaceutical industry. Bio-wastes from agriculture activities are post-harvest plants and residual fruit left in the field. Municipal solid wastes contains from 18 to 60% bio-waste (Twardowska et al. 2010). It comprises food wastes and green wastes from private gardening and public park trimming activities in approximate 1.6 ratio, respectively (David et al. 2010; Ricci-Jürgensen 2012). Food waste is a major contributor to total waste production. Food wastes are often not properly treated and recycled, unlike recyclable materials such as papers. Hence, food waste often ends up in landfill along with regular waste. This creates alarming impact on environment and health hazard as methane and bacteria build up from food waste in landfill.

In the last decade, public opinion sensitivity to food waste issues has grown. This involves both fabrication and product distribution (Segre et al. 2009; Segre and Gaiani 2011). Food waste has two opposite faces. On one hand, it represents an economic and environmental burden. On the other hand, it contains valuable chemical exploitable energy. Food waste come from households, restaurants, food manufacturers, and farms (David et al. 2010) at the stages of food production, processing, retailing and consumption. As of 2013, half of all food is wasted worldwide (Huffington Post 2013; FAO 2011). Loss and wastage occurs at all stages of the food supply chain or value chain. In low-income countries, most loss (81–97% of total food waste) occurs during production, while in developed countries much food waste occurs at the consumption stage (about 100 kg per person per day, amounting to 32–60% of total food waste). In Europe, the total 89 million t food loss and waste per year arises 47% from household, 16% from catering, 6% from retail and wholesale and 44% from manufacturing activities (Barilla Center for Nutrition 2012).

The above data points out that, for abundance and easy availability, the organic humid fraction of urban wastes is potentially the most convenient exploitable source of recyclable renewable organic matter. According to various statistics, American families throw out between 14 and 25% of the food and beverages they buy. This can cost the average family $1365–$2275 annually (Plumer 2012). The majority of waste from households consists of food wastes, close to 60% (David et al. 2010). Their environmental impact has grown dramatically, due to the increase of population urbanization and consumption habits. This has generated higher costs for society due to the need to dispose higher amounts of wastes. On the other hand, the population urbanization has resulted in the creation of a low entropy source of chemical energy by concentrating the bio-wastes in confined spaces. As taxpayers have already paid collection costs, municipal bio-wastes are a negative cost source of chemical energy (Sheldon-Coulson 2011).

Several technologies are in principle available to recycle and exploit the potential chemical energy of bio-waste for the production of thermal and electrical energy, and of value added chemicals. However, the removal (Canilha et al. 2012; Parsell et al. 2015; Liew 2011; Arato et al. 2005) and conversion of lignin to benefit products (Clark 2007; Ma et al. 2014) is a critical point and a major issue for the valorisation of dedicated or residual biomass as source of renewable fuels and chemicals. This is because lignin inhibits fermentation microorganisms, and is an insoluble recalcitrant material withstanding biochemical and chemical treatment. Lignin is the second most abundant organic component next to cellulose in the vegetable world. The emerging biomass refinery industry will inevitably generate an enormous amount of lignin. Development of selective biorefinery lignin-to-bioproducts conversion processes will play a pivotal role in significantly improving the economic feasibility and sustainability of biofuel production from renewable biomass.

Current biomass treatment technology (Canilha et al. 2012; van Ree and van Zeeland 2014) mainly focused to the production of biofuel by fermentation, such as biogas and bioethanol, adopts several biomass pretreatment methods to remove lignin from the fermentable fraction and/or processes the residual lignin fraction by combustion, pyrolysis, hydrocracking, or aerobic fermentation. These processes, respectively, convert the chemical energy to thermal and electric energy, produce hydrocarbons and other platform chemicals, and compost for landscaping and/or soil fertilization use (Luque and Clark 2013). Yet, the bio-waste lignin fraction has further potential (Ragauskas et al. 2014) that can be exploited by low energy consumption chemical technology. The valorisation of lignin in this fashion would contribute important economic and environmental improvements to current waste treatment practices.

The present chapter reviews work performed for urban and agriculture bio-waste lignin valorisation (LV) to bio-based chemicals, in connection with the state of art of bio-waste management technology and related economic and environmental aspects. It shows potential process and products sustainability stemming from research developed at technology readiness level 5 (Nasa 2015; European Commission 2014). It points out how integrated biochemical and low energy consumption green chemical processes may contribute to the realization of a sustainable biorefinery producing fuel and chemicals from bio-waste. Hereinafter, the chapter comprises different Sects. (6.2–6.5). Section 6.2 reviews the state of art of bio-waste management technology. It describes also a typical waste treatment plan located in Italy, the Acea Plant. This plant treats municipal bio-waste through the most advanced integrated anaerobic and aerobic fermentation technology. For this reason, in the performed research work, the Acea plant represents a highly relevant empirical case study. Section 6.3 describes low temperature hydrolysis and oxidation processes developed, starting from different streams of the Acea plant, and the chemical nature of the related products. Section 6.4 describes the applications of the products obtained through the above processes. Section 6.5 addresses the problems and perspectives for scaling the developed processes and products to commercial production level. It proposes also a possible stepwise business development strategy with reduced entrepreneurial risk, which may effectively turn a conventional bio-waste management plant into a biorefinery through a virtuous bio-waste cycle. However, the business model is conceived at a very early stage. In addition, the reduced entrepreneurial risk relies mostly on author's assumptions, after considering research results and referenced real cost data.

6.2 State of the Art of Bio-waste Management Technology: Environmental and Economic Aspects

6.2.1 Environmental Problems

In addition to prevention at source, bio-waste management options include collection (separately or with mixed waste), anaerobic digestion and composting, incineration, and landfilling. The environmental and economic impacts of different treatment methods depend significantly on local conditions such as population density, infrastructure and climate as well as on markets for associated products (energy and composts).

Landfilling is still the most used municipal bio-waste disposal method in the EU. The majority of countries still landfilled more than half of their municipal waste in 2010 (European Environmental Agency 2013). Biodegradable waste decomposes in landfills to produce landfill gas and leachate. The landfill gas, if not captured, contributes considerably to the greenhouse effect as it consists mainly of methane, which is 23 times more powerful than carbon dioxide in terms of climate change effects (European Commission 2015). The leachate, if not collected in accordance with the *Landfill Directive* 1999/31/EC, can contaminate groundwater and soil. Landfills may also be a source of nuisance for neighbouring areas as they generate bio-aerosols, odours, and visual disturbance. An additional negative impact of landfilling is the area of land used, which is bigger than for other waste treatment technological methods. Landfills must be constructed and managed in line with the EU *Landfill Directive* (impermeable barriers, methane capturing equipment) to avoid environmental damage from the generation of methane and effluent.

The first step of modern waste treatment practices is separation into recyclables (glass, metals) and inert materials (stones etc.), paper, plastics, textiles, and biodegradable humid matter. Options are separate source collection of municipal solid wastes and/or mechanical separation of unsorted wastes. Wastes' separation is achieved by various mechanical-physical means (CP Manufacturing 2012). There are many types of machinery for municipal solid waste processing from size reduction to separation of different fraction, and many companies offering suitable equipment. Choice of a machinery/technology for waste sorting depends upon various factors including waste characteristics/composition, purpose for separation (e.g. material and energy recovery etc.) and following processing steps.

Incineration can be viewed as energy recovery or as a disposal. Incineration requires previous separation of materials. These are humid degradable bio-waste (Department for environmental food and rural affairs 2013) and inorganic materials, which lower the efficiency of incineration by their water content or are incombustible and do not contribute to the energy content of the waste. Others may be paper, plastics and textiles, if they can be reprocessed and recycled to further use. The mixture of the separated combustible materials is named refuse-derived fuel. Its energy content may run up to over 50% more than that of the pristine raw municipal solid waste. Incineration of bio-waste as a part of mixed municipal waste may be used to recover energy from a carbon-neutral source, providing an alternative to e.g. fossil fuels and contributing to climate change. The environmental impact of incinerating municipal bio-waste arises mainly from greenhouse gas emission, heavy metals, dioxin, loss of organic matter and other resources contained in biomass, and disposal of ashes and slags.

Bio-waste composting and anaerobic digestion may be classified as recycling, when compost (or digestate) is used on land or for the production of growing media. If no such use is envisaged, it should be classified as pre-treatment before landfilling or incineration. In addition, anaerobic digestion (producing biogas for energy purposes) should be seen as energy recovery. Composting is the most common biological treatment option (some 95% of current biological treatment operations). Anaerobic digestion is especially suitable for treating wet bio-waste, including fat (e.g. kitchen waste). It produces a gas mixture (mainly methane-50 to 75%-and carbon dioxide) in controlled reactors. Biogas can reduce greenhouse gas emissions most significantly, if it is used as a biofuel for transport or directly injected into the gas distribution grid. Its use as biofuel could result in significant reductions of greenhouse gas emissions, showing a net advantage with respect to other transport fuels. The residue from the process, the digestate, can be composted and used for similar purpose as compost, thus improving overall resource recovery from waste. The use of compost and digestate as soil improvers and fertilizers offers agronomic benefits such as improvement of soil structure, moisture infiltration, water-holding capacity, soil microorganisms and supply with nutrients.

The environmental impact of composting is mainly limited to some greenhouse gas emissions and volatile organic compounds. The impact on climate change due carbon sequestration is limited and mostly temporary. The agricultural benefits of compost use are evident (European Commission 2015; CalRecycle 2016) but there is debate about their proper quantification (e.g. by comparison to other sources of soil improvers), while the main risk is soil pollution from bad quality compost. As bio-waste is easily contaminated during mixed waste collection, its use on soil can lead to accumulation of hazardous substances in soil and plants. Typical contaminants of compost include heavy metals and impurities (e.g. broken glass). There is also a potential risk of contamination by persistent organic substances such as polychlorinated dibenzo-p-dioxins, dibenzofurans, biphenyls or polycyclic aromatic hydrocarbons (Fiedler 1998; Lerda 2011; Lingle 2008). The use of anaerobic digestate has an additional limitation, which arises from the amount of ammonia produced during anaerobic digestion because of organic N mineralization. The application of high doses of bio-waste sourced fertilizers to soil enhances environmental problems. For example, maintaining agronomic benefits in soil requires compost application rates of 10 t per ha and year (Sortino et al. 2014). Other negative implications arise from significant long-distance product transport.

6.2.2 Economic Aspects

The currently practiced technologies to treat bio-wastes suffer process costs not compensated by the value of the obtained products (Montoneri et al. 2011; Tang 2012). Nevertheless, bio-waste treatment is necessary to reduce mass, volume and chemical reactivity of the large amount of waste components produced by the modern society. As this implies a cost for citizens, at the current technology state of art the issue is assessing, for each case, which technology may have the lowest impact on the overall economics of waste management.

A recent work (Tang 2012) for instance, has compared two scenarios for managing wastes in Guanghan. Scenario I assumes a waste management system with source separation and separate collection of all types of recyclable materials and that the rest waste flows directly to the landfill. Scenario II differs from Scenario I in that metals are not separated at source, but flows with the rest waste to an incinerator before landfilling, where advanced technologies are applied to control air quality and to recovery energy, ferrous metal and non-ferrous metals. The result is that the benefit outweighs the cost by two million euro when comparing Scenario II to Scenario I, indicating a higher efficiency in resource allocation.

However, the result is highly sensitive to variations in the borrowing cost and the investment cost of equipment and technology. For the specific Guanghan case study, the following conclusions are drawn. The result of the cost benefit analysis indicates potential economic savings for the waste management system in Guanghan as a whole. It is therefore worthwhile for the policy makers to consider adding waste incineration to their agenda of improving the city's waste management system for environmental protection and for economic efficiency.

For the fermentation technologies, a cost benefits analysis has been published for the Acea waste treatment plant, taken as case study (Montoneri et al. 2011). The plant operates according to the ultimate trend to optimize the economy and to reduce the environmental impact of municipal bio-waste treatment. This consists in integrating anaerobic and aerobic fermentation. The Acea plant (Fig. 6.1) processes municipal bio-waste collected from an area of 2200 km^2 populated by 800,000 inhabitants distributed over 100 municipalities. These bio-wastes amount to about 50,000 t $year^{-1}$. The published (Montoneri et al. 2011) cost revenue analysis indicates a process cost of 156 € bio-waste t^{-1}, which is compensated by the

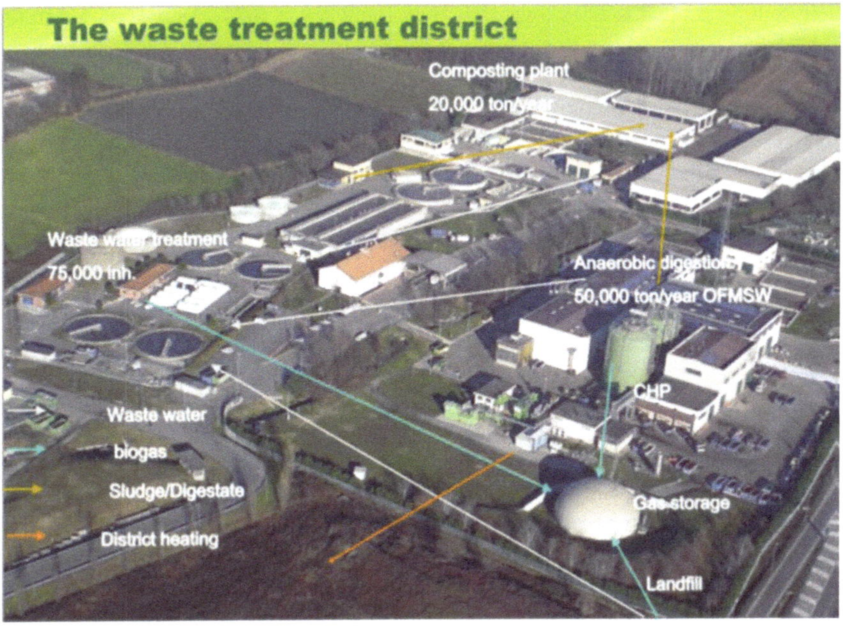

Fig. 6.1 Aerial view of Acea Pinerolese Industriale municipal waste management plant

revenue of 66 and 1 € t^{-1} from biogas and compost sales, respectively, and 90 € t^{-1} from tipping fee and energy recovery incentives. There are a few other similar plants in Italy (Ispra 2012). The composting and anaerobic digestion plants operating in Italy (Centemero 2015) are 240 and 43, respectively. The current trend is to increase the number of anaerobic digestion facilities, in order to integrate compost plants as in the Acea example.

By comparison, 980 and 650 composting and anaerobic digestion plants, respectively, operate in Germany (European Compost Network 2010). Throughout Europe, the potential of organic waste is estimated at 115 Mt $year^{-1}$ (Bart 2010a). There are about 2000 composting sites, with processing capacities ranging from 200 to 70,000 t $year^{-1}$.

A recent comprehensive report has been published in 2015 on the distribution of the bio-waste production and treatment facilities throughout Europe (European Environmental Agency 2013). In spite of the claimed benefits for agriculture (European Commission 2015; CalRecycle 2016) the compost marketability is poor. This implies that the product requires alternative uses. The current alternative is the use for land restoration or landfill cover. The market value of compost for use in agriculture, based on its content and market value of the key N, P and K nutrients, is calculated 4–6 € t^{-1} (WRAP 2016). Real EU market prices in 2005/2006 are reported in the 0–30 € t^{-1} range (Barth 2010b).

Most demand for compost is in advanced countries with mature markets. There is no real demand in starting countries, probably because compost products and their benefits are not well known, yet. Bulk retail prices in USA are reported in the range of 2–50 t^{-1}, depending on the type of sourcing material, location, and use (McEntee 2011, cited in US Environmental Protection Agency 2013). These prices do not include municipal operations that give compost away free of charge.

The Italian case study Acea plant adopts the most advanced bio-waste treatment technology, which is based on the integration of anaerobic and aerobic fermentation (Montoneri et al. 2011). The Acea plant (Fig. 6.1) contains four sections; two for the treatment of solid wastes by anaerobic and aerobic digestion, the third one for treating wastewaters and the last one being a landfill area equipped for biogas collection. The four plant sections are interconnected to maximize biogas and compost yields from bio-waste, thus minimizing bio-refuse disposal to landfill.

The plant allows large operational flexibility to produce different types of compost depending on the nature and relative ratios of the bio-residues constituting the aerobic phase feed. In the plant material balance, most of the plant biogas comes from equal amounts of biogas produced in the bioreactors processing the bioorganic (humid) fraction feed and biogas collected from the landfill area, while the sewage sludge section contributes for only a small part. The total amount of the plant biogas is more than enough for covering the plant energy consumption. Exceeding electrical and thermal energy produced by biogas is sold to the electrical network and to the nearby Pinerolo town residential and commercial districts. In spite of these desirable features, the process economy of the Acea plant, as well as that of all other waste management plants spread in the world, is not profitable due to operational costs exceeding the market value of the energy and/or materials produced (Montoneri et al. 2011).

6.3 Bio-waste Lignin Valorisation (LV) by Low Temperature Green Chemical Processes

The consideration of municipal bio-wastes as source of bio-based chemicals to recycle to the chemical industry is relatively new. Currently practised technology has been developed based mainly on the waste fuel value (see Sect. 6.1). Indeed, incineration producing thermal power is the main option to cope with the large amounts of municipal bio-waste. Anaerobic digestion produces biogas and residual organic matter, which needs disposal. Aerobic digestion although being an exothermal process is not exploited to recover energy, but allows to decrease the waste volume and yields a residual matter which is proposed as soil amending agent.

Current research proceeds along three types of processes, i.e. biochemical, thermal and chemical. Biochemical and thermal processes (pyrolysis and hydrocracking) disrupt the proximates of natural organic matter to obtain small platform molecules. Biochemical processes must cope with the recalcitrant lignin fraction which resists both chemical and biochemical treatment. Thermal processes disrupt all organic matter into simple molecules, but consume a relatively high amount of energy. Low temperature chemical processes are options, to use alone or combined with biochemical and/or thermal processes. Chemical processes require green solvents. No solvent is greener and more available than water.

Recent work (Montoneri et al. 2011; Rosso et al. 2015), shows that low temperature hydrolysis allows obtaining useful lignin-like soluble polymeric products from biomass. Contrary to biochemical and thermo-chemical processes, low temperature hydrolysis does not disrupt the natural molecular structures, but converts them in soluble fragments saving the original C types and functional groups as much as possible and, in doing so, requires low energy consumption and/or amount of equipment needed. Oxidation at room temperature of these polymeric products in water is a further option, which may allow widening the range of obtainable value added biopolymers and simple molecules (Montoneri et al. 2016). Table 6.1 summarizes the main features distinguishing the three types of processes.

Table 6.1 Main features of bio-waste treatment processes, and possible options/integration

	Biochemical	Thermal (no combustion)	Chemical	
			Hydrolysis	Oxidation
Main products	Small molecules: methane, ethanol, lactic acid, platform molecules in general	Small molecules: hydrocarbons	Soluble biopolymers keeping the pristine original C types and functional groups	Biosurfactants and soluble aliphatic poly hydroxy acids
Problems	– Inhibition from recalcitrant lignin fraction – Product recovery from diluted water solution – Residual lignin	High energy consumption	Recalcitrant residual insoluble lignin needs disposal or upgrading to marketable product	Recalcitrant residual insoluble lignin needs disposal or upgrading to marketable product
Possible option/integration	Bio-waste feed pretreatment by chemical hydrolysis to separate soluble lignin from insoluble fermentable matter (Fig. 6.2, III)	Apply to residual lignin from biochemical and/or chemical processes (Fig. 6.2)	Treatment of residual insoluble lignin by pyrolysis or hydrocracking, or combustion (Fig. 6.2, III)	Treatment of residual insoluble lignin by pyrolysis or hydrocracking, or combustion (Fig. 6.2)
	Composting residual lignin, followed by the compost chemical treatment (Fig. 6.2, II)			

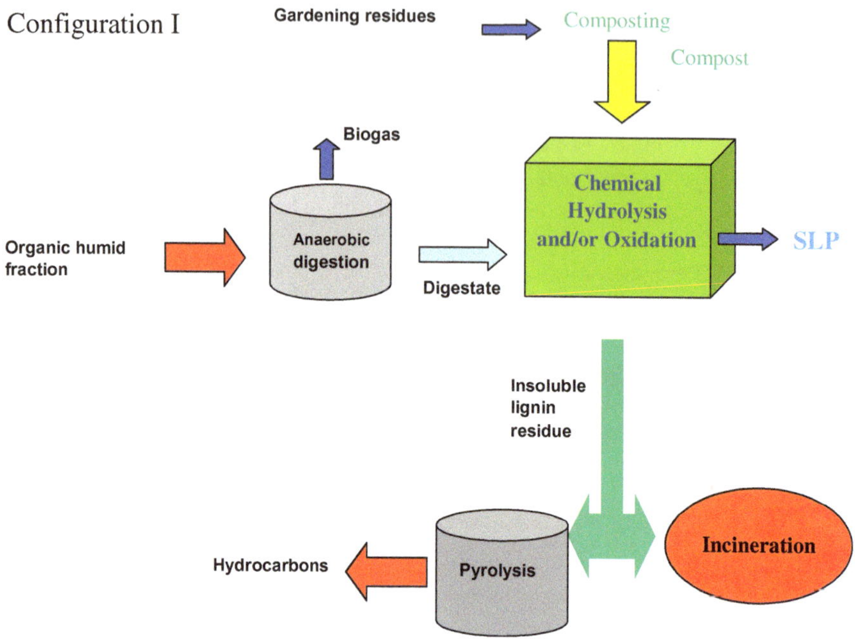

Fig. 6.2 Possible configuration of biochemical, chemical and thermal processes for the treatment of municipal bio-waste, featuring the chemical hydrolysis of the anaerobic fermentation digestate or of composted gardening residue for the production of soluble lignin biopolymers (SLP), followed by pyrolysis or combustion of the insoluble lignin residue

No process alone allows represents the optimum treatment. Figures 6.2, 6.3 and 6.4 show some hypothetical plausible plant configurations, which integrate the biochemical, chemical, thermal processes, and eventually incineration, for the treatment of municipal bio-waste. Obviously, many other configurations are possible. The optimum one should be worked out for each bio-waste type.

The plant configurations in Figs. 6.2, 6.3 and 6.4 are hypothetical, since the bio-waste chemical hydrolysis and/or oxidation process are not operating in real environment, yet. However, at laboratory and 500 L capacity pilot plant scale, several soluble bio-based polymeric substances with molecular weight ranging from 14 to several hundred kDa have been obtained by acid and/or alkaline hydrolysis at 60–100 °C of different urban (Rosso et al. 2015; Franzoso et al. 2015a, b, c; Nisticò et al. 2016) and agriculture (Franzoso et al. 2015b, c, 2016; Baglieri et al. 2014) residues, as collected and after anaerobic and/or aerobic biodegradation. The acid hydrolysates have at least one order of magnitude lower molecular weight than the alkaline hydrolysates. All products contain aliphatic and aromatic C types, and several acid and basic functional groups. The acid hydrolysates contain mainly polysaccharide moieties. The alkali hydrolysates contain mainly lignin-like aromatic moieties. Hereinafter, these polymers will be referred to according to their

Configuration II

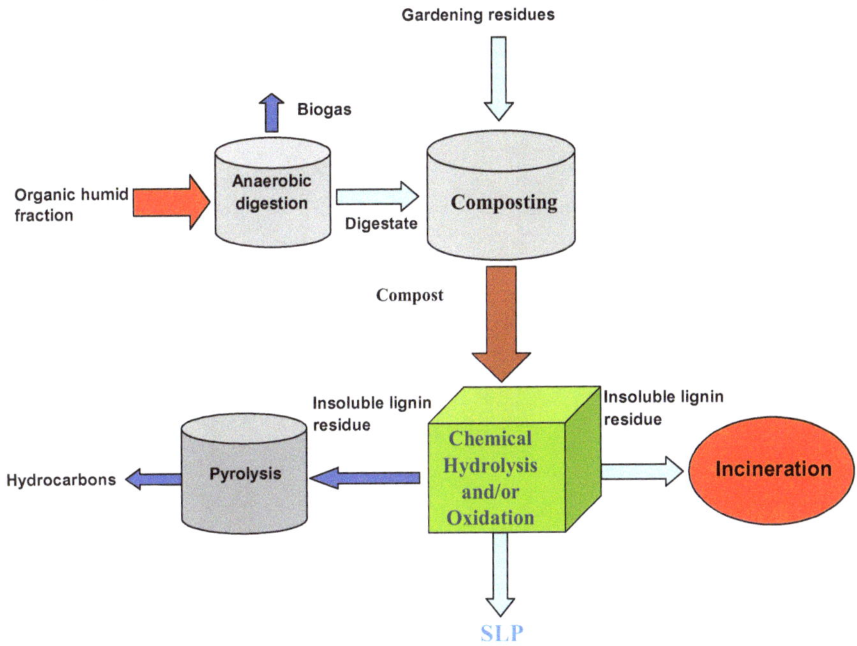

Fig. 6.3 Possible configuration of biochemical, chemical and thermal processes for the treatment of municipal bio-waste: composting of the anaerobic digestate mixed with gardening residues (as the Acea process in Fig. 6.1), which is followed by the chemical hydrolysis of the compost for the production of soluble lignin biopolymers (SLP), and ultimately by pyrolysis or combustion of the insoluble lignin residue

solubility properties and main constituents, i.e. soluble saccharide polymers (SSP) and soluble lignin-like polymers (SLP). The above chemical features are reported to be associated to surface-active properties. The higher molecular weight SLP exhibit similar behaviour as small molecule surfactants at concentration lower than 2 g L^{-1} in water, whereas they behave more likely polyelectrolytes at higher concentration (Montoneri et al. 2010). Composted urban bio-wastes contain more lignin-like matter than the as collected wastes. This is the likely reflection of microbial biodegradation converting the pristine polysaccharide matter to carbon dioxide and water, but not capable to metabolize lignin as well.

For the production of the SLP, a completely green process has been developed at pilot scale (Sortino et al. 2014). In the process, the reaction of the bio-waste feed and water at alkaline pH is performed at relatively mild temperature. The liquid hydrolysate is separated from the insoluble solid. The former is fed to a 5 kDa cut off ultrafiltration membrane. The membrane retentate is dried to yield the SLP. The permeate is recycled to the hydrolysis reactor for further use. Both the SLP and the insoluble product have been proven useful in multiple applications in the chemical industry and/or agriculture (see next subsection). Thus, in the above process,

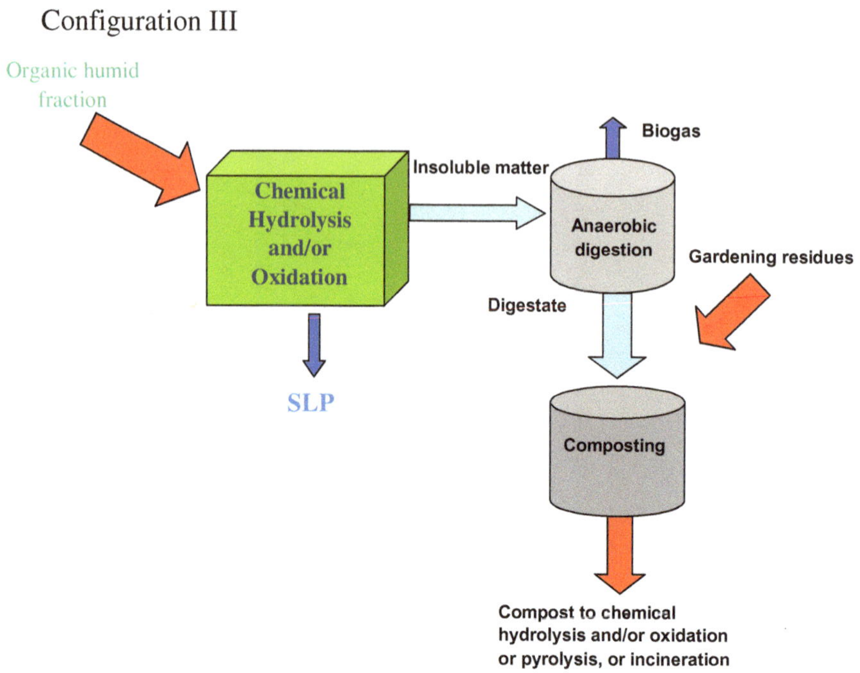

Fig. 6.4 Possible configuration of biochemical, chemical and thermal processes for the treatment of municipal bio-waste, featuring the chemical treatment of the as collected organic humid fraction of municipal bio-waste to produce the soluble lignin biopolymers (SLP), and thus to reduce the lignin content of the feed to the anaerobic digestion reactor. The successive steps are similar to those in Fig. 6.2

solvent and reagent are completely recycled, and no waste is produced, which requires a secondary treatment. Furthermore, most recent work (Rosso et al. 2015) has investigated an alternative hydrolysis process carried out by microwave heating, compared to conventional heating. Non-conventional energy sources such as microwave, compared to conventional heating, can dramatically enhance reaction rates in organic synthesis (Tabasso et al. 2014) and thus allow reducing reagents contact time (Hu and Wen 2008) and reactor volume. This, in principle, is an important step toward the construction of plants that are more compact, cost-effective and safer (Sanders et al. 2012). Therefore, the hydrolysis of a composted municipal bio-waste, taken as case study, has been investigated (Rosso et al. 2015) as a function of solid-liquid contact time, temperature, liquid/solid ratio, and pH. It has been found that similar product yield and type are obtained by conventional and microwave heating. However, by microwave heating, the required solid/liquid contact time is over two order of magnitude shorter than by conventional heating. These findings offer worthwhile scope for further work aimed to compare microwave and conventional heating options for operational and capital costs.

wave heating,

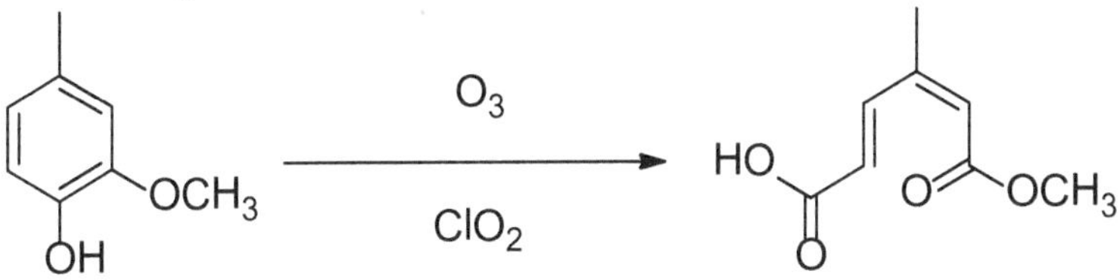

Fig. 6.5 Oxidation of lignin

Room temperature oxidation of SLP has been reported very recently (Montoneri et al. 2016). According to literature (Brunow et al. 1998; Niemel et al. 1985), ozonisation of native lignin destroys double bonds and aromatic rings, leaving the side chains intact in the form of carboxylic acids (Fig. 6.5). In this fashion, the reaction in Fig. 6.5 can ultimately lead to the formation of four C atoms dicarboxylic moieties, –(HOOC)CH–CH(COOH). Application of this reaction to SLP has allowed converting the pristine lignin aromatic ring to aliphatic carboxylic moieties. Thus, new oxidized, and/or more hydrophilic biopolymers have been obtained. These are high molecular weight biosurfactants, with improved colour and surface activity properties, and lower molecular weight aliphatic polycarboxylic macromolecules. Compared to the biosurfactants, the aliphatic polycarboxylic acids are higher oxidation products. These have no surface activity. However, they are potentially valuable for the manufacture of biodegradable polymers (Chiellini and Solaro 2003) and/or value added small platform molecules (Quesada et al. 1999; Clark et al. 2014) to recycle to the chemical industry.

The SSP production process (Franzoso et al. 2015a, b) was not investigated as well as the SLP process (Rosso et al. 2015). The SSP were mainly products of the acid hydrolysis of the bio-waste polysaccharide fraction, and therefore were out of the scope focused on the valorisation of lignin. In the treatment of bio-waste, polysaccharides constitute the fermentable organic fractions. Thus, the SSP are mainly regarded as intermediates for the production of small molecules, such as methane or bioethanol, by fermentation. In the context of the work focused on the valorisation of lignin, the SSP products were obtained to assess, by comparison with SLP, their performance for the manufacture of composite biodegradable polymers. The results, described hereinafter, did not warrant further development work dedicated to SSP.

6.4 Bio-waste Lignin Valorisation (LV): Applications of Products Derived from Urban Food and Agriculture Wastes

Several papers have reported the performance of the above SLP, and/or SSP, in diversified fields; e.g. in the formulation of detergents, textile dyeing baths, flocculants, dispersants and binding agents for ceramics manufacture

Fig. 6.6 Schematic representation of the sourcing urban bio-wastes, materials obtained by anaerobic and aerobic digestions, and virtual molecular fragment and application fields of soluble bio-based substances obtained by chemical processing of digestate and compost materials by hydrolysis

(Montoneri et al. 2011), emulsifiers (Vargas et al. 2014), auxiliaries for soil/water remediation (Montoneri et al. 2014; Avetta et al. 2013; Gomis et al. 2014; Mostofa et al. 2013), and enhanced oil recovery (Baxter et al. 2014), nanostructured materials for chemical (Boffa et al. 2014; Deganello et al. 2015; Testa et al. 2015) and biochemical catalysis (Magnacca et al. 2012), plastic materials (Franzoso et al. 2015a, b, c, 2016; Nisticò et al. 2016), soil fertilizers and plant biostimulants for agriculture (Sortino et al. 2014; Baglieri et al. 2014; Sortino et al. 2013; Fascella et al. 2015; Rovero et al. 2015; Mozzetti Monterumici et al. 2015; Massa et al. 2016), animal feed supplements (Dinuccio et al. 2013; Biagini et al. 2016; Montoneri et al. 2013), and auxiliaries for eco-friendly anaerobic fermentation of urban bio-wastes (Francavilla et al. 2016a, b) and manure (Riggio et al. 2016).

The wide range of applications arises from the fact that these bio-based soluble products are constituted by a mix of polymeric molecules containing organic C and N distributed over a variety of aliphatic and aromatic C moieties substituted by acid and basic functional groups which are bonded to several mineral elements such as Na, K, Ca, Mg, Al and Fe. Figure 6.6 shows a virtual molecular fragment representing the SLP organic chemical features. The C moieties in this fragment are the memories of the protein, fats, polysaccharide and lignin proximate constituting the pristine bio-waste.

They are associated to the products properties as surfactants, agents for sequestering or carrying small molecules and mineral ions in solution, photosensitizers and reactive biopolymers.

One main drawback of the bio-based listed in Fig. 6.6 is the black colour. This is a critical feature in surfactant assisted fabric detergency and textile dyeing. Fabric yellowing has been found to be the critical deficiency in SLP assisted washing (Savarino et al. 2010) or dyeing (Savarino et al. 2009). The results of previous work on the ozonisation of native lignin (Brunow et al. 1998; Niemel et al. 1985) allowed expecting that ozonisation of SLP would destruct the lignin-like aromatic chromophores' moieties and thus yield lighter coloured products. Montoneri et al. (2016) have reported that this is true. The high molecular weight oxidized biosurfactants are light coloured and have improved surface activity properties, compared to the pristine SLP. Thus, wider marketability opportunities are expected for the oxidized SLP biosurfactants.

Based on product performance and/or potential marketability for the tested applications, at the current state of product development, the short term most feasible and appealing uses of SSP and SLP seem to be in agriculture, plastics manufacture, and in anaerobic digestion processes. In a few cases, the performance of SLP in agriculture was studied in comparison with the insoluble hydrolysate product recovered together with SLP. Similarly, to SLP, the insoluble product was proven a valid fertilizer for the cultivation of several food (Sortino et al. 2014; Rovero et al. 2015) and hornamental (Massa et al. 2016) plants.

6.4.1 Performance and Perspectives for SLP Used in Agriculture

Several universities (Sortino et al. 2014; Baglieri et al. 2014; Jindřichová et al. 2016), the Italian center of agriculture research (Fascella et al. 2015; Massa et al. 2016) and Isagro SpA (2014), a company operating on a global level, in about 70 countries, in the market of agropharmaceuticals, have studied effects of SLP as auxiliaries for agriculture. The addition of SLP to soil at 145 kg ha^{-1} dose has been found to increase significantly tomato and red pepper plant photosynthetic activity, growth and productivity, and/or product quality, by 10–20% relatively to farm routine practice with no SLP soil treatment (Sortino et al. 2014). Compared with most biofertilizers used at 20–30 t ha^{-1} dose, the effect/dose ratio exhibited by SLP appears quite remarkable. It prospects several economic and environmental benefits for farmers and soil, respectively. Isagro (2014) has proven the beneficial effects of SLP on tomato, as well as on wheat and tobacco. Other workers have demonstrated that the hydrolysates of urban bio-wastes and post-harvest agriculture residues are capable to enhance growth, productivity, and/or plant photosynthetic activity and protein production, for a variety of other plants such as maize (Fascella et al. 2015), beans (Franzoso et al. 2016), radish (Monzetti Monterumeci et al. 2015),

Table 6.2 Increments (w/w%) of crop production (unless otherwise specified) for different plants cultivated in the presence of SLP from different urban and agriculture source,[a] relative to control plants with no added SLP

Plant		CVDF	CVD	CV	D	TP
Tomato Lycopersicon[b]		20	20	20		
Tomato Micro Tom[c]		46		1	16	
Pepper[d]			66			
Maize[e]		120				
Bean[f]						77–278[j]
Radish[g]						0
Wheat[c]		10		9	9	
Tobacco[c]		6		0	0	
Euphorbia[h]		233			117	
Hibiscus[i]				15[k]	25[k]	

[a]Digestate (D) from anaerobic fermentation of urban organic humid waste; compost made from gardening residues (CV), and from their mixes with D (CVD) and with D and sewage sludge (CVDF); post-harvest tomato plant (TP)
[b]Sortino et al. (2014)
[c]Isagro (2014)
[d]Sortino et al. (2013)
[e]Rovero et al. (2015)
[f]Baglieri et al. (2014)
[g]Mozzetti et al. (2015)
[h]Fascella et al. 2015
[i]Massa et al. (2016)
[j]Biomass protein production
[k]Total biomass weight

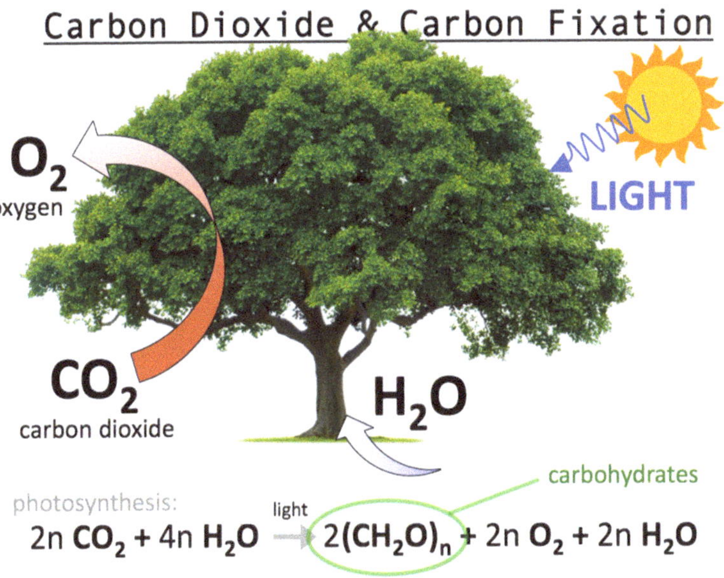

Fig. 6.7 Photosynthesis reaction promoting plant growth

euphorbia x lomi (Fascella et al. 2015), and hibiscus (Massa et al. 2016). The SLP have also been proven efficient plant disease suppressants (Jindřichová et al. 2016).

All performance tests have compared the alkaline soluble hydrolysate product and/or the insoluble hydrolysate product with known commercial products for their effects on several plant response indicators. Some tests have compared the SLP with their pristine sourcing materials, and/or with the insoluble hydrolysate product. In all cases, the SLP effects have been found equal or better than the products investigated for comparison. Table 6.2 summarizes the effects of the SLP obtained from different urban and agriculture wastes on the productivity of investigated plants. The effects clearly depend on the type of SLP and the tested plant. This fact implies that a waste treatment plant may produce a wide variety of SLP tailored for the cultivation of specific plants, depending on the variety of bio-waste sources, treatments types and processes configuration (Figs. 6.2, 6.3 and 6.4).

In this context, particularly interesting is the work by Sortino et al. (2013, 2014) pointing out a possible role of SLP as interphase between agriculture and human activities. These authors discuss the effects of SLP on plant photosynthetic activity in relation to other work (Gomis et al. 2014) that reports SLP as photosensitizer promoting C mineralization of organic pollutants. They propose that, depending upon the experimental conditions, SLP could promote C fixation or mineralization. Both processes occur in nature (Mostofa et al. 2013). The first promotes plant growth through photosynthesis (Fig. 6.7), starting from carbon dioxide and water. The second is the opposite process. It occurs in soil and water, and converts natural organic matter into carbon dioxide and water. These processes are possible since natural organic matter displays physical properties such as the absorption of energy form ultraviolet and photosynthetically available radiation. By their chemical nature (Montoneri et al. 2011) and properties (Sortino et al. 2014; Montoneri et al. 2010) the SLP have been shown to resemble natural organic matter. The idea that SLP may promote either C fixation or mineralization is rather intriguing. It proposes a virtuous role of SLP entering the C cycle to generate benefits for agriculture and the environment.

6.4.2 *Performance and Perspectives for SSP and SLP Used for Plastics Manufacture*

Both the soluble saccharide (SSP) and lignin-like (SLP) bio-polymers, isolated from municipal bio-waste and/or agriculture residues, have been tested as components of blended films with synthetic polyethylene copolymers (Franzoso et al. 2015a, b, c, 2016; Nisticò et al. 2016), i.e. polyvinyl alcohol-co-ethylene (EVOH) and polyethylene-co-acrylic acid (PEEA). These commercial polymers derived from fossil sources are used for the manufacture of a great variety of articles of every-day life. Research for substituting synthetic polymers with bio-waste derived polymers is justified by a number of presumed economic and environmental benefits for the chemical industry and the management of wastes. They include lower consumption of fossil sources, availability of articles that are more compatible with the environment, and the valorisation of the bio-waste sourced products as speciality chemicals.

The realization of these perspectives implies the indispensable and essential conditions that the performance was not lower, and the cost of the biobased article was not higher, compared to the current commercial articles. For the SSP and SLP blends, the reported results (Franzoso et al. 2015a, b, c, 2016) have demonstrated that the above performance condition is satisfied by the blends containing not more than 10% SSP or SLP. Blends with higher biopolymers content have poor unacceptable mechanical properties.

Table 6.3 summarizes the mechanical properties of the best performing SSP and SLP blends. An important property of polymers, which conditions their end-uses, is the response to the application of a force, as indicated by two main types of behaviour: elastic and plastic. Elastic materials will return to their original shape once the force is removed. Plastic materials will not regain their shape. Most materials demonstrate a combination of elastic and plastic behaviour, showing plastic behaviour after the elastic limit has been exceeded. For plastic films, the most common indicators of mechanical properties are the Young's modulus and the strain at break (The engineering toolbox 2013). The Young's Modulus or Modulus of Elasticity is a measure of stiffness of an elastic material. It is used to describe the elastic properties of the material when it is stretched. It can be used to predict the

Table 6.3 Mechanical data for different SSL and SLP blend films with synthetic polymers

Sample	Young's modulus (MPa)	Stress at yield point or maximum stress (MPa)	Strain at break (%)	Stress at break (MPa)
EVOH	352	33	86	26
EVOH-SSP$_T$ 0.1	747	43	35 ± 2	32
EVOH-SLP$_T$ 0.1	389	44	14 ± 3	41
EVOH-SSP$_{IR}$ 0.1	1043	39	70 ± 10	32
EVOH-SLP$_M$ 0.06	352	53	40	0
EVOH-SLP$_D$ 0.06	1082	62	42.3	44
PEAA	30.4	4.7	>300	10.6
PEAA-SLP$_M$ 0.1	76.1	9.3	280	13.5
PEAA-SLP$_M$ 0.2	135.6	13.7	257	19.3
PEAA-SLP$_M$ 0.3	172.2	17.1	55	16.1
PEAA-SSP$_{IR}$ 0.1	23.9	4.4	>300	9.9
PEAA- SSP$_{IR}$ 0.2	43.4	6.7	>300	15.1
PEAA- SSP$_{IR}$ 0.3	23.8	4.6	>300	9.1
	Extruded rods			
	Bending modulus (MPa)	Flexural strength (MPa)		
EVOH	3900	118		
EVOH-SLP$_D$ 0.05	3100	102		
EVOH-SLP$_D$ 0.1	3000	108		
EVOH-SLP$_V$ 0.05	3300	113		
EVOH-SLP$_V$ 0.1	3000	101		

Mechanical data for different SSL and SLP blend films with synthetic polymers obtained by solvent casting (unless otherwise indicated): *EVOH* neat polyvinyl alcohol-co-ethylene; *PEAA* neat polyethylene-co-acrylic acid; A–B ρ blends: *A* EVOH or PEAA; *B* soluble saccharide polymer by acid hydrolysis of tomato plant powder (SSP$_T$) or of the insoluble residue after alkaline hydrolysis (SSP$_{IR}$), or *B* soluble lignin polymer by alkaline hydrolysis of urban bio-waste digestate (SLP$_D$), of compost of urban vegetable gardening residues (SLP$_V$), and of compost of mixed urban vegetable gardening residues, digestate and sewage sludge (SLP$_M$); ρ = B/A w/w ratio in the blend. Data selected from published work for best performing SLP$_M$ and SLP$_D$ blends (Franzoso et al. 2015a), PEAA blends (Franzoso et al. 2015b), SSP$_T$ and SLP$_T$ blends (Franzoso et al. 2015c) and extruded blends

elongation of the object as long as the stress is less than the yield strength of the material. The stress at yield point is the applied force above which the material acquires a permanent deformation. Above this point, further applied stress causes elongation of the material until fracture (Key to Metals AG 2014). The strain at break is the % elongation of the material until it breaks. The stress at break is the tensile stress when the test specimen tears (Ensinger Gmbh n.d.).

For rigid objects, indicators of their end-use performance are the bending modulus and the flexural strength. The bending modulus indicates the tendency for a material to bend (wiseGEEK n.d.). It is a measure of how a certain material will strain when weight or force is applied to bend it, before a permanent deformation occurs. The flexural stress is the maximum amount of bending stress that can be applied before rupture or failure of the material occurs. The data reported in Table 6.3 show in most cases that the poly vinyl alcohol-co-ethylene blends, which contain 6% SLP obtained from the Acea urban food wastes anaerobic digestate (EVOH-SLP_D0.06) and 10% SSP obtained from post-harvest tomato plant (EVOH-SSP_{IR}0.1) exhibit up to three times higher stiffness, but lower strain at break than the neat synthetic polymer (Franzoso et al. 2015a). The EVOH blends containing the soluble lignin-like polymers SLP biopolymers obtained from the bio-waste digestate (EVOH-SLP_D0.06) and from the compost (EVOH-SLP_M0.06) exhibit higher maximum stress than the EVOH blends containing the soluble saccharide or lignin-like polymers obtained from post-harvest tomato plants. This means that the blends containing the soluble lignin-like polymers obtained from urban bio-wastes, compared to the neat synthetic polymer and to the blends containing the tomato plant sourced biopolymers, can bear higher load before undergoing permanent deformation. However, their elongation before breaking is significantly reduced, compared to the neat synthetic polymer. Similar trends are shown by the polyethylene-co-acrylic acids blends, although it is much less evident the relative decrease of the strain at break for the blends, compared to the neat synthetic polymer. Also in the case of the polyethylene-co-acrylic acid blends, the films containing the soluble lignin-like polymers (PEAA-SLP) compared to the neat synthetic polymer and to the films containing the soluble saccharide biopolymers, exhibit higher Young's modulus and stress at yield point, but lower strain at break. This behaviour is the likely reflection of the different chemical nature of the two biopolymers. The soluble lignin-like polymers save the memory of the parent lignin, a though not ductile material, compared to polysaccharides. The melt extruded poly vinyl alcohol-co-ethylene blends could be obtained with SLP, but not with SSP. The latter did not withstand the 200 °C processing temperature of the melt extrusion. The extruded EVOH blends containing the SLP exhibited lower bending and flexural strength, compared to the neat synthetic polymer. The results indicate that it is possible to enhance the mechanical strength of the tested synthetic polymers. However, this occurs with some loss of the elongation capacity. The blend properties appear to depend strongly, not only on the type of and content of biopolymers (i.e. soluble lignin-like versus soluble saccharide biopolymers), but also on the processing technology.

The SLP blends mechanical data, coupled to the effects of the SLP in agriculture, prospect a virtuous scenario where mulch film fabricated with the above reinforced blends, at end of their service life, might contribute their beneficial properties to plants. From the economic point of view, substitution of 5–10% of synthetic polymers with SLP is expected to be cost effective (see Sect. 6.5.2).

6.4.3 Performance and Perspectives for SLP Used in Anaerobic Digestion Processes

The improvement of current municipal bio-waste anaerobic fermentation processes is pending upon the achievement of two main objectives (Al Seadi et al. 2008; Sereno 2010): i.e. enhancing the biogas CH_4/CO_2 ratio and reducing the mineralization of organic N. The former is directly related to the biogas heat value. The other has relevance for the environmental impact of the process digestate. In both cases, ammonia has an important role. Ammonia inhibits methanogenic bacteria, which are especially sensitive to this compound. Ammonia is collected with the digestate. This is normally recycled to farmland. Ammonia emission and/or nitrate leaching can occur due to inappropriate handling, storage and application of digestate as fertiliser (Martins 1992).

The European Nitrate Directive (91/676/EEC) restricts the input of mineral nitrogen on farmland, aiming to protect the ground and surface water from pollution. Downstream technology is available for separating CH_4 from CO_2 and for removing excess inorganic N from the digestate (Zhao et al. 2010; Mirbagheri et al. 2010; Provolo and Riva 2009). This however requires additional process costs, which affect negatively the waste treatment economy and in turn on taxpayers. The separation of CH_4 from CO_2 in the biogas and the abatement of ammonia from the digestate can run 0.13–0.44 € N m^{-3} biogas (Zhao et al. 2010) and 1.6 \$ kg^{-1} N (Burke 2013), respectively. The problem of organic N mineralization is well known also in animal husbandry, which in turn affects agriculture. These two activities are strongly interrelated in as much as agriculture provides feed for animals and these provide manure to recycle to soil as fertilizer for agriculture. Proteins are the main source of ammonia during animal feed digestion, due to proteolytic bacteria activity. Indeed, deficient intestinal fermentation results in increased proteolysis and release of toxic substances, such as ammonia and amines. In this fashion, manure may have negative environmental impact, because of greenhouse gases emission and leaching of mineral nitrogen through soil and ground water. For example, the typical levels of aerial ammonia in a pig farm facility vary from 5 to 35 ppm (Ji et al. 2006), while suggested threshold limit values of ammonia concentration are at 25 ppm level. Greater aerial ammonia level not only reduces the pig growth, but also is harmful to human health.

Recently, various SLP have been tested (Montoneri et al. 2013) as diet supplements to modulate pigs ileal fermentation of a protein feed. These were isolated from the soluble hydrolysates of different streams from the Acea waste treatment plant: i.e. the digestate, and several composts made from the digestate, gardening wastes, and sewage sludge, alone or mixed in different relative weight ratios. The study was carried out in vitro, using the cecal content collected from slaughtered pigs as incubation liquor. The reported results clearly point out reduced proteolysis and N mineralization by 7–17% caused by the compost derived SLP added to the fermentation liquor at 0.1–0.2% concentration. The digestate derived SLP has opposite effect. Consistently with these findings, in vivo animal study were performed (Dinuccio et al. 2013; Biagini et al. 2016), by feeding rabbits with diet supplemented containing 0.05 and 0.25% of SLP isolated from the composted gardening wastes. The reported results demonstrate 25% reduction of ammonia emission from freshly produced manure of rabbits fed with diet containing 0.25% SLP (Dinuccio et al. 2013; Montoneri et al. 2013) and no toxicity for animals by SLP (Biagini et al. 2016). Following these findings, four materials were sampled from the case study Acea plant (Fig. 6.1). These were the as collected bio-waste organic humid fraction, the digestate sampled from the anaerobic fermentation reactor, the compost made from a digestate-gardening waste mix (Francavilla et al. 2016b) and the compost made from gardening waste only (Francavilla et al. 2016a). The SLP were then obtained from the two compost and added in separate lab experiments to 6 L bioreactors containing a mix of the as collected organic humid fraction and digestate, similar to that contained in the plant bioreactors in routine operation. The intent was to assess whether the same ammonia reduction effects by SLP, which had been found in the above animal studies (Mozzetti Monterumeci et al. 2015; Montoneri et al. 2013) occurred in the anaerobic digestion of the organic humid fraction of urban waste, which was normally processed in the Acea plant. In the lab experiments (Francavilla et al. 2016a, b), the SLP obtained from the two different composts were added at 0.05 and 0.20% concentration to the fermentation liquor sampled from the Acea anaerobic digestion reactor. The anaerobic digestion of the control (no added SLP) and the treated (added SLP) fermentation liquor was carried out in parallel reactors operated at 55 °C for 12 days, until the biogas production became negligible. During this time, the control and treated fermentation liquor produced the same amount of biogas. However, the ammonia content of the control fermentation increased by 24%. On the contrary, the fermentation liquors, containing 0.05 and/or 0.20% added SLP, produced no ammonia. In addition, the methane/carbon dioxide mol/mol ratio in the biogas produced by the fermentation liquor containing 0.20% SLP was apparently about 9% higher than that of the control liquor. The results obtained with the two different SLP indicated that the above effects were different upon changing the bio-waste feed and the type of added SLP. This finding offers worthwhile scope for further research scope in order to optimize the type and amount of SLP as a function of the bio-waste feed. Aside from this, for a municipal waste treatment plant as the case study Acea plant (Fig. 6.1), the lab scale results obtained in the SLP assisted anaerobic digestion of the urban organic humid fraction (Francavilla et al. 2016a, b) prospect the realization of a virtuous in-house material cycle with attracting potential economic and environmental benefits. In essence, the plant biogas digestate would be exploited as source of SLP to be added to the biogas reactor, in order to produce digestate with reduced ammonia content and biogas with enhanced methane content. This scenario offers the highest potential economic benefits to the case study Acea plant (see Sect. 6.5.1), as well as to any other similar plant that integrated its anaerobic and aerobic fermentation processes with the compost chemical hydrolysis facility producing the SLP. The plant, in fact, could take advantage from the lower in-house SLP production cost in place of the ammonia abatement costs with the current technology (Table 6.3; Sect. 6.5.1).

The above results obtained for the anaerobic digestion of urban bio-waste are highly relevant also in other contexts. A very important one is animal manure, which is a strong source of ammonia emission (Ji et al. 2006). The entire manure production in the EU is estimated 1.4 billion t yr^{-1} (Foged et al. 2012). This production results from a myriad of farms spread over large areas (Eurostat 2016). Thousands of anaerobic digestion installations throughout EU member states are currently processing only about 8% of the manure production, equal to 108 million t yr^{-1}, containing 556,000 t nitrogen. These range from small on-farm to large centralized facilities. The available technology for the secondary treatment of the manure digestate (Burke 2013; Zarebska 2015) requires high capital cost investments, which cannot be borne by small farms. It is obvious that these circumstances require a simple economically sustainable solution to the problem of ammonia production in the anaerobic fermentation of manure. This solution should be applied locally, in on-farm installations of any size, thus avoiding collection and transportation costs of the digestate to larger centralized plants for secondary treatment. The results obtained in the SLP assisted anaerobic digestion of urban bio-wastes (Francavilla et al. 2016a, b) prospect that anaerobic digestion in the presence of SLP does not require secondary treatment of the digestate for reducing the ammonia content, and that therefore no capital cost for digestate processing facilities is necessary. Based on these perspectives, Riggio et al. (2016) have investigated the anaerobic digestion of cow manure in the presence of 0.2% SLP. They found that the addition of SLP to the fermentation slurry inhibited the production of ammonia during the manure fermentation. These findings offer further environmental and economic argument for scaling the SLP production to commercial level.

6.5 Scaling Promising Technology to Commercial Production Level

Assessing the feasibility of transferring new technology to the market place must take in consideration its promising aspects as well as the impeding factors. For the above-described LV technology, promising features are the following ones. Firstly, urban bio-wastes are negative cost sources (Sheldon-Coulson 2011). Secondly, the hydrolysis process has been tested at prototype level by conventional heating, and a preliminary operational cost evaluation has been obtained (Montoneri et al. 2011). Thirdly, products have been tested for their performance as speciality chemicals at laboratory level and in representative environment (Rosso et al. 2015). Fourthly, the above described research has involved cooperation of the author at the University of Torino and several other public research institutions (see references in Sect. 6.4), and potential end-users of the collected research results, such as Acea (Montoneri et al. 2011; Francavilla et al. 2016a, b) and Isagro (2014). By these features, according to the Nasa (2015) and the European Commission technology readiness level (TRL) definitions (European Commission 2014), the LV technology TRL may be estimated 4–5. The assignment of TRL numbers to processes and products allows a common understanding of technology status, helping decisions that concern the development and transitioning of technology, and evaluating the connected risk (United States Department of Defense 2011). Guidelines for converting research into commercial success have been published (European Commission 2013, 2016). There are some conditions for a successful scale up of process/product from research (TRL 5) to industrial/commercial (TRL 6-9) level:

- The production process must be economically and environmentally viable
- A sizable market exists
- The product is accepted by the consumer
- All players, with the proper knowledge on resource acquisition, production technology and product marketability, are actively involved.

In the specific case of the LV technology, the SLP products have been obtained by assessed biochemical processes (i.e. anaerobic digesting and composting) coupled to pilot chemical hydrolysis. The chemical process is a green process. It involves the hydrolysis of the bio-waste in water at relatively low temperature (60 °C), low energy consumption, complete recycle of solvent and reagents, no formation of unusable product to dispose, and no need of secondary process effluent treatment. The above biochemical and chemical processes have been shown to convert natural cellulose, lignin, protein and fat matter present in urban food and/or gardening wastes, and/or in post-harvest plants, to several SSP and SLP products. As cellulose, lignin, proteins and fats are the major organic constituents of living matter, the above biochemical and chemical processes can be applied to most dedicated crops and bio-organic residues. Thus, an even wider range of products can be obtained, depending on the starting bio-organic material. These features are basic requirements for a successful market-oriented exploitation of the developed bio-waste processes and products (European Commission 2013).

Fulfilling all conditions for successful scale up of the LV technology to commercial production level requires the active participation of different operators, which have proven know how in diversified fields, such as waste collection and management, process engineering, and product technological development and marketing. A recent study (Morone et al. 2015), taking the Italian bioplastics market as case study, has performed a social network analysis to assess the potential of bio-waste for bioplastics production. It shows that the Italian bioplastics producers' network offers great opportunities for the development of a technological niche based on bio-waste valorisation. However, the system is weak especially as far as expectations are concerned, as these are generally low and, more critically, are low for those actors occupying central positions in the network. This shortcoming could jeopardize the niche development process, if no appropriate policy actions are undertaken. More specifically, this study could support decision makers in developing specific strategies to unlock the enormous potential of bio-waste as well of the bioplastics sector by empowering knowledge creation and its diffusion, and by supporting strategic collaboration schemes. For instance, policy measures could be introduced to stimulate social learning as a driver of expectations.

Another study (Sheldon-Coulson 2011) has evaluated the commercial viability of cellulosic bio-refineries in the urban corridor linking New York and Philadelphia. A mature technology such as the non-commercialized Biofine process was taken as case study. This process has been realized at demo scale to obtain the well-known multipurpose platform chemical levulinic acid from carbohydrate feedstock. The results of the study indicate extremely healthy economic returns by scaling up the demo plant to commercial production. However, these returns are not high enough to convince private entrepreneurship to undertake the technology and integration risk of an early stage-venture, without public participation to share the relevant risks.

While the above levulinic acid is a well-known product, risks are even higher, and particularly critical, in the case of new products for several reasons:

- Product not known in the market.
- Difficulty to identify the right market type, and assess the product market desirability, sale value and saleable amount.
- Allocation of the product in government approved marketed product categories and/or development, implementation, assessment, monitoring and evaluation of environmental policy and legislation dedicated to a new product category.

Products such as thermal and/or electrical energy, and fuel, hydrocarbons or platform chemicals with well-defined chemical structure and composition, obtained by bio-waste combustion and pyrolysis, respectively, have a well-assessed market. The SLP do not fall into such categories. They are new products, need market assessment, and therefore imply higher risks for private entrepreneurship. On the other hand, their social impact is much lower than that of energy and fuel. Thus, for politicians, chemical products are less relevant, due to their lower social impact compared to fuel, and thermal and electrical energy. This fact, for the specific case of the LV technology, implies unlikely assumptions of new venture risks by public bodies.

Notwithstanding the above limitations, the finding that SLP may be used as regulator of the bio-waste anaerobic digestion process (Francavilla et al. 2016a, b) offers an attracting opportunity to undertake innovation business with reduced entrepreneurial risk. Indeed, considering the Acea waste management plant (Fig. 6.1) as case study, the integration of the SLP production unit into the existing Acea facility would allow benefiting from the on-site availability of bio-wastes, all necessary utilities and land space, at minimum cost. If SLP was produced only for in-house use (i.e. addition to the biogas production reactors to obtain digestate with reduced ammonia content), no market product allocation would be necessary. Therefore, the above listed risks would not apply. The availability of the SLP production unit for in-house use would allow proceeding cautiously and efficiently to a second scale up stage. Indeed, upon completion of the first stage, enough SLP quantities would be available to involve other players operating in other fields where the SLP could be used (see Sect. 6.4.1). Under this condition, further product development and market assessment tests would be carried on in real operational environment, in order to proceed to full-scale production and product commercialization.

The above strategy would allow optimizing risk management. With reference to the EU proposed types of pathways of market-oriented exploitation (European Commission 2016), the first stage, not necessarily requiring additional research activity, implies a linear and prompt conversion of research outcome into a ready for use technology. The success of this stage does not depend on technology commercialization. Nevertheless, upon completion, it has great potential for optimization and commercialization. The second stage certainly needs additional research activity for product development and market allocation. It sets out as the deferred transformation of research outcome into a product or service available to or ready for the market. Thus, the success of the second stage ends up for depending strongly on product commercialization. The following sustainability evaluation regards the above two acts play, involving one actor in the first act and more actors in the second act. With reference to the range of demonstrated applications (Fig. 6.6), the applications in agriculture and in plastics manufacture seem the most promising ones, based on product maturity and market size. Hereinafter, the estimates of the potential economic impact of adopting this two acts strategy are given for the Acea plant taken as case study.

6.5.1 Economic Impact of SLP Production

Figure 6.8 shows a possible virtuous bio-waste material flow through the Acea plant. It involves the existing anaerobic and aerobic digestion sections, integrated with the compost hydrolysis facility to be installed. The key feature is that SLP is

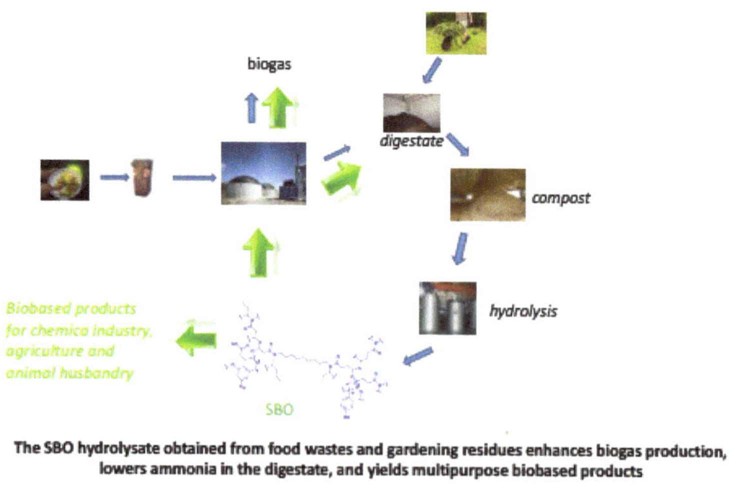

Fig. 6.8 Production of SLP (SBO in the figure) from municipal bio-wastes (*blue arrow*), and (*green arrow*) using SLP as additive to the in-house biogas production reactor for ammonia abatement and for the manufacture of multipurpose products for the external market

Table 6.4 Comparison of ammonia Nitrogen (N_{NH3}) abatement costs by conventional and SLP assisted technology for the Acea anaerobic digestion process

	Amount (t year^{-1})	N_{NH3} abatement cost (€ year^{-1})	Facility capital cost (M€)
Organic humid fraction slurry feed to bioreactors	100,000		
N_{NH3} amounts in feed	126		
N_{NH3} amounts in digestate	156		
N_{NH3} production	30		
N_{NH3} abatement by 0.05% SLP	25		
Required SLP	50	5,000–25,000	0.2–0.3
Cost of N_{NH3} abatement by conventional technology, 1.4 € kg^{-1}		35,000	1–2

produced from the plant compost and used within the same plant to carry on the in-house SLP assisted anaerobic digestion process. The economic impact is expected from the SLP in-house use and from its allocation in the external market of biosurfactants, agriculture auxiliaries and plastics materials. The possibility to allocate the SLP in these markets offers further incentives for implementing SLP production in excess of the amount needed for the plant in-house use.

SLP In-house use

Table 6.4 reports some pertinent figures based on the virtuous cycle depicted in Fig. 6.8. The figures in the Table are extrapolated according to the following current plant feed and on ammonia abatement rate measured in laboratory studies (Francavilla et al. 2016b). The case study Acea plant processes by anaerobic digestion 100,000 t year^{-1} of the municipal bio-waste fermentation slurry. Under normal operational condition, the ammonia nitrogen (N_{NH3}) amounts in the bioreactor feed and digestate liquor at fermentation end are 126 and 156 t year^{-1}, respectively. The process therefore produces 30 N_{NH3} t year^{-1}. The fermentation in the presence of SLP 0.05% will allow abating 25 N_{NH3} t year^{-1}, 83% of the yearly N_{NH3} production in the absence of added SLP. This will require however production of 50 SLP t year^{-1} with a cost of 5–25 $\times$ 10^3 € year^{-1}. The cost of abating 25 N_{NH3} t year^{-1} by the conventional technology (Burke 2013) would be 35 k € year^{-1}. The capital cost of a plant producing 50 SLP t year^{-1} is estimated 200–300 k€. By comparison, the capital cost of the conventional facility for the abatement of the above N_{NH3} amounts is about one order magnitude higher. The data shows that operational cost savings of 5–30 k€ year^{-1} would be obtained by using SLP addition in place of the conventional technology for abating N_{NH3}. However, the capital investment in the SLP option is much lower. To complete this scenario, it should be considered that the SLP production technology is still in the early development stage. Further process development and optimization studies are likely to bring substantial cost reduction for SLP production.

SLP allocation in the agriculture market

The agriculture market comprises mostly mineral and organic products, used as fertilizers and/or plant biostimulants. A minor amount of other agriculture auxiliaries, such as plant disease suppressing agents is marketed. However, their price is very high. Benzothiadiazole is one of the most used plant disease suppressants (Burketova et al. 2015). It is sold at about 820 \$ kg^{-1} (eBiochem 2016). The major fertilizers market comprises mineral and organic products. The world consumption of mineral fertilizers containing N, P and K was estimated 187 million t in 2014, with demand expected to grow at 1.8% per annum from 2014 to 2018 (FAO 2015). Generally, global consumer price inflation is projected to remain subdued as demand weakens, with falling commodity prices. In advanced economies, risks to activities associated with very low inflation have become important, especially in the Euro area, where large output gaps have contributed to low inflation. The international monetary fund considers that there is the possibility of higher real interest rates, an increase in private and public debt burdens, and weaker demand and output. Major agriculture crops are wheat, coarse grain, rice, and sugar and oil crops. The major mineral fertilizers are urea, diammonium phosphate, phosphate rock, potassium chloride, triple superphosphate with production cost ranging from minimum 0.11 € kg^{-1} for phosphate rock to maximum 0.46 € kg^{-1} for diammonium phosphate. Over the last two decades, the market outlook for organic fertilizers has not been bright. An article published in the FAO magazine (Fresco 2003) in 2003 had predicted that non-mineral nutrient sources were unlikely to challenge mineral fertilizer in the future. In 2007, other authors (Kelly and Crawford 2007) reported that organic soil management methods contributed to soil fertility improvement, but were inadequate for meeting the rapid and sustainable growth needed in African countries characterized by low food crop production. Consequently, the only means of both maintaining soil fertility and of achieving the required rate of agricultural growth was to increase significantly the quantities of mineral fertilizers. However, the combined use of mineral and organic fertilizers was a possible option to increase crop output and the amount of biomass available for transfer to land on which crops are grown.

There is no question that there has been a spurt in domestic and international demand for greening agriculture across the countries as a result of initiatives of multiple actors such as institutions/organizations, industrial and trading firms, farming communities, civil society and their representatives. A survey (Kolanu and Kumar 2007) reports that, although organic agriculture in India is likely to provide economic opportunities for different stakeholders, thanks to a number of drivers, several factors exist which constrain the development of this practice. A large number of these problems are due to the relatively newness of this sector from the point of view of different players, such as products' producers/distributors/traders, users (farmers), promoters (governments). Most problems arise from missing market regulation, poor selection of good quality products, scares product information, and insufficient government incentives' and farmers education policy. Nevertheless, in the last few years, a new class of organic products named biostimulants has emerged (du Jardin 2015). A plant biostimulant is any substance or microorganism applied to plants with the aim to enhance nutrition efficiency, abiotic stress tolerance and/or crop quality traits, regardless of its mineral nutrients content. These products modify the physiology of plants, promoting their growth and enhancing their stress response. Compared with biofertilizers, the capacity of biostimulants to promote plant growth under stressful conditions, and at very low doses, is the main distinguishing factor. Biostimulants are generally classified into three major groups based on their source and content. These groups include humic substances, hormone containing products, and amino acid containing products. The humic substances are natural constituents of the soil organic matter, resulting from the decomposition of plant, animal and microbial residues, but also from the metabolic activity of soil microbes using these substrates. Regardless of their source, they include extracts from naturally humified organic matter (e.g. from peat or volcanic soils), from composts and vermicomposts, or from leonardite fossil deposits. The SLP bear similar origin and chemical features as humic substances (Montoneri et al. 2011). Evidence of bio-stimulant properties for SLP has been published (Fascella et al. 2015; Massa et al. 2016). Further work to assess the full potential of SLP biostimulant properties is in progress. Thus, new perspectives are opening for organic substances in general, and for SLP specifically, to replace and/or decrease mineral fertilizers consumption in agriculture. In such scenario, due to the worldwide cost effective availability of municipal bio-wastes (Sheldon-Coulson 2011), as compared to the other humic substances sources, SLP constitute a potentially more viable alternative, which would also contribute reduced depletion of fossil leonardite deposits.

Undoubtedly, now, organic fertilizers belong to a niche market. Some reports estimate the US fertilizer market to be around $40 billion of which organic fertilizers occupy only about $60 million. The rest of it is the share of the various artificial fertilizers (Diffen 2016). Organic fertilizers wholesale prices (Alibaba 2016a, b; Ebay 2016) range from 140 $ t^{-1} for solid products containing 10% soluble organics, to 1500 $ t^{-1} for products with >90% soluble organics, and to 3000 $ t^{-1} for products in solution containing 35% organics and other mineral elements. Based on personal interviews by the author of the present chapter with major Italian distributors of peat derived organic fertilizers, the European market wholesale value can be estimated 20–25 million €, mainly in Spain 5–6 million €, Italy 4–5 million € and France 3.5–4.5 million. At a minimum sale price of 1000 € t^{-1}, this is equivalent to 20–25 thousands t sale.

A most recent paper (Fascella et al. 2015) reports the effects of SLP isolated from municipal bio-waste compost on Euphorbia x Lomi cultivation, in comparison with a commercial product containing humic substances extracted from Leonardite. The SLP are reported more powerful than the commercial Leonardite product in enhancing plant photosynthesis, growth and aesthetic effect, improving flower quality, and optimizing water use efficiency. Enhancement factors of plant performance indicators by SLP range from 1.3 to 8.6 relatively to the control plants, and from 1.2 to 4.5 relatively to plants treated with the commercial Leonardite based product at equal applied dose. The vis-à-vis performance comparison of SLP with the commercial Leonardite derived product demonstrates that SLP could efficiently replace commercial humic products in the agriculture market. The above commercial Leonardite derived product containing 30% dry matter can be purchased in 1 kg package for 7 € kg^{-1} (Fascella et al. 2015). Based on the dry matter content, this price is equivalent to over 23 € kg^{-1} dry matter. The SLP production cost has been estimated about 0.1–0.5 € kg^{-1} (Montoneri et al. 2011). The figures prospect attracting economic benefits deriving from the allocation of SLP in the organic fertilizer market.

Further commercial opportunity may derive from the growth of the biostimulants market, currently estimated in 200–400 million euros in Europe, probably 800 million worldwide (New Ag International journal 2012; Natale 2012). The latter figure is expected to reach over 2900 million euros in 2019 (Marketsand markets n.d.).

To evaluate the economic perspectives deriving from the allocation of SLP in the above context of products categories, types, and market size and prices, it should be taken in consideration that SLP contain all mineral nutrients needed by plants. These are bonded to the soluble lignocellulosic matter. The research results (see Sect. 6.4.1) point out that the observed effects on plant growth and productivity are because SLP supply the plants the mineral nutrient in readily available soluble form, thus facilitating the nutrients uptake by the plant. Based on their organics and minerals content, the SLP would fall into the high price organic fertilizers' category. It should also be considered that SLP are obtained from composted urban bio-wastes, and that the yearly production of Italian organic humid bio-waste is 4.2 million t (Bastioli 2013), which can potentially yield 300–400 thousand t of SLP. This potential production exceeds the above estimated organic fertilizers market size. It is evident that this market cannot absorb all organic fertilizers that can be obtained from the produced compost. It should also be considered that the SLP have been proven efficient plant disease suppressants (see Sect. 6.4.1). The capacity to induce plant protection against pathogens adds 1–3 order magnitude higher market value (eBiochem 2016) to the product, in comparison with fertilizers for enhancing plant growth (Alibaba 2016a, b; Ebay 2016). Yet, since plant disease suppressants are given at low doses, the market size of these products' category is small. The above literature survey however points out that the organic fertilizers market is in the early stage. Under these perspectives, the SLP might be favoured for their capability to provide an integrated complete plant nourishment, which contains both mineral and organic matter of renewable source. In principle, these products could replace current commercial mineral and organic fertilizers. To appreciate the full potential of SLP uses in agriculture, it should be taken also in consideration the work (Franzoso et al. 2015a, b, c) reporting SLP as potential components of new composite mulch films. Used in agriculture, these films might have multiple function, i.e. protecting plants against negative external influences, creating an ideal microclimate, and slowly releasing the SLP into the soil to stimulate plant and crop growth.

SLP allocation in the plastics market

The plastics market is more challenging than the agriculture market. In the recent years, bio-based polymers have raised great interest since sustainable development policies tend to expand with the decreasing reserve of fossil fuel and the growing concern for the environment. These polymers bring a significant contribution to the sustainable development in view of the wider range of disposal options with minor environmental impact. As a result, the market of these environmentally friendly materials is in rapid expansion. However, until now, bio-based polymers have not found extensive applications in industries to replace conventional plastic materials, reasons being their high production costs and sometimes their underperformed properties. Compared to traditional resins costs, which run below 2 € kg^{-1} (Kanellos 2009), current biopolymers are from about 2.0 to 7.0 times more expensive (Roland-Holst et al. 2013). The difference depends on the fluctuation of oil prices and on the type of bioplastics, whether they are from dedicated crops, such as starch, or from fermentation, such as polyhydroxylalkanoates. The following cost and production figures for major biopolymers, compared to synthetic polymers, may be found in literature (Roland-Holst et al. 2013). Depending on which bacterial producer is used to generate polyhydroxylalkanoates, the cost of production can range from 4 to 16 \$ kg^{-1}. However, to be commercially viable, these products should be sold 3–5 \$ kg^{-1}. By comparison, polylacticacid is more cost-competitive. This polymer was selling in bulk at approximately 0.90 \$ lb^{-1} in the last quarter of 2011, against polystyrene and polyethylene terephthalate selling at 1.00 and 0.80 \$ lb^{-1}, respectively. Packaging is one of the fastest growing sectors for bioplastic consumption. Bioplastic packaging consumption has been estimated to be 125,000 t in 2010 with an estimated market value of 454 million \$ (Roland-Holst et al. 2013). These figures correspond to an average 3.6 \$ kg^{-1} sale price.

A vast number of biopolymers (e.g. cellulose, chitin, starch, polyhydroxyalkanoates, polylactide, polycaprolactone, collagen and other polypeptides) have been synthesized or are formed in natural environment during the growth cycles of organisms. Most biopolymers are obtained from dedicated crops. The use of land to cultivate plants for energy or chemicals production raises much socio-environmental and moral concern. Negative impacts on land, water and biodiversity, and food production count among the most discussed side effects of this practice (Green Facts 2015; FAO 2008). Using corn as non-food feedstock may cause food price increase and thus can be controversial. Bio-wastes, as sources of biopolymers, have not been much investigated so far. Yet, their valorisation for this scope can potentially overcome the socio-environmental and moral criticalities connected to the exploitation of dedicated crop for producing chemicals.

In this context, the manufacture of the composite plastic films (see Sect. 6.4.1), which contain SLP blended with poly vinyl alcohol-co-ethylene (Franzoso et al. 2015a, c, 2016) and poly ethylene-co-acrylic acid (Franzoso 2015b), is a praiseworthy approach. Poly vinyl alcohol-co-ethylene is a special high cost polymer, with a market price (Alibaba n.d.) of about 5.5 € kg^{-1}. The production cost of SLP is estimated 0.1–0.5 € kg^{-1} (Montoneri et al. 2011). This figure is rather attracting, upon considering that for the production of polylacticacid and other bioplastics, food crops are a major input, and that the incidence of the cost of corn is 0.26 $ per kg of plastic produced (Roland-Holst et al. 2013). The relatively low SLP production cost prospects that the substitution of a fraction of poly vinyl alcohol-co-ethylene with the cheaper SLP biopolymer should yield a blend with a cost per kg lower or not higher than that for the neat synthetic polymer. Thus, based on the enhanced mechanical strength exhibited by the poly vinyl alcohol-co -ethylene-SLP blend, it seems possible to make articles with the bio-waste based blend, which were more eco-friendly, and equally or better performing than those made by the neat synthetic polymer, without increasing the product final market price.

The potential stake of the above perspective can be appreciated considering that bioplastics are currently only in a small portion (under 1%) of global market share of plastics (European Bioplastics 2016; Storz and Vorlop 2013; Plastermat n.d., Nova Institute 2015), which should be around 1–2 million t year^{-1}. Worldwide bioplastics demand has grown tremendously over the past several years, albeit still representing a small fraction of global plastics demand. As of 2007, it was estimated that worldwide production of bioplastics amounted to approximately 360,000 metric t (890,000 metric t by 2012). It would reach 1.5–4.4 million metric t by 2020 (Roland-Holst et al. 2013). An average size waste treatment plant (see Sect. 6.5.2) can produce 900 t year^{-1} SLP, which in turn would allow obtaining 5000 poly vinyl alcohol-co-ethylene-SLP blend ton year^{-1}. Thus, there are interesting revenue and market share expectations for allocating urban refuse sourced biopolymers in the current bioplastic market.

SLP allocation in the surfactants market

Further economic and environmental benefits may potentially derive from allocating the SLP in the surfactants' market. In the case of the Acea plant, the production and market allocation of SLP has been calculated yielding six times higher earnings than the current plant selling biogas and compost (Montoneri et al. 2011). The result stems from the likely sale value of SLP in the surfactant market at 1000 € per ton against 11 € per ton for compost. The global market for surfactants is currently estimated $32.6 billion per year and is projected to reach a volume of 24 million t and a value of $42.1 billion by 2020. These figures correspond to an average price of $1750 per ton (Marketsandmarkets 2015). In principle, this market could absorb the SLP production from 27,000 MBW treatment plants of Acea similar size. Under these circumstances, 75 million t per year of CO_2 emission from fossil C would be saved and all European composting plants could benefit from the added revenue deriving from selling the SLP at about 1750 € t^{-1} instead of maximum 30 t^{-1} from the sale of the pristine compost. The use of SLP, in place of fossil derived chemicals consumed in EU, would allow 15% (55 Mt year^{-1}) CO_2 lower emission.

There is a wide range of surfactants, which cover a wide range of applications. Small molecule surfactants, obtained through chemical synthesis from fossil sourced hydrocarbons have an average market price of 1 € kg^{-1}. Market prices per product kg may be even much higher for some high performance biodegradable surfactant molecules, such as rhamnolipids (Montoneri et al. 2016). These products are produced by specialized bacterias. They lower the surface tension of water from 70 down to 28 mN m^{-1} at the critical micellar concentration of 0.8–2 g L^{-1}. Their market price may run from 30 to 150 € per kg. The oxidized high molecular weight SLP have shown chemical features similarities with rhamnolipids. However, the former ones do not reach yet the high performance of bacterial surfactants. A real breakthrough would be improving the surface activity properties of the high molecular weight ozonized SLP to match the rhamnolipid biosurfactant properties.

6.5.2 Replicability and Transferability of the LV Technology in Real Operational Environment and Expected Benefits

Bio-wastes contain mainly polysaccharides and lignin. Thus, in principle the above-described LV technology is applicable to all kind of bio-wastes. One main consideration related to the scale up of bio-based products to commercial level is the management of the entire supply chains. This is not straightforward, and transitioning from the development stage to commercialization of a material requires an immense amount of coordination. It implies the availability, collection and processing of the sourcing feedstock, and the product distribution. Another challenge facing bio-based producers is securing funding for the difficult transition from research and development to commercialization. In the case of the SLP products, a waste management plant such as the case study Acea plant should be encouraged to invest in the production of SLP, due to presumed benefits (Table 6.4) obtainable from their in-house use. Once available, the SLP production facility would allow producing the product marketability in the agriculture market. This will require establishing joint venture with other companies operating in the production, marketing and distribution of agrochemicals. The successful demonstration of this approach is expected to be a strong drive for the replicability and transferability of the results. Certainly, feedstock for the production of SLP is not a problem worldwide. Moreover, there is large availability of waste management plant, running composting and anaerobic digestion facilities (see Sect. 6.2.1), which would be interested in the economic benefits deriving from integrating the LV technology into the existing fermentation facilities. On the other hand, another important source of bio-waste of the LV technology are farms, which process agriculture residues such as straw and manure by anaerobic digestion. As anticipated in Sect. 6.4.2, these make up another important large market sections, which may benefit from technological transfer of the SLP assisted anaerobic digestion technology. This scenario depicts the following virtuous renewable C cycle. Urban bio-wastes are used to obtain products (SLP) in large urban waste treatment plants. The SLP producer will use the product for its needs, and will sell the produced excess to the farmer customer. In this fashion, producer and customer, regardless of the operational capacity and the required product amount by the latter one, would share the environmental and economic benefits of the SLP.

The SLP benefits are not limited at the use of the product for reducing the ammonia content in anaerobic digestates. The use of SLP in agriculture would allow reducing the use of commercial fertilizers, and the environmental impact caused by applying high fertilizer doses to soil (see Sect. 6.4.1). The use of SLP to make bio-based plastic blend articles would allow reducing the consumption of synthetic polymers derived from fossil source. The economic and environmental benefits stemming from all these applications may encourage communities worldwide to start or implement further dismissing bio-waste landfilling, and/or to reduce waste incineration practices, in the outlook of more economically rewarding and eco-friendly technology.

Notwithstanding the amount of research carried out, to enter our everyday life the SLP must be assessed for their performance, marketability and sale value in real operational environment, and need life cycle analysis and certification (European Commission 2016). The technological transfer of experimental to industrial and commercial scale may be relatively easy for technologies producing energy or products that are known and assessed in the market. It is more complicated in the case of new technologies, such as the LV technology producing bio-based substances as the SLP, which are not known in the market. These products need not only market assessment, but also legislative acceptance.

According to Italian legislative decree DL 29 April 2010, n. 75, humic extracts, in solid, aqueous suspension or liquid form, obtained from soil, fossil deposits such as peat, and generally from natural humification processes, may be permitted for use as soil improvers or fertilizers for agriculture, if they meet some compositional requirements. The SLP have been demonstrated having chemical composition similar to natural humic substances (Montoneri et al. 2011), and biostimulant performance similar or better than Leonardite sourced humic substances (Fascella et al. 2015). Thus, based on chemical composition and performance, the SLP fall well into the category of fertilizers according to the above Italian legislation.

In addition to these facts, Biagini et al. (2016) have demonstrated that the SLP are not toxic for rabbits. The lack of toxicity has been shown also in pigs fed with a diet containing the same SLP (Montoneri 2012).

A formal request of legislative acceptance for the SLP has not been filed yet with any national authority. Legislative acceptance of new products requires production scale up in order to perform products' on field and laboratory studies, and support by producers' association. For example, recently the Italian Biochar Association (www.ichar.org) has succeeded in getting inclusion of biochar in the list of agriculture soil improvers, which are authorized for use in Italy (see Gazzetta Ufficiale, Serie Generale n° 186 12-8-2015). This result has been accomplished after eight research on biochar, which was started in Italy in 2008. Recently, Mozzetti et al. (2015) have compared the SLP obtained from post-harvest tomato plants and biochar obtained from the pyrolysis of poultry manure and miscanthus. The products were compared for chemical composition and performance in the cultivation of radish. The results show that the SLP and the pristine tomato plants contain lignin, hemicellulose, protein, peptide and/or amino acids moieties, and several mineral elements. The biochar samples contain also similar mineral elements, but the organic fraction is characterized mainly by fused aromatic rings. All materials had positive effects on radish growth.

The comparison of SLP and biochar by Mozzetti et al. (2015), as well as the comparison of SLP and Leonardite sourced humic acids (Fascella et al. 2015), and the lack of toxicity of SLP for plants (Sortino et al. 2014) and for animals (Biagini et al. 2016; Montoneri 2012), offer sound arguments supporting the potential legislative acceptability of SLP for use in agriculture and other sectors.

For the SLP, the task required for the definite market assessment and legislative acceptance may only be accomplished by the construction and operation of a demo plant, which had production capacity adequate for carrying on field tests in near operational environment. This implies risking capital investment. In the case of SLP, this step may be achieved with much reduced entrepreneurial risk by starting with the production of SLP for in-house use. Construction and operation of the SLP production facility, finalized at producing digestate with reduced ammonia content, would also allow producing at very limited risk excess SLP for assessing the perspective to allocate the products in the market for the other uses in the chemical industry and in agriculture (Fig. 6.6). During this time, the capital investment would be paid off by the savings resulting from adopting the SLP technology for abating the ammonia content in the digestate, which is produced routinely in by the plant anaerobic fermentation section. In this fashion, a municipal waste treatment plant, as the Acea plant (Fig. 6.1), might be turned out into a biorefinery with different product lines (Fig. 6.6), where the production output of the different product lines was modulated according to the specific markets' demands. The availability of multipurpose products, such as SLP, is a further benefit, in as much as it allows reducing the risk of saturating a specific market sector. This is a very attracting perspective. Taking the Acea plant as example, the following is a possible stepwise business development strategy in order to maximize the possibility of success and minimize the risk:

- Plan facility for the production of SLP to be integrated into the current waste treatment plant in order to produce SLP in excess of the amount needed for the in-house anaerobic digestion process
- Run performance and marketability tests in real operational environment for SLP use in other market sectors. Produce product formulations for specific uses
- Assess the legislative acceptance of the SLP
- Scale up SLP production facility according to market assessment results
- Invest in new process/products research and development.

Starting with the SLP production for in-house use seems a viable sustainable route toward the construction of a biorefinery fed with municipal bio-wastes. In the long term, this business strategy will allow collecting data in real operational environment, which will be useful for planning a reliable technology transfer to other waste management plant throughout the world. Such scenario implies positive impact in many sectors. First, under this perspective, waste management facilities would become also producers of value added chemicals. Second, the economic benefits of the integrated plant selling biogas and SLP should in principle results in lower taxes for citizens. Third, the chemical market would consume less non-renewable sources for the manufacture of chemicals and benefit processes cost reduction. Fourth, the distributed nature of the wastes source might align geographically with areas of the region where development of new business opportunities and jobs is of vital interest.

Multiple environmental benefits are expected for agriculture, the use of compost, the anaerobic digestion processes, and the chemical sector. These concern main important issues, such as the reduction of fossil sources consumption and of GHG emission, deriving from substituting synthetic chemicals with bio-waste based products. Multiple diversified investors are likely to act. This scenario will contribute to assess two new technological and socio-economic concepts:

- a municipal bio-waste treatment plant may be turned into a biorefinery for the production of fuel and valuable bio-waste based products with friendly environmental impact;
- municipal bio-wastes may be a source revenue, and not merely a burden for society.

Undoubtedly, the appealing economic benefits of the integrated municipal bio-waste plant producing biogas and SLP might be an important socio economic driver for diverting waste from landfill or from incinerating valuable organic matter. In densely populated areas, the value of land is generally high, which makes it critically costly to use areas for landfill. On the other hand, incinerators need densely populated areas to be operated with enough feed to be operated at optimum efficient capacity. For the same reason, in densely populated areas, an integrated municipal bio-waste plant producing biogas and SLP could be operated at optimum cost-effective size. Thus, in densely populated areas, the integrated municipal bio-waste plant producing biogas and SLP might may become a real viable option to reduce landfills and incineration. Perhaps, the best socio-economic message of such scenario is 'before burning everything up, consider saving what is valuable. Municipal bio-waste contains valuable organic matter and the real waste is dismissing to landfills or burning it'. The present chapter shows that the realization of this scenario may start from a typical waste management plant, as the Acea plant, which started producing SLP for internal use. This would allow developing new business innovation models for municipal bio-waste plants, as well as the definition of guidelines and recommendations for a European Common Policy to support the adoption of business innovation models in the new European bio-based chemical industry.

References

Al Seadi T, Rutz D, Prassl H, Köttner M, Finsterwalder T, Volk S, Janssen R (2008) Biogas handbook. University of Southern Denmark Esbjerg (Niels Bohrs Vej 9-10, DK-6700 Esbjerg, Denmark, R). http://www.lemvigbiogas.com/BiogasHandbook.pdf. Accessed 7 Mar 2016

Alibaba (2016a) Leonardite humic acid fulvic acid fertilizer. http://www.alibaba.com/trade/search?fsb=y&IndexArea=product_en&CatId=&SearchText=leonardite+humic+acid+fulvic+acid+fertilizer. Accessed 7 Mar 2016

Alibaba (2016b) Organic fertilizer market. http://www.alibaba.com/showroom/organic-fertilizer-market.html. Accessed 7 Mar 2016

Alibaba (n.d.) EVOH—SOARNOL SG-654-NG. http://www.alibaba.com/product-detail/EVOH-SOARNOL-SG-654-NG_169861914.html. Accessed 6 Mar 2016

Arato C, Kendal Pye E, Gjennestad G (2005) Appl Biochem Biotechnol 871:121–124

Avetta P, Bella F, Bianco Prevot A, Laurenti E, Montoneri E, Arques A, Carlos L (2013) ACS Sustain Chem Eng 1:1545–1550

Baglieria A, Cadilia V, Mozzetti Monterumici C, Gennari M, Tabasso S, Montoneri E, Nardi S, Negre M (2014) Sci Horticulturae 176:194–199

Barilla Center for Food Nutrition (2012) Food wastes: causes, im-pacts and proposals, https://www.barillacfn.com/media/publications/food-waste-causes-impact-proposals.pdf. Accessed 7 Mar 2016

Barth J (2010a) Bio-waste in the context of EU legislation. http://www.google.it/url?sa=t&rct=j&q=&esrc=s&source=web&cd=8&ved=0CF4QFjAHahUKEwjRgsvkx7XIAhWCPxQKHSCvCqg&url=http%3A%2F%2Fwww.act-clean.eu%2Fdownloads%2F3_uba_ifat_bio-waste_barth_final.ppt&usg=AFQjCNELaOtN7OulrqHkwfNxFGoQ9Q0ZGA&sig2=gw-HGCgRZMdR-1Fg6rU-rgg. Accessed 7 Mar 2016

Barth J (2010b) Markets for compost and digestate in Europe. http://compost.it/biblio/2010_beacon_conference_perugia/2nd_day/5.c%20-%20Barth.pdf. Accessed 7 Mar 2016

Bastioli C (2013) Bioplastics: an Italian case study of bioeconomy in Italy. http://www.kyotoclub.org/docs/Bruxelles_060313_Bioplastics_Bastioli.pdf. Accessed 7 Mar 2016

Baxter MD, Acosta E, Montoneri E, Tabasso S (2014) Ind Eng Chem Res 53:3612–3621

Biagini D, Gasco L, Rosato R, Peiretti PG, Gai F, Lazzaronia C, Montoneri C, Ginepro C (2016) Anim Feed Sci Technol. doi:10.1016/j.anifeedsci.2016.02.005

Boffa V, Perrone DG, Magnacca G, Montoneri E (2014) Ceram Int 40:12161–12169

Brunow G, Lundquist K, Gellerstedt G (1998) In: Sjostro E, Alen R (eds) Analytical methods in wood chemistry, pulping and papermaking. Springer Science and Business Media, Berlin. https://books.google.it/books?id=U-0k4NCbbNQC&pg=PR6&lpg=PR6&dq=Eero+Sj%C3%B6str%C3%B6m,+Raimo+Alen,+1998&source=bl&ots=38EM_u6Nd_&sig=i74LLwJL2FKKarHUP-5FLQ_JHng&hl=it&sa=X&ei=sV-YVNjNOYbkUuiZgPAO&ved=0CCMQ6AEwAA. Accessed 7 Mar 2016

Burke DA (2013) Economical recovery of ammonia from anaerobic digestate. In: Waste to worth, spreading science and solutions. Denver, CO. http://www.extension.org/pages/67578/economical-recovery-of-ammonia-from-anaerobic-digestate#.VEYae51H5b0. Accessed 7 Mar 2016)

Burketova L, Trda L, Ott PG, Valentova O (2015) Biotechnol Adv 33:994–1004

CalRecycle (2016) Organic materials management, compost use in agriculture, Jan 15. http://www.calrecycle.ca.gov/organics/farming/; http://ec.europa.eu/environment/waste/compost/…/econanalysis_finalreport.pdf; http://ec.europa.eu/…/waste/…/econanalysis_execsum.pdf. Accessed 3 Mar 2016

Canilha L, Chandel AK, dos Santos Milessi TS, Antunes FAF, da Costa Freitas WL, das Gracas Almeida Felipe M, da Silva SS (2012) J Biomed Biotechnol, Article ID 989572. doi:10.1155/2012/989572

Centemero M, Ricci M, Giavini M, Dall'Anna D, Campagnol J, Longu G (2015) Annual report of the Italian composting and biogas association. http://www.renewablematter.eu/partners/CIC/CIC%20annual_report2015eng.pdf. Accessed 7 Mar 2016

Chiellini E, Solaro R (2003) Biodegradable polymers and plastic. Springer Science+Business Media, New York. doi:10.1007/978-1-4419-9240-6. https://books.google.it/books?id=CdLiBwAAQBAJ&pg=PA386&lpg=PA386&dq=biodegradable+polymers+based+on+polyhydroxy+acids&source=bl&ots=-z098-XF69&sig=JFr0gSZjp1uHN74niYdHm4GY1HM&hl=it&sa=X&ved=0ahUKEwjpl8OqwajJAhXEKw8KHUuiDOIQ6AEIS-zAG#v=onepage&q=biodegradable%20polymers%20based%20on%20polyhydroxy%20acids&f=false. Accessed 7 Mar, 2016

Clark J (2007) J Chem Technol Biotechnol 82:603–609

Clark J, Deswarte F, Farmer TJ, Mascal M (2014) In: Introduction to chemicals from biomass, 2nd ed. Copyright © 2015 Wiley, NJ. Published online: 26 Dec 2014. doi:10.1002/9781118714478.ch4. http://onlinelibrary.wiley.com/doi/10.1002/9781118714478.ch4/summary. Accessed 7 Mar 2016

CP Manufacturing (2012) Solid waste management equipment. http://www.cpmfg.com/material-recovery-facility/municipal-solid-waste-recycling/solid-waste-management-equipment/. Accessed 3 Mar 2016

David M, Joshi T, Negin D, Tan KY, Fagan JM (2010) Indus-trial biodegradable waste with primary focus on food waste. https://webcache.googleusercontent.com/search?q=cache:MT3tHA7_Oa4J;https://rucore.libraries.rutgers.edu/rutgers-lib/37176/pdf/1/+&cd=1&hl=it&ct=clnk&gl=it. Accessed 6 Mar 2016

Deganello F, Tummino ML, Calabrese C, Testa ML, Avetta P, Fabbri D, Bianco Prevot A, Montoneri E, Magnacca G (2015) New J Chem 39:877–885

Department for environmental food and rural affairs (2013) Incineration of municipal solid wastes. https://www.gov.uk/government/uploads/system/uploads/attachment_data/file/221036/pb13889-incineration-municipal-waste.pdf. Accessed 3 Mar 2016

Diffen (2016) Chemical fertilizer vs. organic fertilizer. http://www.diffen.com/difference/Chemical_Fertilizer_vs_Organic_Fertilizer#Cost. Accessed 7 Mar 2016

Dinuccio E, Biagini D, Rosato R, Balsari P, Lazzaroni C, Montoneri E (2013) Adv Anim Biosci 4 (Special Issue 2):515

du Jardin P (2015) Sci Hortic 196:3–14

Ebay (2016) Leonardite. http://www.ebay.it/itm/LEONARDITE-5-Kg-CONCIME-PIANTE-SIEPI-FIORI-ORTO-GIARDINO-CACTUS-BONSAI-/270841276887. Accessed 7 Mar 2016

eBiochem (2016) Wholesale 2,1,3-benzothiadiazole, http://www.ebiochem.com/product/2-1-3-benzothiadiazole-71535. Accessed 7 Mar 2016

Ensinger Gmbh (n.d.) Tensile strength at break. http://www.ensinger-online.com/ru/technical-information/properties-of-plastics/mechanical-properties/tensile-strength-at-break/. Accessed 11 Oct 2016

European Bioplastics (2016) Bioplastics facts and figures. http://www.google.it/url?sa=t&rct=j&q=&esrc=s&source=web&cd=2&ved=0ahUKEwiymt7lrq7LAhXCbRQKHSBs-BMMQFggnMAE&url=http%3A%2F%2Fwww.muoviteollisuus.fi%2Fdocument.php%2F1%2F156%2Feuropean_bioplastics_-_facts_and_figures%2Fb4313531b6f32c56e3b4a4d3d5fdd15c&usg=AFQjCNGRZsjgqlXbzlwgl2n_uTKZx51RfA&sig2=d_G6fqLgWyJL82H-GWbVsw. Accessed 7 Mar 2016

European Commission (2013) Innovation, how to convert research into commercial success story? http://ec.europa.eu/research/industrial_technologies/pdf/how-to-convert-research-into-commercial-story_en.pdf. doi:10.2777/10737. Accessed 7 Mar 2016

European Commission (2014) Horizon 2020—Work programme 2014–2015, general annexes G, technology readiness lev-els (TRL), extract from part 19—Commission Decision C(2014) 4995, http://ec.europa.eu/research/participants/data/ref/h2020/wp/2014_2015/annexes/h2020-wp1415-annex-g-trl_en.pdf. Accessed 7 Mar 2016

European Commission (2015) Environment, biodegradable waste. http://ec.europa.eu/environment/waste/compost/. Accessed 3 Mar 2016

European Commission (2016) REACH, http://ec.europa.eu/growth/sectors/chemicals/reach/index_en.htm. Accessed 7 Mar 2016
European Compost Network (2010) Country report of Germany. http://www.compostnetwork.info/germany.html. Accessed 7 Mar 2016
European Environmental Agency (2013) Managing municipal solid wastes. EEA report. http://www.google.it/url?sa=t&rct=j&q=&esrc=s&source=web&cd=2&ved=0ahUKEwimrrm7hK7LAhVD8RQKHdJdATgQFggrMA-E&url=http%3A%2F%2Fwww.eea.europa.eu%2Fpublications%2Fmanaging-municipal-solid-waste%2Fdownload&usg=AFQjCNG4-8WUDJj9wZU0HMBKSSzxm4upCw&sig2=uSgUDij0O-FDP08ub1mV3g. Accessed 7 Mar 2016
Eurostat (2016) Farm structure statistics. http://ec.europa.eu/eurostat/statistics-explained/index.php/Farm_structure_statistics. Accessed 7 Mar 2016
FAO (2008) The state of food and agriculture, Part I: biofuels: prospects, risks and opportunities. http://www.fao.org/docrep/011/i0100e/i0100e00.htm. Accessed 7 Mar 2016
FAO (2011) Global food losses and food waste—extent, causes and prevention, Rome. http://www.fao.org/docrep/014/mb060e/mb060e00.pdf. Accessed 3 Mar 2016
FAO (2015) World fertilizer trends and outlook to 2018. http://www.fao.org/3/a-i4324e.pdf. Accessed 7 Mar 2016
Fascella G, Montoneri E, Ginepro M, Francavilla M (2015) Sci Hortic 197:90–98
Fiedler H (1998) Sources and environmental impact of PCDD/PCDF. In: Proceedings of the subregional awareness raising workshop on persistent organic pollutants (pops), Kranjska Gora, Slovenia, 11–14 May 1998. http://www.chem.unep.ch/pops/POPs_Inc/proceedings/slovenia/CONTKG.html. Accessed 7 Mar 2016
Foged H, Flotats Ripoll X, Bonmatí Blasi A, Palatsi Civit J, Magrí Aloy A, Schelde KM (2012) Inventory of manure processing activities in Europe. http://agro-technologyatlas.eu/docs/21010_technical_report_I_inventory.pdf. Accessed 7 Mar 2016
Francavilla M, Beneduce L, Gatta G, Montoneri E, Monteleone M, Mainero D (2016a) Biochem Eng J 116:75–84
Francavilla M, Beneduce L, Gatta G, Montoneri E, Monteleone M, Mainero D (2016b) J Chem Technol Biotechnol 91:2679–2687
Franzoso F, Antonioli D, Montoneri E, Persico P, Tabasso S, Laus M, Mendichi R, Negre M, Vaca-Garcia C (2015a) J Appl Polym Sci 132(18):6006
Franzoso F, Causone D, Tabasso S, Antonioli D, Montoneri E, Persico P, Laus M, Mendichi R, Negre M (2015b) J Appl Polym Sci 132(18):5803. doi:10.1002/app.41909
Franzoso F, Tabasso S, Antonioli D, Montoneri E, Persico P, Laus M, Mendichi R, Negre M (2015c) J Appl Polym Sci 132(4):1301. doi:10.1002/app.41359
Franzoso F, Vaca-Garcia C, Rouilly A, Evon P, Montoneri E, Persico P, Mendichi R, Nisticò R, Francavilla M (2016) J Appl Polym Sci 133(9):2110. doi:10.1002/APP.43009
Fresco LO (2003) FAO spotlight/2003, fertilizer and the future. http://www.louiseofresco.com/pdf/FAO/2003%20-%20FAO%20Spotlight%20-%20Fertilizer%20and%20the%20future.pdf. Accessed 7 Mar 2016
Gomis J, Bianco Prevot A, Montoneri E, Gonzalez MC, Amat AM, Martire DO, Arques A, Carlos L (2014) Chem Eng J 235:236–243
Green Facts (2015) Liquid biofuels for transport prospects, risks and opportunities. http://www.greenfacts.org/en/biofuels/l-3/4-environmental-impacts.htm. Accessed 7 Mar 2016
Hu Z, Wen Z (2008) Biochem Eng 38:369–378
Huffington Post (2013) Food waste: Half of all food ends up thrown away, 10 Jan 2013. http://www.huffingtonpost.co.uk/2013/01/10/food-waste-half-of-all-fo_n_2445022.html. Accessed 6 Mar 2016
Isagro SpA (2014) Efficacia biologica di Compost forniti dall'Università di Torino (Unpublished confidential report delivered to the author). www.isagro.com
Ispra (2012) Banche dati, environmental data yearbook. http://annuario.isprambiente.it/pdf/environmental-data-yearbook-2012. Accessed 7 Mar 2016
Ji F, McGlone JJ, Kim SW (2006) J Anim Sci 84:2482–2490
Jindřichová B, Burketová L, Rosso D, Montoneri E (2016) (to be published)

Kanellos M (2009) Bioplastics close in price to regular plastic and trader Joe's coconut packages. https://www.google.it/#q=83.%09Kanellos+M%2C+Bioplastics+Close+in+Price+to+Regular+Plastic+and+Trader+Joe%E2%80%99s+Coconut+Packages. Accessed 7 Mar 2016
Kelly V, Crawford E (2007) Policies and actions to stimulate private sector fertilizer marketing in sub-Saharan Africa. Agricultural management, marketing and finance occasional paper 15. http://www.fao.org/3/a-a1136e.pdf. Accessed 7 Mar 2016
Key to Metals AG (2014) Engineering stress-strain curve. http://www.totalmateria.com/Article107.htm
Kolanu RT, Kumar S (2007) Greening agriculture in India an overview of opportunities and constraints. http://www.fao.org/docrep/article/agrippa/658_en00.htm#TopOfPage. Accessed 7 Mar 2016
Lerda D (2011) Polycyclic aromatic hydrocarbons (PAHs) fact sheet. https://ec.europa.eu/jrc/sites/default/files/Factsheet%20PAH_0.pdf. Accessed 7 Mar 2016
Liew NL (2011) Solid-state anaerobic digestion of lignocellulo-SiC biomass for biogas production. Master thesis, The Ohio State University. https://etd.ohiolink.edu/!etd.send_file?accession=osu1306870552&disposition=inline. Accessed 7 Mar 2016
Lingle J (2008) Examination of the sources of polycyclic aromatic hydrocarbons (PAHs) in urban background soil. http://yosemite.epa.gov/sab/sabproduct.nsf/F334E867D04350FB85257E270059D7F0/$File/EPRI-1015558-Sources+of+PAHs+in+Urban+Background+Soil-2008.pdf. Accessed 7 Mar 2016
Luque T, Clark J (2013) Sustain Chem Process, pp 1–10
Ma R, Xu Y, Zhang (2014), ChemSusChem 7:1–29
Magnacca G, Laurenti E, Vigna E, Franzoso F, Tomasso L, Montoneri E, Boffa V (2012) Process Biochem 47:2025–2031
Marketsandmarkets (2015) Surfactants market by product .... trends and forecast to 2020. http://www.marketsandmarkets.com/Market-Reports/biosurfactants-market-493.html. Accessed 12 Oct 2016
Marketsandmarkets (n.d.) Biostimulants market ... Global trends and forecasts to 2021. http://www.marketsandmarkets.com/PressReleases/biostimulant.asp
Martins O (1992) Bioresour Technol 42:103–111
Massa D, Prisa D, Montoneri E, Battaglini D, Ginepro M, Negre M, Burchi G (2016) Sci Hortic 205:59–69
McEntee K (2011) National compost prices. http://www.recycle.cc/compostprices.pdf. Accessed 11 Oct 2016
Mirbagheri SA, Poshtegal MK, Parisai MS (2010) Desalination 256:70–76
Montoneri E (2012) Efficacy of an extract obtained from gardening trimmings in pigs diet, unpublished study commissioned by Università di Torino to Cerzoo. Piacenza, Italy
Montoneri E, Boffa V, Savarino P, Perrone DG, Montoneri C, Mendichi R, Acosta EJ, Kiran S (2010) Biomacromolecules 11:3036–3042
Montoneri E, Mainero D, Boffa V, Perrone DG, Montoneri C (2011) Int J Glob Environ Issues 11:170–196
Montoneri C, Montoneri E, Tomasso L, Piva A (2013) J Agriculture Sci 13:31–44
Montoneri E, Tomasso L, Colajanni N, Zelano I, Alberi F, Cossa G, Barberis R (2014) Int J Environ Sci Technol 11(2):251–262
Montoneri E, Rosso D, Bucci G, Berto S, Baglieri A, Mendichi R, Quagliotto P, Francavilla M, Mainero D, Negre M (2016) Chem Select 1:1613–1629
Morone PG, Tartiu VE, Falcone P (2015) J Clean Prod 90:43–54
Mostofa KMG, Yoshioka T, Abdul Mottalemb M, Vione D (2013) Photobiogeochemistry of organic matter. Springer, Berlin. http://www.springer.com/br/book/9783642322228. Accessed 7 Mar 2016
Mozzetti Monterumici C, Rosso D, Montoneri E, Ginepro M, Baglieri A, Novotny E, Kwapinski W, Negre M (2015) Int J Mol Sci 16:8826–8843
Nasa (2015) Definition of technology readiness levels. http://esto.nasa.gov/files/trl_definitions.pdf. Accessed 7 Mar 2016

Natale G (2012) The world market for biostimulant in crop production. http://www.biostimulants.eu/wp-con-tent/uploads/2012/12/121128_TheWorldMarketForBiostimulants_Natale_def_LOWRES.pdf. Accessed 4 Mar 2016

New Ag International Journal (2012) Nov/Dec, Biostimulants: What's behind the name? NewAgInternational_excerpt_Biostimulants-Whatsbehindthename.pdf. http://www.lawrieco.com.au/uploaded_files/document_uploads/2012_11-12_. Accessed 4 Mar, 2016

Niemel K, Alen R, Sjçstrçm E (1985) Holzforschung 39:167–172

Nisticò R, Evon P, Labonne L, Vaca-Medina G, Montoneri E, Francavilla M, Vaca-Garcia C, Magnacca G, Franzoso F, Negre M (2016) Chem Select 1:2354–2365

Nova Institute (2015) Bio-based building blocks and polymers in the world. http://bio-based.eu/download/?did=32875&file=0. Accessed 7 Mar 2016

Parsell T, Yohe S, Degenstein J, Jarrell T, Klein I, Gencer E, Hewetson B, Hurt M, Kim JI, Choudhari H, Saha B, Meilan R, Mosier N, Riberio F, Delgass WN, Chapple C, Kenttamaa HI, Agarwal R, Abu-Omar MM (2015) Green Chem 17:1492–1499

Plastermat (n.d.), Bioplastics to grow to 7% of total plastics market by 2020. http://www.plastemart.com/Plastic-Technical-Article.asp?LiteratureID=1903&Paper=bioplastics-to-grow-to-7-per-cent-total-plastics-market-by-2020. Accessed 7 Mar 2016

Plumer B (2012) How the U.S. manages to waste $165 billion in food each year. The Washington Post, 22 Aug. https://www.washingtonpost.com/news/wonk/wp/2012/08/22/how-food-actually-gets-wasted-in-the-united-states/. Accessed 7 Mar 2016

Provolo G, Riva E (2009) Modelli gestionali per l'uso sostenibile degli effluenti di allevamento, GEA Quaderni della ricerca n. 104, Regione Lombardia, http://www.arca.regione.lombardia.it/shared/ccurl/409/896/GEA_q104_web,0.pdf. Accessed 7 Mar 2016

Quesada J, Rubio M, Gómez D (1999) J Wood Chem Technol 19:115–137

Ragauskas AJ, Beckham GT, Biddy MJ, Chandra R, Chen F, Davis MF, Davison BH, Dixon RD, Gilna P, Keller M, Langan P, Naskar, Saddler JN, Tschaplinski TG, Tuskan GH, Wyman CE (2014) Science 344:1246843. doi:10.1126/science.1246843

Ricci-Jürgensen M (2012) Bio-waste management in Italy. http://www.assobioplastiche.org/wp-content/uploads/2012/10/Newmann-David-Bio-waste.pdf. Accessed 6 Mar 2016

Riggio V, Rosso M, Comino E, Biagini D, Montoneri E (2016) J Chem Tech Biotechnol. doi:10.1002/jctb.5106

Roland-Holst D, Triolo R, Heft-Neal S, Bayrami B (2013) Bioplastics in California. http://www.calrecycle.ca.gov/publications/Documents/1469%5C20131469.pdf. Accessed 7 Mar 2016

Rosso D, Fan J, Montoneri E, Negre M, Clark J, Mainero D (2015) Green Chem 17:3424–3435

Rovero A, Vitali M, Rosso D, Montoneri E, Chitarra W, Tabasso S, Ginepro M, Lovisolo C (2015) Int J Agronomy Agricultural Res (IJAAR) 6:75–91

Sanders JPM, Clark JH, Harmsen GJ, Heeres HJ, Heijnen JJ, Kersten SRA, van Swaaij VPM, Moulijn JA (2012) Chem Eng Process 51:117–136

Savarino P, Montoneri E, Bottigliengo S, Boffa V, Guizzetti T, Perrone DG, Mendichi R (2009) Ind Eng Chem Res 48:3738–3749

Savarino P, Montoneri E, Musso G, Boffa V (2010) J Surfactants Detergents 13:59–68

Segre A, Gaiani S (2011) Transforming food waste into a re-source. RSC Books 2013, Royal Society of Chemistry Publishing, UK. http://www.rsc.org/globalassets/05-journals-books-databases/booksellers/portfolio-flyers/rsc-books-food-science-portfolio-flyer.pdf?id=13583. Accessed 6 Mar 2016

Segre A, Falasconi L, Morganti E (2009) Last minute market, increasing the economic, social and environmental value of unsold products in the food chain. Total food sustainability of the agrifood chain. RSC, London, pp 161–167

Sereno CK (2010) Impact of ammonia and long chain fatty acids on thermophilic anaerobic digestion of swine wastes. PhD dissertation, North Carolina State University. http://www.lib.ncsu.edu/resolver/1840.16/6147. Accessed 7 Mar 2016

Sheldon-Coulson GA (2011) Production of levulinic acid in urban biorefineries. Master of Science thesis in technology and policy, Massachusetts Institute of Technology. https://dspace.mit.edu/bitstream/handle/1721.1/68450/769021899-MIT.pdf?sequence=2. Accessed 7 Mar 2016

Sortino O, Dipasquale M, Montoneri E, Tomasso L, Avetta P, Bianco Prevot A (2013) Agron Sustain Dev 33(2):433–441

Sortino O, Montoneri E, Patanè C, Rosato R, Tabasso S, Ginepro M (2014) Sci Total Environ 487C:443–451

Storz H, Vorlop KD (2013) Appl Agric For Res 4:321–332

Tabasso S, Montoneri E, Carnaroglio D, Caporaso M, Cravotto G (2014) Green Chem 16:73–76. doi:10.1039/C3GC41103F

Tang J (2012) A cost benefit analysis of waste incineration with advanced bottom ash separation technology for a Chinese municipality Guanghan. Master of science thesis, Vienna School of International affairs. http://www.seas.columbia.edu/earth/wtert/sofos/PubDat_210340.pdf. Accessed 3 Mar 2016

Testa ML, Tummino ML, Agostini S, Avetta P, Deganello F, Montoneri E, Magnacca G, Bianco Prevot A (2015) RSC Adv 5:47920–47927

The engineering toolbox (2013) Modulus of elasticity or Young's modulus and tensile modulus for some common materials. http://documents.mx/documents/elastic-properties-and-young-modulus-for-some-materialspdf.html; http://www.engineeringtoolbox.com/young-modulus-d_417.html. Accessed 10 Oct 2016

Twardowska I, Janta Koszuta K, Miszcsak E, Stefaniak S (2010) Occurrence and binding strength of metals in composted bio-waste and sewage sludge. In: Pawlowski L, Dudzviska M, Pawlowski A (eds) Environmental engineering III. Taylor and Francis, London. https://books.google.it/books?id=_NXKBQAAQBAJ&pg=PA339&lpg=PA339&dq=food+waste+as+%25+of+total+bio-waste+production&source=bl&ots=tNyLi2_aQm&sig=MbNROoxerzh0XeK3UGPWvOq-Tn8&hl=it&sa=X&ei=FkamVKCvEYboUtLNgOgI&ved=0CEsQ6AEwBQ#v=onepage&q=food%20waste%20as%20%25%20of%20total%20bio-waste%20production&f=false. Accessed 6 Mar 2016

United States Department of Defense (2011) Technology readiness assessment (TRA) guidance. http://www.acq.osd.mil/chieftechnologist/publications/docs/TRA2011.pdf. Accessed 7 Mar 2016

US Environmental Protection Agency (2013) Global mitigation of non-CO_2 greenhouse gases. https://www.epa.gov/sites/production/files/2016-06/documents/mac_report_2013-iii_waste.pdf. Accessed 11 Oct 2016

van Ree R, van Zeeland A (2014) Iea bioenergy task42, biorefining. http://www.iea-bioenergy.task42-biorefineries.com/en/ieabiorefinery.htm. Accessed 7 Mar 2016

Vargas AKN, Savarino P, Montoneri E, Tabasso S, Cavalli R, Bianco Prevot A, Guardani R, Roux GAC (2014) Ind Eng Chem Res 53:8621–8629

wiseGEEK (n.d.) Bending modulus. http://topics.wisegeek.org/topics.htm?bending-modulus#. Accessed 7 Mar 2016

Wrap (2016) Compost calculator. http://www.wrap.org.uk/content/compost-calculator. Accessed 7 Mar 2016

Zarebska A, Romero Nieto D, Christensen KV, Fjerbæk Søtoft L, Norddahl B (2015) Crit Rev Environ Sci Technol 45:1469–1521

Zhao Q, Leonhardt E, MacConnell C, Frear C, Chen S (2010) Purification Technologies for biogas generated by anaerobic digestion. CSANR research report 2010—001. http://csanr.wsu.edu/publications/researchreports/CFF%20Report/CSANR2010-001.Ch09.pdf. Accessed 7 Mar 2016

Chapter 7
Techno-Economic Study and Environmental Assessment of Food Waste Based Biorefinery

Aude Pommeret, Xiaofeng Yang, Tsz Him Kwan, Elias A. Christoforou, Paris A. Fokaides and Carol Sze Ki Lin

Abstract Sustainability consists of three major components, namely economic, ecological and social impacts. The most important driver for food waste based biorefinery is whether the proposed design is profitable. The development of highly efficient and cost-effective biorefineries is a prerequisite for such a bio-based economy. There are many factors that influence the overall costs and returns of the food waste based biorefinery process, and affect the overall economic performance as well. In this chapter, the economic and environmental impacts of food waste based biorefinery is evaluated by using Techno-economic Study and Life Cycle Assessment (LCA) in terms of non-renewable energy use (NREU) and greenhouse gases (GHG) emission. Special focus on the economics of Green Chemistry, and the current status of LCA studies on succinic acid and thermochemical processes for biomass conversion to biofuels are covered.

A. Pommeret · T.H. Kwan · C.S.K. Lin (✉)
School of Energy and Environment, City University of Hong Kong,
Kowloon Tong, Hong Kong
e-mail: carollin@cityu.edu.hk

A. Pommeret
e-mail: aude.pommeret@cityu.edu.hk

T.H. Kwan
e-mail: thkwan-c@my.cityu.edu.hk

X. Yang
School of Bioscience and Bioengineering, South China University
of Technology, Guangzhou 510006, China
e-mail: biyangxf@scut.edu.hk

E.A. Christoforou · P.A. Fokaides
School of Engineering and Applied Sciences,
Frederick University, Limassol, Cyprus
e-mail: res.ce@frederick.ac.cy

P.A. Fokaides
e-mail: eng.fp@frederick.ac.cy

P. Morone et al. (eds.), *Food Waste Reduction and Valorisation*,
DOI 10.1007/978-3-319-50088-1_7

Keywords Food waste · Biorefinery · Life cycling assessment · Techno-economic analysis · Biofuels · Bio-succinic acid

7.1 Introduction

Growing population and the environmental issues combined with the increasing global demand for energy, chemicals and materials for social development have fostered research efforts to develop low environmental impact and sustainability technologies based on renewable raw materials to meet such global targets, contributing to the biorefinery creation. Biorefinery has been described as the process that entails refining of biomass in a commercial context for the production of chemicals, materials, fuels, food and value-added ingredients. In this case, biomass and waste (especially food waste) are selected as renewable feedstocks and converted into valuable marketable products by using a series of sustainable and low environmental impact technologies (Clark et al. 2009). This concept stimulating a great deal of interaction between scientists from different fields including chemistry, biology, environmental sciences and economics in an attempt to utilise renewable resources for a bio-based economy. Recent studies pointed out that the development of a more integrated approach is essential for the resource management based sustainable strategies along the whole food supply chain (Luque et al. 2008; Cherubini 2010).

Food waste (FW) is currently a major issue worldwide, becoming more and more important in both developing and developed countries. The causes of FW are numerous; occurring at almost every stages of the food industry and about 1.3 billion tons of FW is thrown away per year, which roughly accounts for one-third of food produced for human consumption in the world (Gustavsson et al. 2011; 2013, UNIDO n.d.). FW or losses has not only financial consequences, but also a great impact on climate change, because the food production consumes resources, such as water, fertilizers, pesticides, seeds, energy and labour (Zorya et al. 2011). The Food and Agriculture Organization of the United Nations (FAO 2013) estimated that the food produced but not eaten guzzles up huge volumes of water equivalent to the annual flow of the Russian river Volga. It is responsible for adding 3.3 billion tons of carbon dioxide per year as well, which makes food wastage the third top greenhouse gases (GHG) emitter for the US and China (FAO 2013). In some developed countries such as the United States, the quantity of FW accounts for approximately 12% of the total solid waste (Xu et al. 2015). In Hong Kong, the number of FW per day is 3640 tons in 2014 and increasing every year (Hong Kong Government n.d.). The sustainable development strategy and public environmental concern make the utilization of FW necessary and urgent.

The advantages of using FW for fuel and chemical production lies on the available infrastructure and expertise in collection, distribution and processing. Since FW contains rich nutrients e.g. carbohydrates, proteins, oils and fats, it can be a valuable source for carbon and energy use (Kiran et al. 2014; Lin et al. 2014; Pleissner et al. 2013). For example, with proper technology based on the biorefinery concept, FW can act as a renewable feedstock and be converted into valuable products (Lin et al. 2013; Ravindran and Jaiswal 2016; Yang et al. 2015). For the past decades, the production of various value-added bio-based chemicals, biofuels, and renewable energy using FW as feedstock has been demonstrated by implementing biological techniques (Zhang et al. 2013a, b; Ohkouchi and Inoue 2007; He et al. 2012). These bioprocesses can also reduce significantly the organic waste pollution affecting the environment, as well as fossil fuels consumption and CO_2 emissions. In addition, the valorisation of FW into biofuels can reduce dependency on crude oil.

In the light of these comments, this book chapter is aimed to provide a comprehensive and multidisciplinary approach on advanced and innovative food valorisation practices providing a variety of case studies that illustrate the potential of FW valorisation and its contribution to a future bio-based economy.

Concerns on climate change, energy-security, desires to lower waste and greenhouse gases production, depletion of fossil resources and a growing world population that deserves access to basic needs and a good environment are the drivers that promote the transition to a biobased economy. The environmental impact of the biorefinery process and the potential commerciality of the developed products must be considered and evaluated before the investment is launched up. To carry out 'green' business actions, the environmental impact assessment tool or method is one of the key issues. Techno-economic analysis (TEA) and life cycle assessment (LCA) are the most extensive applied assessment tools. LCA is the systematic tool to assess the complete life cycle of a product in terms of its environmental burden from raw materials to its disposal. It offers a boundary of 'cradle to grave' or 'cradle to gate' to view a product or a process, considering its environmental impacts (Williams 2009; Peters 2010). TEA focuses more on the process technologies and their economic impact. There are many factors that influence the overall costs and returns in each parts of the FW based biorefinery process, and affect the overall economic performance and environmental impact.

7.2 Techno-economic Study of FW Based Biorefinery

As a switch to bio-based economy, biorefinery has attracted increasing attention for economic, environmental, and energy security considerations. First-generation biorefineries could lead to food versus fuel dilemma. Many bio-based production goals have been legislated supporting the non-food competing materials. For instance, the Revised Renewable Fuel Standard (RFS2) by US Environmental Protection Agency (EPA) mandates that by the year 2022, at least 16 billion gallons per year of cellulosic biofuels will be produced and consumed in the US (Schnepf and Yacobucci 2011). However, the bio-based products have been significantly below the targets, which is mainly due to technical immaturity and feedstock availability issues (Brown 2015).

Techno-economic analysis comprises the design, synthesis and optimisation of (bio)chemical processes and utility systems covering a wide range of research activities. Although the design of an economically relevant business plan for the production of bio-based products is important, the environmental sustainability should be carefully considered as well. It is possible to be precise about operating costs in process design, but capital costs remain a major area of uncertainty. In response to the special problems encountered in the bioprocess, several models and algorithms have been exploited. A sensitivity analysis has been conducted to appraise the effects of the interactions of various parameters on the economic feasibility of a single pathway and of an integrated pathway with fast pyrolysis and bio-oil gasification, and the comparison between these two paths (Li and Hu 2016). By establishing and modifying the factored estimate method, the designer can optimise various procedures. Many bioprocesses are found to be profitable, but so far always at high risk due to numerous assumptions and uncertainties in the TEA.

The problem for biorefinery designers is that the increasing level of sophistication requires an increasing level of sophistication in the design software. Therefore, the developments in process design research depend on the development of commercial software for inherently safe designs with the minimum environmental impact. Waste minimisation is one of the key factors that should be considered coupled with the optimisation of the batch bioreactor systems, especially when wastes such as FW are used as feedstock.

To date, most of the techno-economic studies regarding bioprocesses focus on the production of bioethanol. The National Renewable Energy Laboratory (NREL) has investigated the production economics of biochemical conversion of lignocellulosic biomass to ethanol since 1999. According to the latest technical report issued by NREL in 2011, the minimum ethanol selling price (MESP) is estimated to be US$ 2.15/gal (US$ 0.72/kg) for a plant designed to process 2000 tons of dry corn stover/day at 76% theoretical ethanol yield (79 gal/dry tonne). The capital investment is estimated to be US$ 422 million while the feedstock used is corn stover with a price of US$ 59/tons of dry material. Major cost contributors were identified to be the feedstock (35%), cellulase (16%), wastewater treatment (16%) and pretreatment (13%) (Humbird et al. 2001). In fact, the high costs of enzymes have been recognised as a significant challenge presented in biorefineries for ethanol production by many literatures. However, the cost contribution of enzymes to the production of lignocellulosic ethanol varies significantly. Klein-Marcuschamer et al. (2012) reported that the cost contribution of enzymes to ethanol produced by the conversion of corn stover was $0.68/gal. Aden and Foust reported $0.10/gal in the TEA of the dilute sulfuric acid and enzymatic hydrolysis process for the conversion of corn stover to ethanol. It is because different assumptions and simulation background such as scale of the plant, estimated cost of raw materials and cost of labour, were used. To order to address the inconsistency in the economic analysis, it is necessary to conduct sensitivity analysis to identify the most critical factor and potential risk in each TEA. From the single-point sensitivity analysis reported by NREL, conversion parameters of cellulose to glucose, xylan to xylose, xylose to ethanol and arabinose to ethanol have the largest impact to the MESP. Compared to the reports issued in the past 6 years by NREL, the MESP gradually decreased from US$ 3.53/gal in 2007 to US$ 2.15/gal in 2012. The price of ethanol as a commodity chemical is ranging from US$ 2.11/gal to US$ 3.90/gal.

Apart from bioethanol, the bio-based commercial production of chemicals (e.g. lactic acid, succinic acid) is also nowadays a reality. Supplementation with expensive nutrient sources (e.g. yeast extract) in the agro-industrial by-product and waste streams can increase the yield as well as the raw material costs significantly. Downstream process of chemicals production usually contributes a high proportion of cost. Tejayadi and Cheryan (n.d.) and Gonzalez et al. (2007) studied the production of lactic acid and indicated recovering lactic acid from fermentation broths has a high cost, i.e. corresponding to more than 40% of the capital cost. Kwan et al. (2015) reported production cost of US$ 944 per metric tonne of lactic acid using glucose and microalgae biomass as carbon and nitrogen source, respectively. There are limited open literature studies on the economics of large-scale production of 1,3-propanediol (PDO). The production cost of PDO from glucose is estimated around US$ 1.3/kg, while US$ 1.4/kg PDO when glycerol is used as raw material, when the annual capacity of the plant has been set to 15,300 tons of PDO (2002). In 2013, Genomatica and DuPont Tate & Lyle Bio Products Company, LLC built up the first successful commercial-scale production of 1,4-butanediol (BDO).

Another outstanding example is the production of bio-succinic acid, since the U. S. Department of Energy reported succinic acid as one of the top platform chemicals derived from biomass (FAO 2013; Xu et al. 2015). Bio-based succinic acid becomes a key building block for deriving both commodity and high-value chemicals, which makes it an attractive compound in the coming bio-based economy. The global production ranges from 30,000 to 50,000 tons in 2014, and the expected market grows quickly and have a potential size of US$ 15 billion (McKinlay et al. 2007). Several companies and industrial consortia, e.g. BioAmber, Reverdia, Myriant and BASF, have begun to develop their industrial production on a large scale in North America, Europe and Asia Pacific. In 2012, Myriant became the first renewable chemicals company to build a pilot plant for producing 13,600 tons of bio-succinic acid annually from grain sorghum and other commercially available sugars. Orjuela et al. (2013) conducted a TEA of bio-succinic acid production process with a novel recovery strategy and estimated that the minimum production cost is US$ 1.85/kg. Vlysidis et al. (2011) conducted a comprehensive TEA of an integrated biorefinery for co-production of biodiesel and succinic acid produced by using crude glycerine from the biodiesel process. It was found that succinic acid co-production can enhance the profit of the biodiesel plant by 60%, as compared to the disposal of crude glycerine as a waste (Vlysidis et al. 2011).

The most important characteristics of a biorefinery that can affect the overall economic performance have been identified (Van Dien 2013):

- Final product concentration (should exceed 50 g/L).
- Productivity (should exceed 2.5 g/L h).
- Yield (should at least be 80% of the theoretical value).

These limits should be treated as general figures of merit for the production of any commodity chemical with a selling price in the range of US$ 0.3–5/kg.

7.2.1 Economics on Green Chemistry

At the macro-level, green chemistry is modelled as a backstop technology with a corresponding cost. As soon as such a technology appears, its cost indicates a maximum price for the fossil fuel substitute. There is then an incentive for fossil fuel producers to lower fossil fuel price to sell their fossil fuel stocks entirely. It is important to note that as long as such a backstop technology is not cost-effective, it does not prevent from using fossil fuels and it can even speed up their use.

It is therefore critical to turn to the micro-level that is concerned with the effect of green chemistry at the level of a plant and considers cost-effectiveness. Cost-effectiveness is typically appraised using cost-benefits analyses. The basic idea is to add all potential discounted costs and benefits generated by the new technology and to compare the net present value obtained with that of the fossil fuel(s) it is supposed to replace. For green chemistry, costs to be considered are in general the following:

- Land;
- Fixed-capital investment including equipment costs (such as grinders, blending tanks, bioreactors, fermenters, bowl centrifuge, absorbers, ion-exchange columns etc.), but also equipment installation, process piping, instrumentation and controls, electrical systems buildings, yard improvements, auxiliary facilities, engineering, construction expenses, legal expenses, contractor's fee …
- Maintenance and repair;
- Utility costs (for electricity, steam, water …);
- Labour costs;
- Chemical products.

Benefits arise from the sale of biofuel or valuable products as plasticisers or succinic acid (sometimes together with other by-products resulting from the process) obtained thanks to the green chemistry plant. They may embed gains in terms of environmental improvements as well, which can be computed using LCA. It is important to note that these costs and benefits occur at different points in time, sometime in the future meaning that they are uncertain, which makes the computation of the costs-benefit analysis not so straightforward. We now consider the following example based on succinic acid production using FW to illustrate how cost-benefits analysis is conducted.

Table 7.1 Equipment cost of FW based refinery for succinic acid production

Equipment	Unit cost (US$)	Quantity	Sub-total (US$)
Shake flask rack	51,000	2	102,000
Grinder	26,282	1	26,282
Blending tank	24,167	1	24,167
Bioreactor	66,921	1	66,921
Bowl centrifuge	34,230	2	68,460
Fermenter	174,720	1	174,720
Freeze dryer	94,500	1	94,500
Solvent recycler	94,050	1	94,050

The techno-economic analysis is made by Yi (2016) and based on the studies of Dr. Carol Lin's group at School of Energy and Environment in City University of Hong Kong (Kwan et al. 2015; Lam et al. 2014). The plant is designed to convert 1 t of FW per day to succinic acid. Fixed capital investment considers the costs of building up the pilot plant, which includes equipment cost, installation cost, and other related cost. The working capital cost is assumed to be 5% of the fixed capital investment (Humbird et al. 2001). Total equipment cost is estimated to be US$ 651,100 (Table 7.1). The processing cost of fixed capital investment is approximately 40% (Kalk and Langlykke 1986). Hence, the FCI is estimated to be US$ 1,627,750 and the corresponding working capital cost is about US$ 81,400. Thus, the estimated capital cost is US$ 1,709,150.

The annual operating cost includes raw materials cost, labour cost, cost of utilities, maintenance cost, and other related cost (Table 7.2). The estimated operating cost is US$ 211,550 annually. Labour cost and cost of raw materials contribute the most on the overall operating cost, which are 47 and 34% respectively.

The optimal yield is 550 kg succinic acid from 1 ton FW (Lam et al. 2014). The biorefinery is designed to perform 260 batches every year that is proximately 5 days a week, so the annual production of succinic acid is 143,000 kg. The unit price of succinic acid is US$ 9/kg. Hence, the annual revenue will be US$ 1,287,000 with a

Table 7.2 Annual operating cost of FW based refinery for succinic acid production

Item	Cost (US$)
Raw material	71,550
H_2SO_4	1200
NaOH	1800
Carbon dioxide	68,000
Magnesium carbonate	350
Enzyme	200
Labour	100,000
Utility	20,000
Maintenance	20,000

net cash-flow of US$ 1,075,450 per year. NPV determines if the plant is profitable for a certain period of time by discounting the future cash flows to the present value (Eq. 7.1).

$$NPV = \sum_{t=1}^{T} \frac{C_t}{(1+d)^t} - C_0 \tag{7.1}$$

Where t is year, C_t is the net cash flow during period t, C_O is the initial investment and d is the discount rate. Assuming the plant will exist for 20 years and the discount rate is 5%: NPV is equal to US$ 11.7 million. As it is positive, it is profitable to construct such a plant. Another critical profitability indicator is the Internal Rate of Return (IRR) defined as (Eq. 7.2):

$$\sum_{t=1}^{T} \frac{C_t}{(1+IRR)^t} - C_0 = 0 \tag{7.2}$$

In our example, it is around 62.92%. Also note that the payback period (PP) is defined as (Eq. 7.3):

$$\sum_{t=1}^{PP} \frac{C_t}{(1+d)^t} - C_0 = 0 \tag{7.3}$$

The PP is 2 in our example, meaning that the cost of the capital investment is recovered after two years.

A standard way to account for the uncertainty is to conduct a sensitivity analysis to evaluate the impact of each variable upon the economic performance. Variables, including raw materials cost, labour cost, utility cost and prices of products, should be independently evaluated and set at certain percentage of variation (e.g. ±20%) at the beginning of the plant's lifetime.

7.3 Environmental Assessment Process Design

As environmental awareness increases, environmental pollution become seriously considered. Nowadays the environmental impact of a business behavior or a commercial product is considered and evaluated before the business is launched up. Many corporates or groups have responded to this environmental protection concept by giving out environmental-friendly products and using non-energy intensive processes. To carry out 'green' business actions, the environmental impact assessment tool or method is one of the key issues. One of the most widely used assessment tools is the Life Cycle Assessment (LCA) methodology. LCA is the systematic tool for viewing the complete life cycle of a product. For a bio-based products, a 'cradle-to-grave' approach covers the extraction and production of all raw materials and energy carriers used in the agricultural phase, as well as in the transformation, use and waste management phases (Williams 2009; Peters 2010).

Herein, several cases using LCA to monitor the environmental assessment of a number of processes are reviewed. Bio-succinic acid is selected as the typical representative of the bio-based chemical production from FW. Also the life cycle impact of biofuels are summarized in the sections below.

7.3.1 LCA in Biobased Succinic Acid Production from FW

FW contains valuable resources. By employing biological techniques we can transform FW into renewable energy sources and chemical products, which can significantly reduce organic waste pollution to the environment as well as decrease the consumption of fossil fuels and the emissions. Our group previously demonstrated that FW could be used as feedstock in fermentative succinic acid production (Sun et al. 2014; Leung et al. 2012). To assess the effect of the succinic acid production from FW on the environment, LCA is carried out in the following section.

Works have been reported on utilizing daily FW as feedstocks to produce chemical products such as succinic acid (Sun et al. 2014; Leung et al. 2012), ferulic acid (Mussatto et al. 2006), apocarotenoid (Chedea et al. 2010), proteins (Tahergorabi et al. 2011), carotenoid pigments (Shahidi and Synowiecki 1991), and so on. In this section, LCA methodology is used as a tool which allows compilation and evaluation of potential environmental impacts of succinic acid production based upon our earlier study (Sun et al. 2014), in which mixed FW directly collected from the restaurants was used as feedstock in succinic acid fermentation. Such a LCA study can act as a guideline for the FW based biorefinery for fermentative succinic acid production in Hong Kong.

7.3.1.1 Research Method

The LCA includes four steps (Williams 2009; Peters 2010):

1. **Goal and scope definition**. In this study, we want to know: what are the impacts of the succinic acid producing process on environment? What are the similarities and differences of these impacts compared with those of petrochemical processes and other LCA works? The scope we consider in this work is 'from cradle to gate', that is from the raw materials to the product. The functional unit is 1 kg of produced succinic acid, the inputs/outputs of materials and energy used are based on this unit.

2. **Life cycle inventory**. The required energy and raw materials for the production of succinic acid are obtained from FW and the quantity converted into the functional unit (Sun et al. 2014).
3. **Life cycle impact assessment**. Here we choose the Non-Renewable Energy Use (NREU) and the GHG as the impact categories. The inventory indicators are calculated with respect to the functional unit.
4. **Interpretation of the results**. A complete LCA will include consistency and completeness check, contribution and comparison analyses of inventories, sensitivity, uncertainty and scenario analyses of the LCA.

7.3.1.2 System Boundary

The flow diagram of the succinic acid production from FW via biorefinery process is displayed in Fig. 7.1. The succinic acid production process is 'from cradle to gate'. FW was the feedstocks for fungal solid state fermentation to produce amylolytic and proteolytic enzymes. These fungal mashes would be considered as crude enzyme sources and added to FW suspensions to hydrolyse the key carbon and nitrogen components. The resulting FW hydrolysate contained 31.9 g/L glucose and 280 mg/L free amino nitrogen, which was then used as feedstock for the subsequent *Escherichia coli* fermentation process to produce SA (Sun et al. 2014).

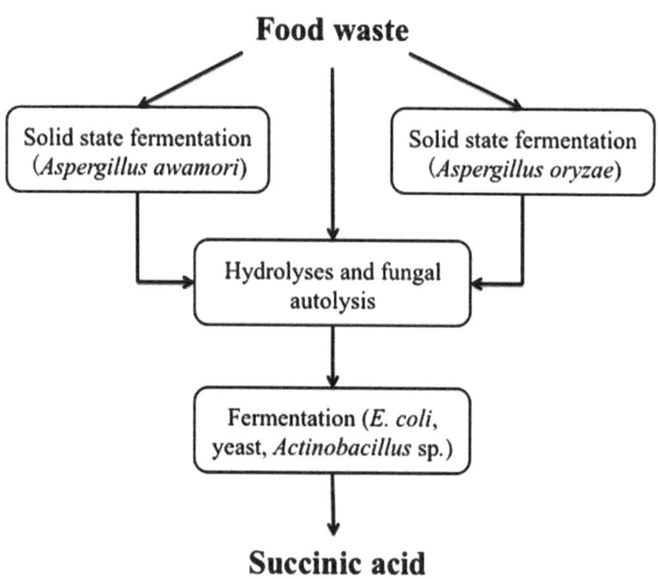

Fig. 7.1 Flow chart of food-waste based biorefinery for succinic acid production processes (Sun et al. 2014)

The SA production in the 2.5 L fermentor was 29.9 g/L with a yield of 0.224 g/g, making 74.8 g SA from 333.7 g FW.

7.3.1.3 Life Cycle Inventory Analysis

The original materials and energy used for the SA production are listed in Table 7.3. The quantities of the input materials and energy were converted with respect to 1 kg of SA. The major nutrients from FW are glucose and free amino nitrogen. Their amounts were calculated by considering that the volume of FW hydrolysate for later fermentation process was 1.5 L, which made the mass percentage of glucose and free amino nitrogen in the FW equal to ~15%. According to the functional unit, the required input of glucose and free amino nitrogen to produce 1 kg SA is 0.64 and 0.0056 kg, respectively. The fermentation required pH control of the solution, which was adjusted with the addition of NaOH and H_2SO_4. The volume of NaOH and H_2SO_4 solution was estimated to be 0.05 and 0.95 L respectively, corresponding to 0.27 kg NaOH and 0.63 kg H_2SO_4 with respect to 1 kg SA.

The amount of process water and fermentation water used were chosen as 2 L respectively, which was estimated from the volume of the processing and fermentation facilities. The wastewater was chosen as the sum of processing and fermentation water. As for the energy source, the major cost was electricity consumption during the process of the facilities. The bioreactor, BioFlo115 (Eppendorf, Operating Manual, Revision E), was used to estimate electricity used. The power of the facility is 1380 W, and the operating hours of hydrolysis was 24 h and 72 h for the *E. coli* fermentation. Therefore, the energy consumption to produce 1 kg SA is 6380.31 MJ.

Table 7.3 Lists of the life-cycle Inventory data for the production of 1 kg succinic acid

Item	Input or output	Input/output amount	kg/kg SA
Glucose	Input	47.85 g	0.64
Free amino nitrogen	Input	0.42 g	0.0056
NaOH	Input	10 M, ~0.05 L (1.5 L hydrolyses was added, then the remaining 1 L should be NaOH and H_2SO_4)	0.27
H_2SO_4	Input	0.5 M, ~0.95 L	0.63
Process water	Input	~2 L	26.76
Fermentation water	Input	~2 L	26.76
Waste water treatment	Output	4 L	53.51
Electricity	Input	1380 W, 24 h, 72 h	6380.31 MJ

Table 7.4 Characterization factors for each input/output items for the production of succinic acid (Sun et al. 2011)

Items	Input/output	Characterization factors	
		NREU (MJ/kg SA)	GHG (kg CO_2, eq/kg SA)
Glucose	Input	0.721	0.234
Free amino nitrogen	Input	41.672	2.0962
NaOH	Input	21.359	1.0964
H_2SO_4	Input	2.020	0.1237
Process water	Input	0.006	0.0003
Fermentation water	Input	0.017	0.0008
Waste water treatment	Output	0.321	0.1810
Electricity[a]	Input	0.421	0.0638

[a]Electricity, units in (MJ/MJ and kg CO_2, eq/MJ)

7.3.1.4 Life Cycle Impact Assessment

In this work, GHG and the NREU were chosen as characterization factors, whose values for each category were obtained from an online course titled 'Industrial Biotechnology' by Delft University of Technology (TUDelft) and are listed in Table 7.4 (Sun et al. 2011). The characterization factors were in the unit of MJ/kg and kg CO_2, eq/kg. To obtain the environmental impact of each category, the characterization factors had to be multiplied by the amount of inventory in kg unit. An equation (Eq. 7.4) was used to calculate the category impact indicator:

$$Category\,indicator = \sum_{1}^{s} (Characterization\,Factors(s) * Emission\,Invertory(s)) \tag{7.4}$$

(s denotes the component)

By multiplying the inventory data and characterization factor, we obtained the category indicator results of each item, which are listed in Table 7.5 major NREU and GHG were from the electricity use, accounting for ~97% of the NREU and GHG emissions of the whole SA production process. This was because the batch operation time for hydrolysis and fermentation process together took up one week. Other large consumption parts were the use of wastewater treatment and NaOH.

7.3.1.5 Interpretation of Results

Comparison and sensitivity analysis. It was clearly displayed from Table 7.5 that the major contribution of NREU and GHG emissions came from electricity consumption, which accounted for ~99 and ~97% of the total respectively.

Table 7.5 Category indicator results for each input/output items for the production of 1 kg succinic acid (Sun et al. 2011)

Items	Input/output	Characterization factors	
		NREU (MJ/kg)	GHG (kg CO_2, eq/kg)
Glucose	Input	0.46	0.15
Free amino nitrogen	Input	0.23	0.012
NaOH	Input	5.72	0.30
H_2SO_4	Input	1.27	0.078
Process water	Input	0.16	0.0080
Fermentation water	Input	0.45	0.021
Wastewater treatment	Output	17.18	9.69
Electricity	Input	2686.11	407.06
Total		2711.58	417.32

This means that the environmental impact of SA production was very sensitive for hydrolysis and fermentation processes. Other categories which have large effect were the wastewater treatment and the addition of NaOH in fermentation process.

To test whether the method was applicable for the production of SA from FW in a large scale, we compared our LCA study with that by Cok et al. (2014), who carried out the NREU and GHG emission assessment of three downstream recovery processes for SA recovery, direct crystallization (BioSA-DC), succinic salt production through anaerobic fermentation at neutral pH and then DSP by electrodialysis (BioSA-ED), and a process similar to BioSA-ED, to produce succinic acid from carbohydrates, and compared the biological methods with several petrochemical processes, petrochemical maleic anhydride, petrochemical succinic acid, and petrochemical adipic acid.

The NREU and GHG emission of these methods are displayed in Fig. 7.2. The NREU and GHG emissions of bio-based processes were smaller than the ones derived from traditional petrochemical methods. This indicated that producing SA from biomass was more environmental friendly. Among these three bio-based methods, the NREU and GHG of the BioSA-DC process were 32.7 kg per kg of SA and 0.88 kg CO_2, eq per kg of SA, respectively, which were the smallest among the methods considered, meaning that such SA production was the most environmental friendly one. On the other hand, the NREU and GHG strategies assessed in this work turned out to generate 2686.11 kg/kg SA and 407.06 kg CO_2, eq/kg SA, which is much higher than what is obtained using petrochemical strategies (Cok et al. 2014). However, the large contribution parts, wastewater treatment and electricity in this study, were estimated from the fermentation and downstream recovery. For those bio-based methods, the electricity used for the production of 1 kg of SA was only ~3 kWh. Also, the amount of NREU and GHG related to wastewater treatment is negligible. If the effect of waste water treatment is ignored and electricity consumption is 3 kWh (i.e. same in our previous work using mixed FW for SA production (Sun et al. 2014), the total amount of NREU and GHG of the SA producing method would be 12.84 MJ and 1.26 kg CO_2 eq per kg SA,

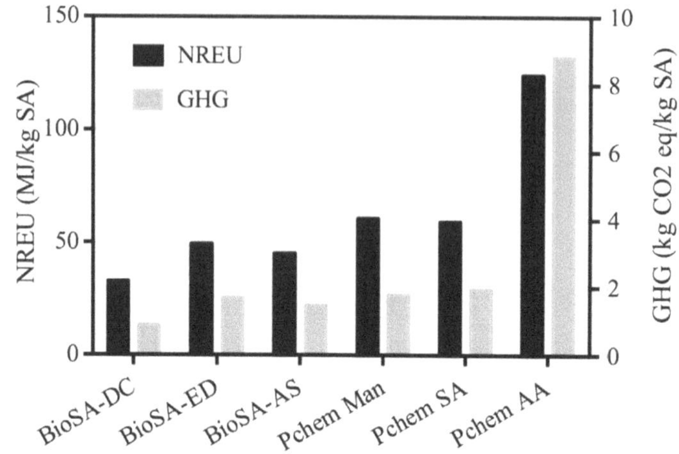

Fig. 7.2 Illustration of the NREU and GHG emissions of various methods for succinic acid production (Cok et al. 2014)

respectively, which significantly reduced the impact of the SA production on the environment, and the NREU was even reduced by 60% with respect to that of BioSA-DC method.

Completeness check. The LCA included data of all the necessary input/output material streams. However, other factors have not been taken into consideration in this study. The NREU and GHG data related to the collection and transportation of FW had not been included and assessed. Here we only considered the environmental impact of NREU and GHG, which could be categorized into the resource depletion and climate change respectively. Other aspects like eutrophication, land-use, and ecotoxicity were not included. On the other hand, the system boundary of the current study was only from 'cradle to gate', while a more extensive analysis should be from 'cradle to grave', which include the consumption and disposal treatment process of the product.

Consistency check. The method described in this work has been testified to be effective to produce succinic acid. By carrying out the environmental impact assessment study, the primary energy use (NREU) and GHG related to the succinic acid production were determined. The assumptions, data and methods used throughout the LCA process are generally consistent with the scope and goal of the study.

Uncertainty check. The major uncertainty arises from the fact that raw materials and energy consumption are not clearly stated in our former study (Sun et al. 2014). Several raw material data were obtained through estimation. Among the data, the amounts of NaOH and H_2SO_4 were estimated by assuming that the volume of input hydrolysate for fermentation was 1.5 L. The total volume of added NaOH and H_2SO_4 solution was 1 L to make the total volume equal to 2.5 L. The ratio of NaOH:H_2SO_4 is 1:20 (the inverse of the concentration ratio). The process water and fermentation water were estimated according to the volume of the bioreactor used, which may not be very accurate. The amount of wastewater is supposed to be the sum of process and fermentation water. Another major error source is the electricity used. The actual energy use of the fermentor was assumed to be the same as that of the facility used for hydrolysis. The fermentation time was 72 h (Leung et al. 2012). Among the estimated data, NaOH, waste water, and electricity contribute the highest to primary energy consumption and GHG emission.

7.3.1.6 Summary of LCA Study of Fermentative SA Production from FW

Due to technological limitations, a large amount of investment has to be spent on the setting up and operation of the facilities. However, with the development of technology, the production price will be lower and the bio-based refinery process can be very competitive in the industrial areas. Besides, the awareness of human society about the importance and necessity of protecting our environment makes this attempt worthy and accessible in the future.

By carrying out the LCA of succinic acid production, we identify the categories that result in large NREU and GHG emissions. To make the succinic acid production applicable, wastewater treatment and electricity consumption have to be modified. The long processing time of SA production from FW was compared to that reported by Cok et al. (2014) and may be due to the difficult treatment of FW into useful substances. Hong Kong has a good opportunity to turn this FW based biorefinery into reality due to its strong sense of environmental protection and related encouraging policies having been set up by the Government in recent years.

Apart from sugar and protein, FW contains a significant amount of lipids with a range of 6.4–30% (Kiran et al. 2014). This lipids-rich FW could be converted into fatty acids and biodiesel either by direct transesterification using alkaline or acid catalysts, or by the transesterification of microbial oils produced by various oleaginous microorganisms. Nowadays, transesterification of lipid into biodiesel is a hot topic. Recently, various scales of plants have been built for biodiesel production in countries, and the feasibility of valorization of FW into biodiesel has been demonstrated in some of the pilot plants, for instance SENECA Green Catalyst S. L. in Spain, ASB Biodiesel Ltd in Hong Kong and Brocklesby Ltd in the UK (Lin et al. 2014). To promote the industrial scale of biodiesel production from FW, LCA has been introduced in this field.

7.3.2 *LCA in Biomass Biofuels*

7.3.2.1 Introduction

The management of the significant potential of FW remains a challenge for the society, but at the same time creates an opportunity for energy recovery and

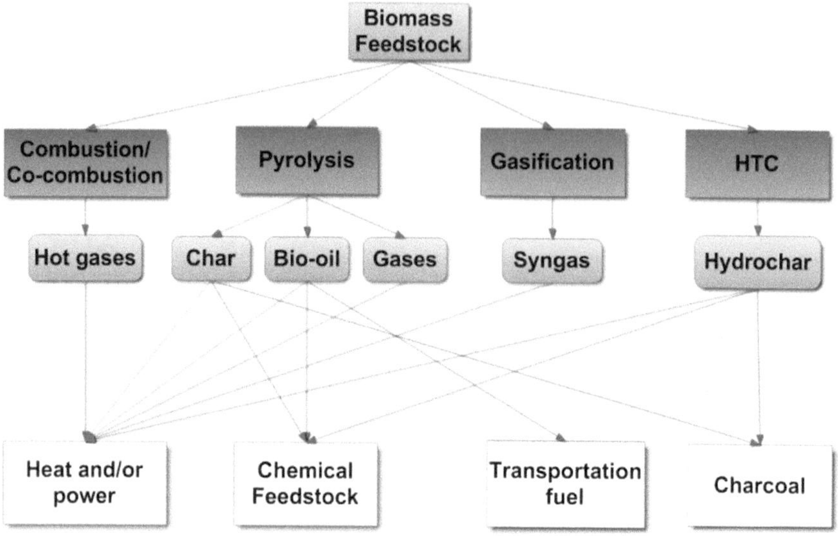

Fig. 7.3 Thermochemical biomass conversion technologies

valuable fuel production. This fact has been recognized during the last decades, and it has motivated numerous research efforts towards the exploitation of this potential waste-to-energy conversion routes. At the same time, it contributes to the reduction of the environmental burden of its disposal and to the minimization of natural resources depletion.

Various technologies can be used for the thermochemical conversion of waste biomass streams, including FW to energy via combustion/incineration, co-combustion or co-firing, gasification, pyrolysis and hydrothermal carbonization (Fig. 7.3). Significant work has been undertaken in the last decade to evaluate the environmental impact of those technologies, as well as the impact of pre-treatment methods, using LCA methodology (Standardization IOf 2006) with an effort to improve their environmental performance and minimise the effects of the processes (Astrup et al. 2015; Kylili et al. 2016).

7.3.2.2 Thermal Conversion Technologies

Combustion/Incineration

Combustion is defined as the complete oxidation of the fuel (Van Loo and Koppejan 2012). The hot gases produced from the combustion process can be used for heating purposes, directly or indirectly. Power production is achieved using secondary conversion technologies (e.g. by exploiting the produced heat to operate steam turbines) (Autret et al. 2007). Biomass combustion depends on several parameters such as the properties of the feedstock, the temperature, and the combustion atmosphere. During the process, volatilization of combustible vapours from the biomass occurs. New technologies and improved air emission control systems can create favourable conditions for environmentally friendly application of direct combustion as a waste to energy method.

According to a review study conducted by Astrup et al. (2015), waste incineration was found to be the most frequently assessed waste-to-energy (WtE) technology in terms of its environmental impact.

Several studies focus on the environmental performance of waste incineration (Autret et al. 2007; Assamoi and Lawryshyn 2012; Fruergaard et al. 2010; Liamsanguan and Gheewala 2007; Wang et al. 2012; Tyskeng and Finnveden 2010; Riber et al. 2008). Recently conducted LCA studies for biomass combustion have shown that significant reduction of GHG emissions can potentially be achieved by replacing conventional fuels with biomass, in combustion systems. LCA approach was applied by Caserini et al. (2010) to investigate the environmental impacts of biomass combustion in small domestic appliances and in two types of centralized combined heat and power plants. A GHG emissions reduction in the range of 0.08–1.08 tons of CO_2 eq. per ton of dry biomass was found, under different scenarios, in case biomass was used instead of conventional fuels. In the study conducted by Kim et al., three FW disposal methods, namely anaerobic digestion, co-digestion, and incineration of dried feedstock, were evaluated from the perspective of global warming and energy/resource recovery, using LCA (Kim et al. 2013). With regard to the dryer-incineration, a GWP of 342 kg of CO_2-eq from 1 ton of FW was estimated. The environmental credit for dryer-incineration of 1 ton of FW was found to be 315 kg of CO_2-eq. based on electricity and thermal energy production as well as primary materials avoidance.

Co-firing/Co-combustion

Biomass co-combustion (or co-firing) involves the addition of biomass to existing fossil fuel-based systems, the main fuel (e.g. coal-fired) and their simultaneous firing in a high efficiency boiler. Alternative options for biomass-coal co-firing are: the direct co-firing of the fuels, indirect co-firing, which involves the gasification of the biomass and the combustion of the product fuel gas in the furnace; and parallel combustion, which involves the combustion of the biomass in a separate combustor and boiler, and the utilization of the steam produced within the coal plant and power generation systems (Van Loo and Koppejan 2012).

Co-combustion of biomass with coal gains significant research interest since it has an advantage in the disposal of waste products and reduces the cost of fuel (Atimtay and Topal 2004, Li et al. 2015). Co-firing method is one of the ways that can be used in large coal-fired plants to limit the use of the non-renewable resources. It is regarded as one of the attractive short-term options for biomass utilization in the power generation industry (Agbor et al. 2014; Dzikuc and Piwowar 2016).

LCA studies have concluded that significant carbon reduction can be achieved by employing co-combustion of biomass in conventional-fueled heat and power systems, whereas biomass co-firing can positively contribute to the improvement of the efficiency of existing systems from an environmental perspective (Huang et al. 2013). LCA is applied to evaluate the environmental impact and benefits of a biochar co-firing (i.e. produced by rice straw torrefaction) for electricity generation. A carbon reduction of 4.32 and 4.68 metric tons CO_2, eq/ha/year was estimated at 10 and 20% co-firing ratios, respectively. GHG emissions of co-firing and biomass-fired power plants were evaluated by Sebastian et al. (2011), concluding that a 29% net electric efficiency biomass-fired power plant would be required to achieve the same global GHG emissions decrease as biomass co-firing. In the study conducted by Lu and Zhang (2010), combined conventional process-based LCA with economic input–output LCA was used to evaluate the ecological and economic performance of 13 crop residues conversion technologies. Co-firing with coal was found to be the best technology for crop residues utilisation in terms of four factors, namely the environmental impact, GHG, net energy value, and economic viability of the technology. Andric et al. (2015) assessed the environmental performance of biomass and coal co-firing in power plants and showed that the addition of approximately 20% biomass to the mass of the combustion mixture causes the decrease in carbon dioxide emissions by nearly 11–25%. It was also highlighted that the co-firing process is environmentally acceptable if the biomass supply stocks are within the area determined by maximum supply distances. Zuwala (2012) conducted LCA analysis of biomass and coal co-firing in combined heat and power (CHP) plants. It was revealed that the partial substitution of coal with biomass leads to a decline of the total life-cycle non-renewable energy resources depletion.

Pyrolysis

Pyrolysis refers to the thermal degradation of biomass in the absence of an externally supplied oxidizing agent (Van Loo and Koppejan 2012; McKendry 2002). The pyrolysis products are mainly tar and carbonaceous charcoal, and low molecular weight gases. Product yields and their properties are affected by many parameters and variables such as the fuel type, temperature, pressure, heating rate, and the reaction time (Van Loo and Koppejan 2012).

Depending on the operating conditions (i.e. temperature and heating rate), the pyrolysis process can be categorized as slow, fast, or flash. Solid biofuels are mainly derived from slow pyrolysis, whereas fast and flash pyrolysis are used for liquid (bio-oil) and gaseous biofuels production (Encinar et al. 2008; Jahirul et al. 2012).

Biomass pyrolysis LCA studies have shown that the process is environmentally friendly with little impact to the environment. Significant GHG savings can be achieved by utilizing pyrolysis products for energy production, whereas the production of biofuels using pyrolysis presented significantly lower environmental impact compared to fossil-fuel production.

Dang et al. (2014) applied LCA method to liquid biofuel production from corn stover using fast pyrolysis, and subsequent upgrading using the Greenhouse gases Regulated Emissions and Energy use in Transportation (GREET) model. LCA methodology was used in an effort to assess the environmental impact of flash pyrolysis of wasted wood for biofuel production and power generation (Zhong et al. 2010). The obtained results showed that flash pyrolysis of wood waste is in fact environmentally friendly, and the process has little contribution to the environment. Fan et al. carried out an LCA of the GHG emissions associated with energy generation from forest resources through pyrolysis-based processing (Fan et al. 2011). GHG savings of 77–99% were estimated for power generation from pyrolysis oil combustion relative to fossil fuels combustion, depending on the processed biomass feedstock and the combustion technology used. Iribarren et al. (2012) used LCA approach to evaluate the environmental performance of a biofuel production system based on the fast pyrolysis of short-rotation poplar biomass, under different impact categories. Focusing on GHG emissions, savings of 72% were calculated using fast pyrolysis compared to the conventional fossil fuels production. Moreover, LCA was used to identify the process with the highest impact to the environment in a biofuels production system which employs a fast pyrolysis plant and hydro-upgrading of biofuel (Peters et al. 2015; Corti and Lombardi 2004). The results indicated potential GHG savings in the order of 54.5% for the produced fuel mix compared to conventional gasoline and diesel production. Roberts et al. (2009) used LCA to estimate the full life-cycle energy, GHG emissions balance, and economic feasibility of biochar produced from a slow pyrolysis system. The results indicated that the switchgrass biochar-pyrolysis system could be a net GHG emitter.

Gasification

Gasification is the conversion of biomass into a combustible gaseous fuel mixture called syngas, by the partial oxidation of the carbon contained in the biomass at high temperatures (i.e. 800–900 °C) in the presence of a gasifying agent such as air, oxygen or steam (McKendry 2002; Van Loo and Koppejan 2012; Wu and Chein 2015; Bridgwater 2003). The properties of the final product depend on several parameters such as the processed feedstock, the operation conditions (i.e. temperature, residence time, oxidant agent) and the type of gasifier. Syngas with a heating value in the range of 4–7 MJ/m^3 can be obtained when air is used as the gasification medium, whereas using pure oxygen or steam as an oxidant leads to significantly higher heating values of the gas in the range of 10–18 MJ/m^3 (Heidenreich and Foscolo 2015). The process has some advantages over traditional incineration technology, mainly related to the availability of coupling the operating conditions and the features of the used reactor to obtain a syngas suitable for use in different applications.

Biomass utilization in gasification systems has been found to lead to considerable reduction in GHG emissions compared to fossil-fuel based systems, offering significant environmental benefits in terms of global warming, NREU and other environmental impact categories. Corti and Lombardi (2004) applied LCA to investigate the environmental performance of a biomass-fed integrated gasification combined cycle (IGCC). The results of the study, with respect to CO_2 emissions, were found significantly better (i.e. 167 kg CO_2/MWh) than in a conventional coal IGCC (i.e. 700–800 kg CO_2/MWh). Kimming et al. (2011) examined three different CHP production systems with organically produced biomass, and a scenario based on natural gas from the consequential LCA perspective. The results indicated a considerable reduction of GHG emissions when biomass-based systems were used, compared to the fossil fuel-based system. The scenario considering gasification of raw biomass and combustion of the produced biogas in an internal combustion (IC) engine was found to be the most efficient in terms of primary energy and fossil energy inputs. Nguyen et al. (2013) aimed to investigate the environmental performance of biomass gasification for electricity production based on wheat straw and to compare the results with alternative power production solutions, straw-fired and fossil-fired production. The study concluded that the production of electricity from straw based on gasification appears to be more environmentally friendly than the direct combustion of straw in all impact categories considered, while straw use instead of coal and natural gas for electricity production would offers significant environmental benefits in terms of global warming, NREU and eutrophication. Yang and Chen (2014) investigated the entire lifetime GHG emissions of a crop residue gasification project, using static and dynamic LCA approaches, and concluded that the largest contributions in terms of emissions are the operation and construction stages, due to the consumption of crop residue, electricity and steel. In the study conducted by Arafat et al. (2013), the environmental impact of different treatment methods for municipal solid waste (MSW), including incineration and gasification, was assessed by means of LCA. The methods were compared with recycling, where applicable, and the results indicated gasification as the best solution for textile wastes management for energy recovery purposes.

Hydrothermal Carbonisation (HTC)

Hydrothermal carbonisation gains significant research interest due to the fact that it allows the processing of waste streams with high moisture content (i.e. 80-90%). The application of HTC has been reported for several types of biomass feedstocks (Basso et al. 2015; Funke and Ziegler 2010; Heilmann et al. 2011; Huff et al. 2014; Hwang et al. 2012; Liu and Balasubramanian 2012; Pala et al. 2014; Xiao et al. 2012). During HTC, biomass is heated with hot compressed subcritical water (200–260 °C) under autogenous pressures and relatively low temperatures (180–350 °C) (Berge et al. 2011; Libra et al. 2011; Reza et al. 2014). The wet raw material is decomposed by a series of simultaneous reactions including hydrolysis, condensation, dehydration, decarboxylation, polymerization and aromatization resulting in a reduction of the oxygen and hydrogen content of the feedstock (Lu and Zhang 2010; Oliveira et al. 2013). The products of HTC are gases (mainly CO_2, CO, H, CH_4, C_2H_6, C_3H_6), a liquid fraction which contains the solvent applied in the reaction and solubilised organic products, and a solid mixture (i.e. hydrochar) which retains 55–90% of the mass and 80–95% of the fuel value from the processed feedstock (Reza et al. 2014).

Although there are many studies suggesting HTC as a biomass conversion process with potential environmental benefits such as a reduction in GHG emissions and lower energy requirements for the conversion of wet feedstocks over other thermal conversion methods (Titirici et al. 2007), there is a lack of published works and data. This was also highlighted by Berge et al. (2011). LCA was applied by Christoforou and Fokaides (2015) to investigate the environmental impact of HTC and torrefaction of two phase olive mill wastes (2POMW). It showed that HTC is more energetically feasible compared to torrefaction for wet biomass feedstock, such as 2POMW. The study conducted by Berge et al. (2015) highlighted the benefits associated with the carbonization of FWs and the potential utilization of the produced biomass-derived hydrochar for power generation.

7.4 Conclusions and Prospect

The awareness on FW based biorefinery as a mean to achieve the illusive target of sustainability is rising in the current society that involves governments, policies, regulations, stakeholders, companies, products, and most importantly consumers and public opinion. The recent laboratory and pilot scale studies have successfully demonstrated the potential of utilizing waste and by-product streams from various industries for the production of platform chemicals and biopolymers. Sustainability consists of three major components, namely economic, environmental and social impacts. The most important driver for FW based biorefinery from laboratory study to commercial production is whether the proposed design is profitable. There are many factors that influence the overall costs and returns and choices made at each stage of the FW based biorefinery process have an effect on the overall economic performance. Apart from the techno-economic view, the environmental and social impacts are major components of the sustainability as well.

This chapter has been aimed to demonstrate the techno-economic feasibility and the improved ecological impact of advanced FW valorization and biorefinery for the coming bio-based economy. The studies on both the economic and environmental impacts of FW based biorefinery were summarized. From the idea to the design and establishment, integrated FW valorisation and biorefinery are discussed extensively, and case-specific examples are analysed, which are key drivers for achieving commercialization. Low environmental impact chemical technologies, conversion of waste into value-added products, zero-waste process should be focused on to attain a more sustainable bio-based society. TEA and LCA have been applied to appraise green chemistry, traditional refinery and biorefinery, in particular for the succinic acid production, and thermochemical processes for biomass conversion to biofuels. To maximize the profitability and meet environmental requirements, production of biofuels or special chemical product, the designer's vision should not be limited to one or two particular processes, but the whole supply chains should be considered, using for instance mathematical programming conceptually-based approach and drawn from pinch analysis of the whole supply chains. Currently, it demonstrates not only the breadth of problems being addressed in the general area of techno-economic analysis, but also demonstrates the wide range of design techniques now available for application in process design.

Acknowledgements C.S.K. Lin, Elias A. Christoforou and Paris A. Fokaides gratefully acknowledge the contribution of the COST Action FP1306. C.S.K. Lin also acknowledges the contribution of the COST Action TD1203-EUBis. Mr. Huaimin Wang (Research assistant from C. S.K Lin's research group) is acknowledged for his assistance in compiling the references.

References

Agbor E, Zhang X, Kumar A (2014) A review of biomass co-firing in North America. Renew Sustain Energy Rev 40:930–943

Andric I, Jamali-Zghal N, Santarelli M, Lacarriere B, Le Corre O (2015) Environmental performance assessment of retrofitting existing coal fired power plants to co-firing with biomass: carbon footprint and energy approach. J Clean Prod 103:13–27

Arafat HA, Jijakli K, Ahsan A (2013) Environmental performance and energy recovery potential of five processes for municipal solid waste treatment. J Clean Prod

Assamoi B, Lawryshyn Y (2012) The environmental comparison of landfilling versus incineration of MSW accounting for waste diversion. Waste Manag 32(5):1019–1030

Astrup TF, Tonini D, Turconi R, Boldrin A (2015) Life cycle assessment of thermal Waste-to-Energy technologies: review and recommendations. Waste Manag 37:104–115

Atimtay AT, Topal H (2004) Co-combustion of olive cake with lignite coal in a circulating fluidized bed. Fuel 83(7–8):859–867

Autret E, Berthier F, Luszezanec A, Nicolas F (2007) Incineration of municipal and assimilated wastes in France: assessment of latest energy and material recovery performances. J Hazard Mater 139(3):569–574

Basso D, Patuzzi F, Castello D, Baratieri M, Rada EC, Weiss-Hortala E, Fiori L (2015) Agro-industrial waste to solid biofuel through hydrothermal carbonization. Waste Manag 2015

Berge ND, Li L, Flora JR, Ro KS (2015) Assessing the environmental impact of energy production from hydrochar generated via hydrothermal carbonization of food wastes. Waste Manag 43:203–217

Berge ND, Ro KS, Mao J, Flora JR, Chappell MA, Bae S (2011) Hydrothermal carbonization of municipal waste streams. Environ Sci Technol 45(13):5696–5703

Bridgwater AV (2003) Renewable fuels and chemicals by thermal processing of biomass. Chem Eng J 91(2–3):87–102

Brown TR (2015) A techno-economic review of thermochemical cellulosic biofuel pathways. Bioresour Technol 178:166–176

Caserini S, Livio S, Giugliano M, Grosso M, Rigamonti L (2010) LCA of domestic and centralized biomass combustion: the case of Lombardy (Italy). Biomass Bioenergy 34(4): 474–482

Chedea VS, Kefalas P, Socaciu C (2010) Patterns of carotenoid pigments extracted from two orange peel wastes (Valencia and navel var.). J Food Biochem 34(1):101–110
Cherubini F (2010) The biorefinery concept: using biomass instead of oil for producing energy and chemicals. Energy Convers Manag 51(7):1412–1421
Christoforou EA, Fokaides PA (2015) A review of quantification practices for plant-derived biomass potential. Int J Green Energy 12(4):368–378
Clark JH, Deswarte FEI, Farmer TJ (2009) The integration of green chemistry into future biorefineries. Biofuels Bioproducts and Biorefining-Biofpr 3(1):72–90
Cok B, Tsiropoulos I, Roes AL, Patel MK (2014) Succinic acid production derived from carbohydrates: an energy and greenhouse gas assessment of a platform chemical toward a bio-based economy. Biofuels Bioproducts and Biorefining-Biofpr 8(1):16–29
Corti A, Lombardi L (2004) Biomass integrated gasification combined cycle with reduced CO_2 emissions: performance analysis and life cycle assessment (LCA). Energy 29(12):2109–2124
Dang Q, Yu C, Luo Z (2014) Environmental life cycle assessment of bio-fuel production via fast pyrolysis of corn stover and hydroprocessing. Fuel 131:36–42
Dzikuc M, Piwowar A (2016) Ecological and economic aspects of electric energy production using the biomass co-firing method: the case of Poland. Renew Sustain Energy Rev 55:856–862
Encinar JM, Gonzalez JF, Martinez G, Gonzalez JM (2008) Two stages catalytic pyrolysis of olive oil waste. Fuel Process Technol 89(12):1448–1455
Fan J, Kalnes TN, Alward M, Klinger J, Sadehvandi A, Shonnard DR (2011) Life cycle assessment of electricity generation using fast pyrolysis bio-oil. Renewable Energy 36(2): 632–641
Food wastage footprint: Impacts on natural resources (2013) http://www.fao.org/docrep/018/i3347e/i3347e.pdf
Food waste. http://www.unido.org/fileadmin/import/32068_35FoodWastes
Fruergaard T, Hyks J, Astrup T (2010) Life-cycle assessment of selected management options for air pollution control residues from waste incineration. Sci Total Environ 408(20):4672–4680
Funke A, Ziegler F (2010) Hydrothermal carbonization of biomass: a summary and discussion of chemical mechanisms for process engineering. Biofuels, Bioprod Biorefin 4:160–177
Gonzalez MI, Alvarez S, Riera F, Alvarez R (2007) Economic evaluation of an integrated process for lactic acid production from ultrafiltered whey. J Food Eng 80(2):553–561
Gustavsson J, Cederberg C, Sonesson U, Otterdijk RV, Meybeck A (2011) Global food losses and food waste. In: Extent, causes and prevention, vol 38. Food and Agriculture Organization of the United Nations, Rome, Italy
Gustavsson JCC, Sonesson U, Emanuelsson A (2013) The methodology of the FOA study: global food losses and food waste-extent, causes and prevention, vol 38. The Swedish Institute for Food and Biotechnology, Gothenburg, Sweden
He Y, Bagley DM, Leung KT, Liss SN, Liao B-Q (2012) Recent advances in membrane technologies for biorefining and bioenergy production. Biotechnol Adv 30(4):817–858
Heidenreich S, Foscolo PU (2015) New concepts in biomass gasification. Prog Energy Combust Sci 46:72–95
Heilmann SM, Jader LR, Sadowsky MJ, Schendel FJ, Von Keitz MG, Valentas KJ (2011) Hydrothermal carbonization of distiller's grains. Biomass Bioenergy 35(7):2526–2533
Huang Y, Syu F, Chiueh P, Lo S (2013) Life cycle assessment of biochar cofiring with coal. Bioresour Technol 131:166–171
Huff MD, Kumar S, Lee JW (2014) Comparative analysis of pinewood, peanut shell, and bamboo biomass derived biochars produced via hydrothermal conversion and pyrolysis. J Environ Manage 146:303–308
Humbird D, Davis R, Tao L, Kinchin C, Hsu D, Aden A, Schoen P, Lukas J, Olthof B, Worley M et al (2001) Process design and economics for biochemical conversion of lignocellulosic biomass to ethanol. Colorado, U.S
Hwang I-H, Aoyama H, Matsuto T, Nakagishi T, Matsuo T (2012) Recovery of solid fuel from municipal solid waste by hydrothermal treatment using subcritical water. Waste Manag 32(3):410–416

Indicative Chemical Prices A-Z. http://www.icis.com/chemicals/channel-info-chemicals-a-z/
Iribarren D, Peters JF, Dufour J (2012) Life cycle assessment of transportation fuels from biomass pyrolysis. Fuel 97:812–821
Jahirul MI, Rasul MG, Chowdhury AA, Ashwath N (2012) Biofuels production through biomass pyrolysis-a technological review. Energies 5(12):4952–5001
Kalk J, Langlykke A (1986) ASM manual of industrial microbiology and biotechnology. In: Li Q, Hu G (eds) Techno-economic analysis of biofuel production considering logistic configurations, vol 206. Bioresource Technology 2016, pp 195–203
Kim MH, Song HB, Song Y, Jeong IT, Kim JW (2013) Evaluation of food waste disposal options in terms of global warming and energy recovery: Korea. Int J Energy Environ Eng 4(1):1–12
Kimming M, Sundberg C, Nordberg Å, Baky A, Bernesson S, Norén O, Hansson P-A (2011) Biomass from agriculture in small-scale combined heat and power plants—a comparative life cycle assessment. Biomass Bioenergy 35(4):1572–1581
Kiran EU, Trzcinski AP, Ng WJ, Liu Y (2014) Bioconversion of food waste to energy: a review. Fuel 134:389–399
Klein-Marcuschamer D, Oleskowicz-Popiel P, Simmons BA, Blanch HW (2012) The challenge of enzyme cost in the production of lignocellulosic biofuels. Biotechnol Bioeng 109(4):1083–1087
Kwan TH, Pleissner D, Lau KY, Venus J, Pommeret A, Lin CSK (2015) Techno-economic analysis of a food waste valorization process via microalgae cultivation and co-production of plasticizer, lactic acid and animal feed from algal biomass and food waste. Bioresour Technol 198:292–299
Kylili A, Christoforou E, Fokaides PA (2016) Environmental evaluation of biomass pelleting using life cycle assessment. Biomass Bioenergy 84:107–117
Lam KF, Leung CCJ, Lei HM, Lin CSK (2014) Economic feasibility of a pilot-scale fermentative succinic acid production from bakery wastes. Food Bioprod Process 92(C3):282–290
Leung CCJ, Cheung ASY, Zhang AY-Z, Lam KF, Lin CSK (2012) Utilisation of waste bread for fermentative succinic acid production. Biochem Eng J 65:10–15
Li J, Paul MC, Younger PL, Watson I, Hossain M, Welch S (2015) Characterization of biomass combustion at high temperatures based on an upgraded single particle model. Appl Energy 156:749–755
Li Q, Hu G (2016) Techno-economic analysis of biofuel production considering logistic configurations. Bioresour Technol 206:195–203
Liamsanguan C, Gheewala SH (2007) Environmental assessment of energy production from municipal solid waste incineration. Int J Life Cycle Assess 12(7):529–536
Libra JA, Ro KS, Kammann C, Funke A, Berge ND, Neubauer Y, Titirici M-M, Fühner C, Bens O, Kern J (2011) Hydrothermal carbonization of biomass residuals: a comparative review of the chemistry, processes and applications of wet and dry pyrolysis. Biofuels 2(1):71–106
Lin CSK, Koutinas AA, Stamatelatou K, Mubofu EB, Matharu AS, Kopsahelis N, Pfaltzgraff LA, Clark JH, Papanikolaou S, Kwan TH et al (2014) Current and future trends in food waste valorization for the production of chemicals, materials and fuels: a global perspective. Biofuels, Bioprod Biorefin 8(5):686–715
Lin CSK, Pfaltzgraff LA, Herrero-Davila L, Mubofu EB, Abderrahim S, Clark JH, Koutinas AA, Kopsahelis N, Stamatelatou K, Dickson F et al (2013) Food waste as a valuable resource for the production of chemicals, materials and fuels. Current situation and global perspective. Energy Environ Sci 6(2):426–464
Liu Z, Balasubramanian R (2012) Hydrothermal carbonization of waste biomass for energy generation. Proc Environ Sci 16:159–166
Lu W, Zhang T (2010) Life-cycle implications of using crop residues for various energy demands in China. Environ Sci Technol 44(10):4026–4032
Luque R, Herrero-Davila L, Campelo JM, Clark JH, Hidalgo JM, Luna D, Marinas JM, Romero AA (2008) Biofuels: a technological perspective. Energy Environ Sci 1(5):542–564
McKendry P (2002) Energy production from biomass (part 2): conversion technologies. Bioresour Technol 83(1):47–54

McKinlay JB, Vieille C, Zeikus JG (2007) Prospects for a bio-based succinate industry. Appl Microbiol Biotechnol 76(4):727–740
Monitoring of solid waste in Hong Kong—waste statistics for 2013 [http://www.epd.gov.hk/epd/english/environmentinhk/waste/data/stat_treat.html]
Mussatto SI, Dragone G, Roberto IC (2006) Brewers' spent grain: generation, characteristics and potential applications. J Cereal Sci 43(1):1–14
Nguyen TLT, Hermansen JE, Nielsen RG (2013) Environmental assessment of gasification technology for biomass conversion to energy in comparison with other alternatives: the case of wheat straw. J Clean Prod 53:138–148
Ohkouchi Y, Inoue Y (2007) Impact of chemical components of organic wastes on L(+)-lactic acid production. Bioresour Technol 98(3):546–553
Oliveira I, Blöhse D, Ramke H-G (2013) Hydrothermal carbonization of agricultural residues. Bioresour Technol 142:138–146
Orjuela A, Orjuela A, Lira CT, Miller DJ (2013) A novel process for recovery of fermentation-derived succinic acid: Process design and economic analysis. Bioresour Technol 139:235–241
Pala M, Kantarli IC, Buyukisik HB, Yanik J (2014) Hydrothermal carbonization and torrefaction of grape pomace: a comparative evaluation. Bioresour Technol 161:255–262
Peters G (2010) Life cycle assessment: principles, practice and prospects. Australasian J Environ Manage 17(3):190–191
Peters JF, Iribarren D, Dufour J (2015) Simulation and life cycle assessment of biofuel production via fast pyrolysis and hydroupgrading. Fuel 139:441–456
Pleissner D, Lam WC, Sun Z, Lin CSK (2013) Food waste as nutrient source in heterotrophic microalgae cultivation. Bioresour Technol 137:139–146
Ravindran R, Jaiswal AK (2016) Exploitation of food industry waste for high-value products. Trends Biotechnol 34(1):58–69
Riber C, Hander GS, Christensen TH (2008) Environmental assessment of waste incineration in a life-cycle-perspective (EASEWASTE). Waste Manage Res 26(1):96–103
Roberts KG, Gloy BA, Joseph S, Scott NR, Lehmann J (2009) Life cycle assessment of biochar systems: Estimating the energetic, economic, and climate change potential. Environ Sci Technol 44(2):827–833
Reza MT, Uddin MH, Lynam JG, Coronella CJ (2014) Engineered pellets from dry torrefied and HTC biochar blends. Biomass Bioenergy 63:229–238
Schnepf R, Yacobucci BD (2011) Renewable fuel standard (RFS): overview and issues. Library of Congress. Congressional Research Service, Washington
Sebastian F, Royo J, Gomez M (2011) Cofiring versus biomass-fired power plants: GHG (Greenhouse Gases) emissions savings comparison by means of LCA (Life Cycle Assessment) methodology. Energy 36(4):2029–2037
Shahidi F, Synowiecki J (1991) Isolation and characterization of nutrients and value-added products from snow crab (Chionoecetes opilio) and shrimp (Pandalus borealis) processing discards. J Agric Food Chem 39(8):1527–1532
Standardization IOf (2006) Environmental management-life cycle assessment-principles and framework. In: ISO 14040; 01 July 2006. Geneva: ISO
Sun A, Davis R, Starbuck M, Ben-Amotz A, Pate R, Pienkos PT (2011) Comparative cost analysis of algal oil production for biofuels. Energy 36(8):5169–5179
Sun Z, Li M, Qi Q, Gao C, Lin CSK (2014) Mixed food waste as renewable feedstock in succinic acid fermentation. Appl Biochem Biotechnol 174(5):1822–1833
Tahergorabi R, Beamer SK, Matak KE, Jaczynski J (2011) Effect of isoelectric solubilization/precipitation and titanium dioxide on whitening and texture of proteins recovered from dark chicken-meat processing by-products. Lwt-Food Sci Technol 44 (4):896–903
Tejayadi S, Cheryan M (1995) Lactic acid from cheese whey permeate. Productivity and economics of a continuous membrane bioreactor. Appl Microbiol Biotechnol 43(2):242–248

Titirici MM, Thomas A, Antonietti M (2007) Back in the black: hydrothermal carbonization of plant material as an efficient chemical process to treat the CO_2 problem? New J Chem 31 (6):787–789
Tyskeng S, Finnveden G (2010) Comparing energy use and environmental impacts of recycling and waste incineration. J Environ Eng Asce 136(8):744–748
Van Dien S (2013) From the first drop to the first truckload: commercialization of microbial processes for renewable chemicals. Curr Opin Biotechnol 24(6):1061–1068
Van Loo S, Koppejan J (2012) The handbook of biomass combustion and co-firing. Routledge, Earthscan, London, United Kingdom
Vlysidis A, Binns M, Webb C, Theodoropoulos C (2011) A techno-economic analysis of biodiesel biorefineries: assessment of integrated designs for the co-production of fuels and chemicals. Energy 36(8):4671–4683
Wang L, Templer R, Murphy RJ (2012) A life cycle assessment (LCA) comparison of three management options for waste papers: bioethanol production, recycling and incineration with energy recovery. Bioresour Technol 120:89–98
Williams AR (2009) Life cycle analysis: a step by step approach. The authority of the State of Illinois, Illinois
Wu K, Chein R (2015) Modeling of biomass gasification with preheated air at high temperatures. Energy Proc 75:214–219
Xiao L-P, Shi Z-J, Xu F, Sun R-C (2012) Hydrothermal carbonization of lignocellulosic biomass. Bioresour Technol 118:619–623
Xu C, Shi W, Hong J, Zhang F, Chen W (2015) Life cycle assessment of food waste-based biogas generation. Renew Sustain Energy Rev 49:169–177
Yang J, Chen B (2014) Global warming impact assessment of a crop residue gasification project A dynamic LCA perspective. Appl Energy 122:269–279
Yang X, Zhu M, Huang X, Lin CSK, Wang J, Li S (2015) Valorisation of mixed bakery waste in non-sterilized fermentation for L-lactic acid production by an evolved Thermoanaerobacterium sp strain. Bioresour Technol 198:47–54
Yi YM (2016) Life cycle assessment studies of food waste base biorefinery. City University of Hong Kong project, Hong Kong
Zhang AYZ, Sun Z, Leung CCJ, Han W, Lau KY, Li MJ, Lin CSK (2013a) Valorisation of bakery waste for succinic acid production. Green Chem 15(3):690–695
Zhang C, Xiao G, Peng L, Su H, Tan T (2013b) The anaerobic co-digestion of food waste and cattle manure. Bioresour Technol 129:170–176
Zhong Z, Song B, Zaki M (2010) Life-cycle assessment of flash pyrolysis of wood waste. J Clean Prod 18(12):1177–1183
Zorya S, Morgan N, Rios LD, Hodges R, Bennett B (2011) Missing food: the case of postharvest grain losses in sub-Saharan Africa. The World Bank, Washington
Zuwala J (2012) Life cycle approach for energy and environmental analysis of biomass and coal co-firing in CHP plant with backpressure turbine. J Clean Prod 35:164–175

Chapter 8
Techno-Economic Evaluation of Refining of Food Supply Chain Wastes for the Production of Chemicals and Biopolymers

Anestis Vlysidis, Apostolis Koutinas and Ioannis Kookos

Abstract The development of sustainable and efficient refining of food supply chain wastes is dependent on the production of various end-products with diversifying market outlets and the identification of cost-effective processing schemes. Design and costing of proposed biorefinery concepts is essential in order to identify those processes that could be implemented on industrial scale. The successful implementation of microbial bioconversion of renewable resources for the production of chemicals and biopolymers is highly dependent on the development of cost-competitive biorefinery concepts. The recent literature on techno-economic assessment of food supply chain waste biorefining is presented. One detailed case study is presented focusing on the techno-economic evaluation of refining of orange peel wastes.

Keywords Food waste biorefineries · Process design · Techno-economic evaluation · Citrus processing waste

8.1 Introduction

Although, there are uncertainties regarding the percentage of total food production that is currently lost through the whole supply chain (Parfitt et al. 2010), this has been estimated to be around 1.3 billion tones, which is approximately one third of the global production of food for human consumption (Galanakis 2012). Food waste (FW) is generated in the entire supply chain starting from agricultural production and postharvest (upstream process) to the processing of goods, distribution and consumption (downstream process) (Food Wastage Footprint 2013).

A. Vlysidis · A. Koutinas (✉)
Department of Food Science and Human Nutrition, Agricultural University of Athens, Iera Odos 75, 118 55 Athens, Greece
e-mail: akoutinas@aua.gr

I. Kookos
Department of Chemical Engineering, University of Patras, 26504 Patras, Rio, Greece

P. Morone et al. (eds.), *Food Waste Reduction and Valorisation*,
DOI 10.1007/978-3-319-50088-1_8

In developing countries, most of FW is generated during the upstream process due to lack of infrastructure, while in developed countries the vast majority of FW is produced at the consumption stage (Parfitt et al. 2010). In 2006, EU-27 generated approximately 90 million tones of FW from the manufacturing sector and households, not including food losses from fisheries and agriculture (Monier et al. 2010). This amount accounts to 180 kg per person per year and is projected to increase up to 126 million t by 2020. Apart from the obvious economic losses in all relating sectors (agriculture, production, retailers and households) there is a significant environmental impact as it has been estimated that around 1.9 t of CO_2 are produced per t of FW (Monier et al. 2010).

Due to the substantial quantities of FW generated each year worldwide, there is a global need governed from societal and economical features to re-use, re-cycle and/or re-cover these 'losses' under sustainable approaches. In recent years, the research community has focused on the valorization of FW as a renewable resource for the production of various commodity or value-added products. FW has been designated as a renewable resource that can play a significant part in the forthcoming bio-economy era as its chemical complexity fits perfectly to the concept of biorefinery development for the production of energy, chemicals and bio-based polymers (Lin et al. 2013; Koutinas et al. 2014a; Mirabella et al. 2014). This chapter focuses on the techno-economic evaluation of biorefineries using FW as renewable resource in order to assess new designs and the production of diversified end-products. A case study will be presented using food manufacturing wastes from an orange juice production factory.

8.2 Techno-Economic Assessment of Food Waste Biorefining

Biorefineries are facilities analogous to petroleum refineries that use biomass instead of crude oil for the production of various end-products including chemicals, materials, energy, fuels and biopolymers. The economic sustainability of these facilities is questioned as their end-products usually have higher production costs and cannot compete with the corresponding materials produced from petroleum. Hence, the economic evaluation of these new designs and end-products is of critical importance and need to be performed as a first step towards their successful commercialization. Various profitability criteria should be measured, most important of which are the net present value (NPV) and the internal rate of return (IRR) (Vlysidis et al. 2011). When there is not a firm market value for the obtained new products their minimum selling price (MSP), corresponding to zero NPV at the end of the life cycle of the plant, is calculated. The latter is usually assessed for different design parameters such as the capacity of the plant, the interest rate, the prices of raw materials and/or end-products and operational parameters like fermentation yields, alternative downstream processes, alternative raw materials and/or end-products (Koutinas et al. 2014a, b).

Preliminary economic studies should underline key factors that affect the profitability criteria of these new investments. Critical outcomes of these assessments should provide information regarding the stages of the process that should be modified towards the optimization of the profitability of the plant and the identification of the best available technology (Koutinas et al. 2014a). Most of the techno-economic assessments reported in the literature focus on the use of food waste and by-products coming from the manufacturing process, such as the sugarcane and dairy industries (Summers et al. 2015; Koutinas et al. 2016). These industrial streams are nowadays considered as by-products. There are also studies focusing on techno-economic assessment of valorization of food waste produced from restaurants and hotels (Han et al. 2016; Kwan et al. 2015).

Summers et al. (2015) carried out a techno-economic analysis using delactosed whey permeate for the production of renewable diesel via microbial fermentation followed by hydrolthermal liquefaction. The designed facility had a plant capacity of 1.25 million m^3 of dairy liquid waste per year and it was based on lab-scale experimental results. The plant life and interest rate was assumed to be 30 years and 8%, respectively. The depreciation schedule was 7 years following the Modified Accelerated Cost Recovery System (MACRS). It was concluded that the MSP of renewable diesel production is 1.26 $/L which is higher than the average prices for soybean-derived biodiesel (1.15 $/L) and diesel (0.90 $/L). The higher MSP value was mainly attributed to the operational requirements of the yeast fermentation, the preparation of the inoculum and the intense conditions of hydrothermal liquefaction. It was estimated that the MSP of renewable diesel production could be reduced to 1.15 $/L, if the capacity of the process is increased approximately thirty times. A further reduction of up to 0.76 $/L on the MSP could be achieved via process optimization (i.e. fermentation yield and productivity as well as performance of hydrotreatment performance) (Summers et al. 2015).

Kwan et al. (2015) developed a techno-economic study on FW biorefining for the production of a spectrum of end-products such as plasticizers, lactic acid and animal feed. The FW was collected from restaurants and bakeries and was hydrolysed after grinding to small particles using enzymes produced via solid state fermentation. The FW hydrolysates were then fermented for algae production using the microalgae *Chlorella pyrenoidosa*. At the end of the fermentation, the lipid content of the algae was extracted so as to be used in the production of plasticizers, while lipid-free algae biomass rich in nitrogen source was used as substrate in lactic acid production via fermentation or as animal feed. Both scenarios were based on experimental results from previous studies and were designed in the software SuperPro Designer (Intelligen Inc.). The operational capacity of the plant was 1 t of food waste per day. Results from the techno-economic analysis indicated that NPV, IRR and payback period were 3.03 M$, 19% and 7.6 years, respectively, for a plant lifetime of 30 years and a discount rate of 5%. According to the sensitivity analysis, the market price of lactic acid had the most significant impact on the NPV compared to other raw-materials and end-products, accounting for approximately 30% reduction in NPV based on 10% variation in its market price (Kwan et al. 2015).

Han et al. (2016) developed a techno-economic analysis for the valorization of FW collected from a University canteen. The processing capacity of the plant was 1095 tones per year. The process included a hydrolysis stage converting FW into fermentable nutrients followed by microbial fermentation for the production of hydrogen. The hydrolysis process was carried out using crude enzymes produced via solid state fermentation. The mass and energy balances were computed using the design software Aspen Plus. The NPV of the plant was calculated for different interest rates and lifespans of the plant and it was above zero for an interest rate of 10% and a lifetime higher than 6.2 years. Although the low scale of the process decreased the profitability potential of the plant as the NPV of the investment after 15 years was around 0.44 M$, the IRR was considerably high accounting to 24.1%. Apart from H_2, the pilot plant also co-produced solid biomass as animal feed (Han et al. 2016).

Koutinas et al. (2014b) have presented a techno-economic analysis for the production of microbial oil from glucose-based media. Waste or by-product streams from confectionary industries and bakeries could be employed. The capacity of the plant was 10,000 t of microbial oil production per year, while the plant operation was set at 8300 h/y. Once the microbial oil is produced by oleaginous yeast, the microbial mass is separated and dried. Cells are then disrupted mechanically and oil is separated from cell debris via a centrifugation unit using hexane. The latter is then recovered and recycled through a one-step evaporation unit. The microbial oil was then used to produce biodiesel via either direct or indirect transesterification. The mass and energy balances were calculated using the process simulation software, SuperPro Designer and UniSim (Honeywell). The cost of manufacture was largely affected by the cost of the bioreactors and was equal to $3.41 per kg microbial oil considering a zero market price of glucose. If the price of the raw material increases to $0.4/t the total production cost rises to $5.48 per kg microbial oil. Koutinas et al. (2014b) mentioned that the unitary production cost of microbial oil is significantly affected by the productivity of the microbial fermentation and the market price of the raw material used. In order to drastically decrease the manufacturing cost to around $1.76 per kg microbial oil, the productivity of the fermentation stage should be increased to 2.5 g/L/h for a zero glucose price (Koutinas et al. 2014b).

Koutinas et al. (2016) carried out techno-economic evaluation for the production of 2,3-butanediol using three different raw materials, one of which was sugarcane molasses. The process simulation software UniSim was used. Koutinas et al. (2016) conducted a sensitivity analysis for different market prices of the raw material and different plant capacities calculating each time the MSP of 2,3-butanediol production. The capacity of the plant was set at 10,000 t per year, while the plant operated for 8300 h/y. The MSP of 2,3-butanediol production was higher than 1 $/kg, which is generally regarded as the target in order to characterize a chemical production as basic or platform chemical. The MSP ranged from 2.6 to 4.8 $/kg for all raw material prices and fixed capital investments. It was stressed that the plant capacity of 10,000–40,000 t/y can be crucial as the MSP gradually drops by approximately 14%. Further capacity increase results in insignificant MSP reduction (Koutinas et al. 2016).

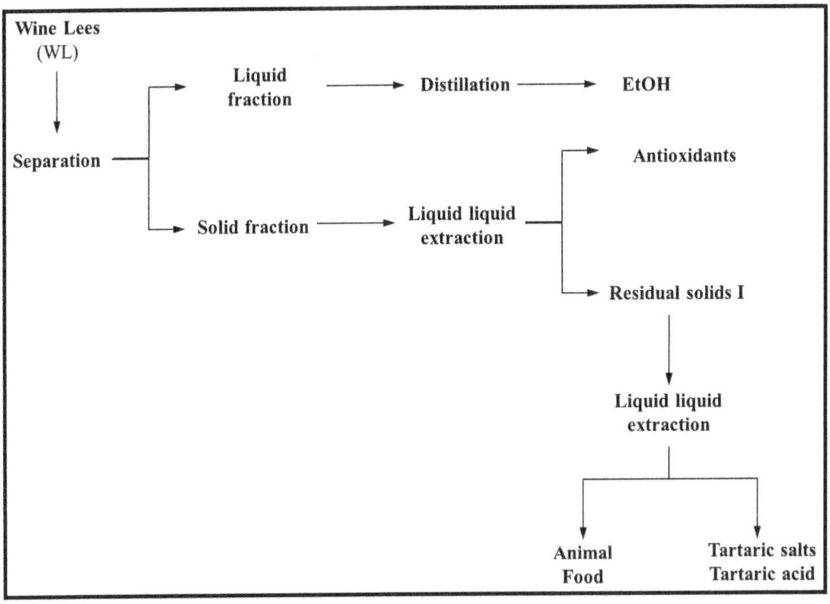

Fig. 8.1 Schematic diagram of a biorefinery using wine lees for the production of ethanol, antioxidants, tartaric acid and animal food (Dimou et al. 2015)

Another industrial FW that presents considerable interest is wine lees produced from the vinification process. Dimou et al. (2015) showed that wine lees could be used for the production of ethanol, an antioxidant-rich extract, tartaric acid and yeast cells (Fig. 8.1). The latter can be used as animal feed or for the production of nutrient supplements for microbial fermentations (Dimou et al. 2015). A techno-economic assessment conducted for a biorefinery using wine lees as renewable resource has been developed (results not published yet). The profitability of the plant utilizing the wine lees was dependent on the MSP of the antioxidant-rich fraction with respect to the plant capacity as this end-product does not currently have a firm market. A ten-fold increase from 500 to 5000 kg/h of processed wine lees can result in a significant drop of the MSP of antioxidants from 122 to 11 $/kg.

8.3 Case Study—Techno-Economic Evaluation of Biorefining Citrus Waste

The development of biorefineries focusing on the valorization of citrus waste first gained attention in 1940s and 1950s in the USA where juice industries were evaluating ways to give value to the huge amounts of citrus waste generated each year after the juice extraction process (Anonymous 1956; Hull et al. 1953; Van Antwerpen 1941). These studies proposed technologies to recover added-value compounds from citrus wastes such as essential oils, flavonoids and pectin as well as the production of a liquid stream called *citrus molasses* rich in soluble sugars (Anonymous 1956). According to FAO statistics, the year 2013 approximately 71.3 million tones of oranges were produced worldwide. Around 40% of this amount was processed in juice production and 50% of this amount was discarded as citrus peel waste (Pfaltzgraff 2014). This leads to a total annual amount of orange peel wastes equal to 14.3 million tones. The main constituents of orange peels are cellulose, hemicellulose and pectin which account to 50–70% of the dry orange peel. It also contains a fraction of soluble sugars such as xylose, glucose sucrose and fructose, 3–4% of D-limonene and 4–5% of flavonoids (Pfaltzgraff 2014). Due to the prospect of producing valuable compounds from citrus wastes, a number of techno-economic studies have been developed (Lohrasbi et al. 2010; Grohmann 2007; Zhou et al. 2007). The extraction of D-limonene is a process already employed in large scale citrus processing plants as it is used in the pharmaceutical, food and cosmetic industry. Pectin extraction is a more complicated process and it is hardly applied to orange juice factories. Pectin is used as a gelling agent in foods (Lopez et al. 2010). Flavonoids are chemical substances of low molecular weight that contain more than three phenolic hydroxides. They are abundant in nature as secondary metabolites and they are one of the most interesting groups having biological active compounds. They are used as antioxidants mainly in the pharmaceutical and cosmetic industry but they also have application in the food industry (Anagnostopoulou 2005). Figure 8.2 presents the main end-products derived from citrus wastes that have been widely investigated. In this chapter, a case study has been developed evaluating the development of a biorefinery concept using citrus waste for the production of D-limonene, energy and bioethanol. Outcomes are compared with results from similar literature-cited studies.

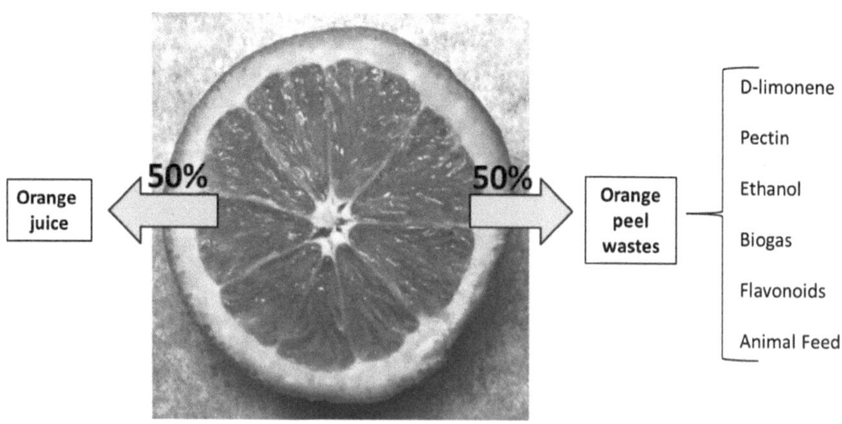

Fig. 8.2 Current valorization options for citrus wastes

8.3.1 Process Design

The design of the biorefinery processing orange peel waste into various end-products was performed using literature-cited results (Lohrasbi et al. 2010; Pourbafrani et al. 2010; Humbird et al. 2011), while the energy and material balances were determined using the process simulation software Unisim. The plant processes 50,000 t of orange peel waste per year which leads to an hourly flowrate of 12.5 t/h as the plant operates seasonally for 4000 h/y (approximately 5.5 months). The developed plant covers only a base case scenario for the valorization of orange peel wastes extracting the D-limonene and producing bioethanol via fermentation of free sugars as well as the hydrolysates of cellulose and hemicellulose. However, this base case scenario can be compared to more advanced biorefinery designs that extract also pectin and flavonoids. The process flow diagrams (PFDs) developed in this study for the extraction of D-limonene and the production of ethanol are shown in Figs. 8.3 and 8.4, respectively. The composition of orange peel waste (Table 8.1) has been obtained from Lohrasbi et al. (2010) and Pourbafrani et al. (2010).

The PFD for the production of D-limonene is presented in Fig. 8.3. The orange peel waste enter in a rotary cutter (M-101) through a belt conveyor (C-101) where the size of orange peel waste is reduced and the surface area available to acid hydrolysis that follows is increased. The shredded orange peel waste enter via stream 1 (12.5 t/h) to the hydrolysis reactor (R-101) where the partial hydrolysis of hemicellulose and cellulose takes place. The conversion yields achieved from cellulose and hemicellulose to the respective sugars are 50 and 60%, respectively. The composition of orange peel waste in cellulose and hemicellulose is given by Aravantinos et al. (1994) where hemicellulose is composed mainly of hexoses (60.6%). Therefore, it was considered that the hydrolysis of hemicellulose gives 60% (w/w) hexoses and 40% (w/w) pentoses according to the following stoichiometric equation:

$$-(C_6H_{10}O_5)_1-(C_5H_8O_4)_{0.8} + 1.8\,H_2O \rightarrow C_6H_{12}O_6 + 0.8\,C_5H_{10}O_5 \qquad (8.1)$$

Hence, 100 kg of hemicellulose could be hydrolysed to 67.26 kg of hexoses and 44.84 kg of pentoses. Cellulose is hydrolysed into glucose. Besides polysaccharides, the orange peel waste contains also free sugars including fructose, sucrose and glucose. During the hydrolysis process, the sucrose contained in the orange peel waste is also hydrolysed to give one molecule of glucose and one molecule of fructose according to the following stoichiometric reaction:

$$C_{12}H_{22}O_{11} + H_2O \rightarrow C_6H_{12}O_6 + C_6H_{12}O_6 \qquad (8.2)$$

Hence, 100 kg of sucrose are hydrolysed to 52.63 kg of glucose and 52.63 kg of fructose. The differences in masses both in (8.1) and (8.2) are due to the addition of water.

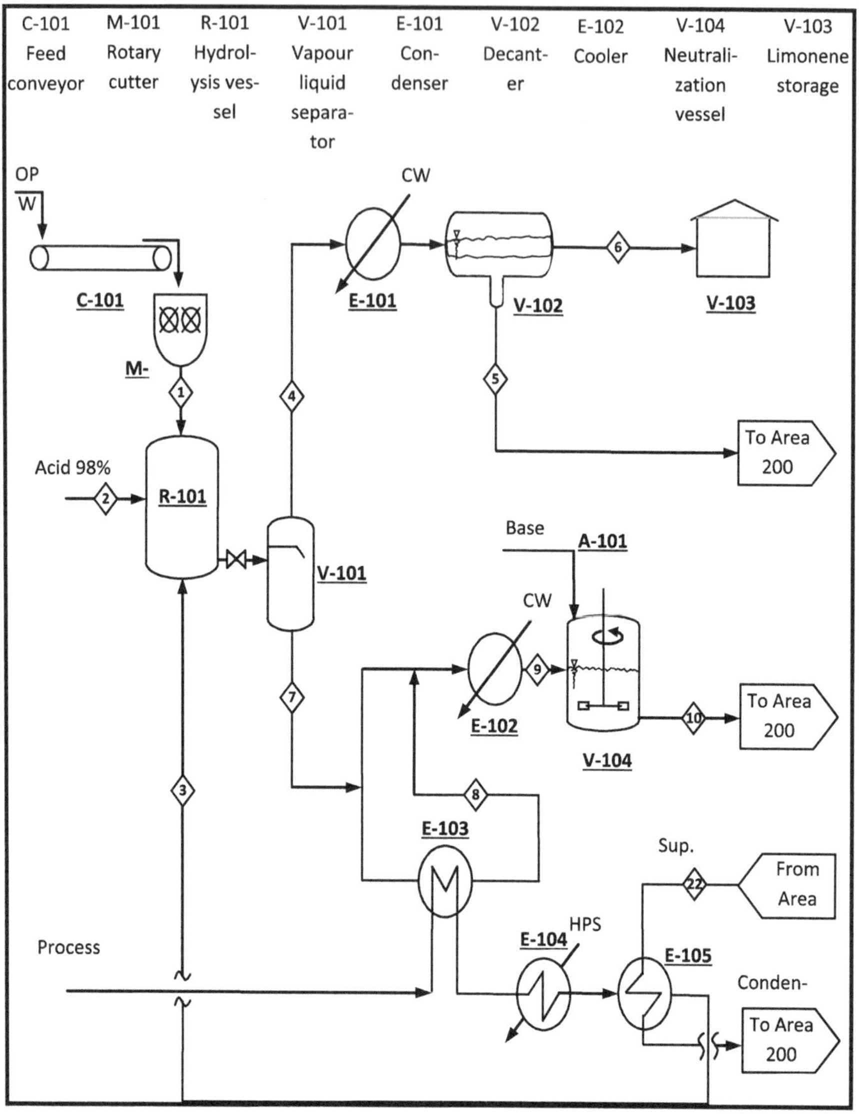

Fig. 8.3 PFD for the hydrolysis of orange peel waste and the extraction of D-limonene

The hydrolysis takes place at 150 °C by steam explosion under mild acidic conditions using 0.25% (w/w) of sulfuric acid solution. Steam explosion is carried out by providing 2.6 t/h of high pressure steam (HPS). The hydrolysis reaction lasts for 10 min and another 5 min are needed for loading and uploading the reactor. Due to fact that the hydrolysis is a batch process, a train of four reactors has been assumed. The steam is produced in a series of heat exchangers (E-103, E-104 and E-105) where process water is transformed into HPS. The superheated steam from

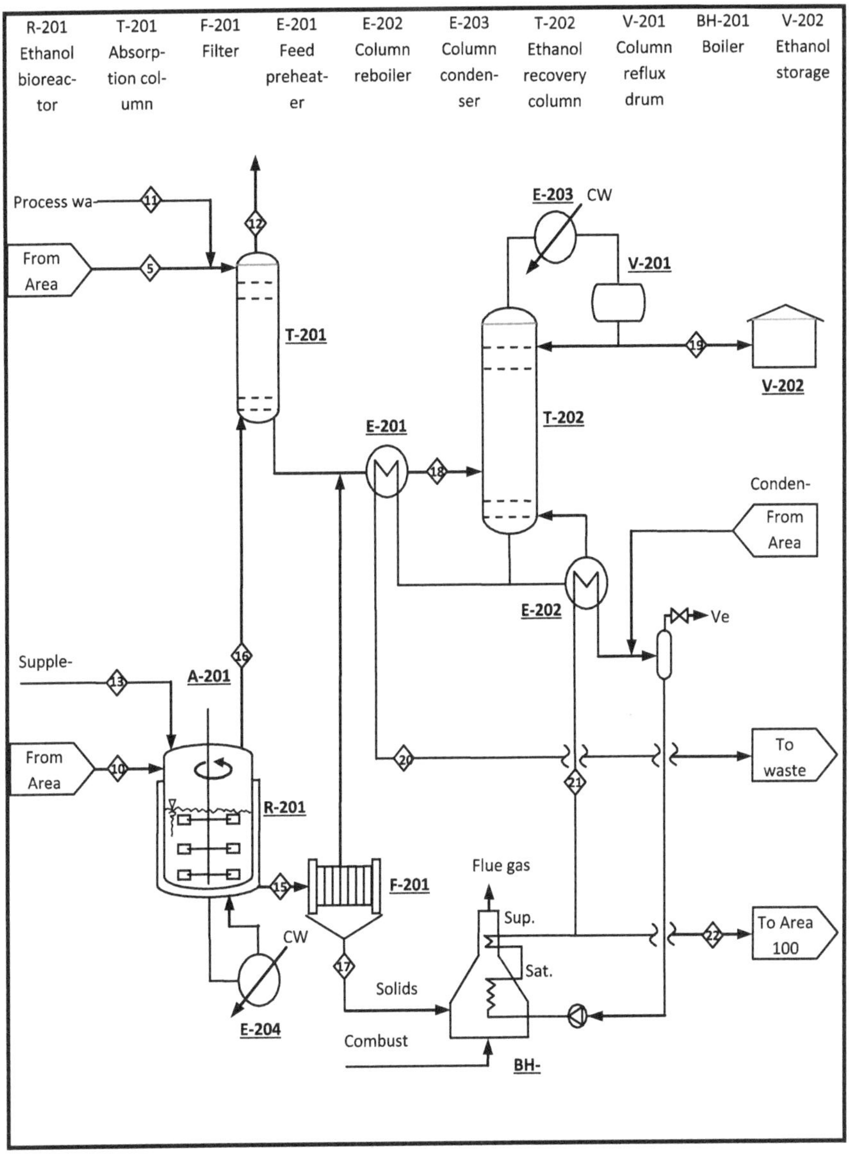

Fig. 8.4 PFD for ethanol production and recovery including the production of HPS from orange peel waste

stream 22 is produced in Area 200 by burning all the remaining solids of orange peel waste after the fermentation process. At the end of the hydrolysis process, pressure is partially released from 10.0 to 4.9 bar in a vapour-liquid separator (V-101) assuming a constant temperature of 150 °C so as to recover a gas stream comprised by D-limonene and water (stream 4) and a liquid/solid stream (stream 7)

Table 8.1 Composition of orange peel waste (Lohrasbiet al. 2010; Pourbafrani et al. 2010)

Component	Amount in kg/100 kg of orange peel waste
Water	80.4
Glucose	1.6
Fructose	2.4
Sucrose	0.6
Pectin	5.0
Protein	1.2
Cellulose	4.4
Hemicellulose	2.2
lignin	0.4
Limonene	1.0
Ash	0.8

containing the produced sugars, water and the remaining solids (pectin, lignin, protein and ash). Stream 4 passes through a cooler (E-101) that condensates D-limonene and water to 25 °C. The liquid stream then enters into a decanter (V-102) where the D-limonene (stream 6) is separated from water (stream 5) as these two liquids are immiscible, forming two distinct liquid phases. Finally, the D-limonene is stored in a storage tank able to store the weekly production of D-limonene with a mass flowrate of 123.75 kg/h. Stream 5 goes into Area 200. Stream 7 is cooled down in the heat exchanger E-102 and partially in E-103 from 150 to 30 °C. Stream 9 then enters into vessel V-104 where the neutralization of the loading of the bioreactor takes place by adding a base such as caustic soda (31.2 kg/h). This vessel also acts as a holding tank before the operation of the fermentation process.

Stream 10 from Area 100 enters into the bioreactor R-201 (Area 200) together with the necessary amounts of nutrient supplements for the production of ethanol (Fig. 8.4). The fermentation time was 36 h (Humbird et al. 2011) and assuming 12 h for cleaning, preparation and loading and another 12 h for uploading, the total batch cycle time is 60 h. The number of batches per year is 67 that can be calculated by dividing the annual operating time of the plant with the duration of a single batch cycle. A seed train of five bioreactors (not included in the PFD of Fig. 8.4) with total volumes of 100.00, 10.00, 1.00, 0.1 and 0.01 m^3 have been also considered in this process design. The ethanol is produced by the microorganism *Zymomonas mobilis* that can ferment both pentoses and hexoses into ethanol with a yield of 0.34 g/g (Humbird et al. 2011). One train of seed bioreactors will be enough to support the main ethanol bioreactor as the cycle time of each seed bioreactor is 36 h (12 h of turnaround time and 24 h of batch time). The input in each bioreactor is 14.9 t/h which includes the inoculum volume, stream 10 and the supplementary nutrients which in this case are corn steep liquor and diammonium hydrogen phosphate. These nutrients provide the necessary nitrogen and phosphorus to the microorganism. There are two output streams from the bioreactor R-201, one vapor stream (stream 16) that comes out from the tower (0.5 t/h) and passes through an absorption column where ethanol is stripped by water. The other stream is a solid/liquid stream (stream 15) that passes through a filtration unit (F-201) where all solids (yeast cells and the remaining solids of orange peel waste) are removed from the liquid. The liquid stream is then mixed with the output from the absorption column (T-201) to form stream 18. The latter enters into the distillation column (T-202). The distillate (stream 19) consists of 95% (w/w) of ethanol with a flowrate of 518.7 kg/h and the product is stored in the tank V-202 with storage capacities for one week. The solid stream (stream 17) after the filtration unit (F-201) enters into a boiler (BH-201) that can process up to 40% moisture content where the solids are burnt to produce HPS to fulfill most of the steam requirements in the hydrolysis process, but also to supply steam to the reboiler E-202. The heat produced by the boiler is 2344 kW that produces 5016 kg/h of saturated steam at 55 bar.

8.3.2 Cost Estimation

The characteristics of each equipment of the two PFDs were determined based on standard engineering procedures, while the f.o.b. cost of each unit operation was calculated using literature-cited data (Peters et al. 2003; Ulrich 1984; Turton et al. 2009). The bare module cost (C_{BM}) was then determined using the chemical engineering plant cost index (CEPCI) and the material factor for each type of equipment. The fixed capital investment is then estimated by using the equation $FCI \approx 1.2 \times$ *Total Installed Equipment Cost*. The individual cost elements relative to the estimation of the FCI are summarized in Table 8.2. The total installed equipment cost is M$13.8, which leads to a FCI of M$16.6. The most expensive unit operation is the four hydrolysis reactors, which account for the 28.6% of the total C_{BM} followed by the boiler needed to produce 5 t/h HPS, which contributes approximately 20% of the total installed equipment cost. The bioreactor accounts for 12% of the total C_{BM} including the agitator and coil. Finally, the neutralization reactor V-104 and the distillation column T-202 contribute around 7.34 and 7.72% of the total installed equipment cost, respectively.

Apart from the capital investment, the total production cost was also estimated. The cost of utilities (C_{UT}), the labour cost (C_{OL}), the cost of raw materials (C_{RM}) and the waste treatment cost (C_{WT}) were determined. The total production cost without depreciation was calculated based on the following empirical equation (Turton et al. 2009):

$$TPC_{woD} = 0.18FCI + 2.73C_{OL} + 1.23(C_{RM} + C_{UT} + C_{WT}) \tag{8.3}$$

The utilities used in this case study are presented in Table 8.3. The total utilities cost is 0.15 M$/y. Due to the heat integration techniques implemented in this design the requirements in HPS were reduced to only 1.2 t/h. High energy requirements are needed in order to agitate the bioreactor. The C_{WT} is estimated by assuming that the non-toxic wastes have a cost of disposal equal to 50 $ per

1000 m^3. The wastes produced in this biorefinery are mainly from the bottom of the distillation column T-202 that ends up in stream 20 with a flowrate of 12.4 m^3/h. This amount leads to an annual C_{WT} of 2480 \$/y, which is insignificant compared to the C_{UT}. To calculate the labour cost, the number of workers required has been estimated for each unit operation based on well-known methods taken from the literature (Turton et al. 2009) and results are shown in Table 8.4.

The annual operating labor cost is 720,000 \$/y (the annual salary of each worker is 30,000 \$). The cost of raw material is calculated by multiplying the annual requirements of each raw material with its unitary cost (Table 8.5). Orange peel wastes are considered to have null price as their transportation will be minimized as the plant will be constructed in an existing orange juice factory. The rest of the chemicals needed for the hydrolysis, neutralization and fermentation process have an insignificant effect mainly due to the low amounts required. The total C_{RM} is M\$ 0.96. The requirements in corn steep liquor and diammonium hydrogen phosphate were determined so as to have 2.5 kg corn steep liquor per t broth and 0.33 kg diammonium hydrogen phosphate per m^3. By using the above data and implementing (8.3), the TPC_{WoD} was measured at M\$ 5.26 per year. The revenues of the plant were calculated similarly as the C_{RM}. The amount of ethanol produced is 2075 t/y and the amount of D-limonene is 495 t/y. As the unit price of limonene is very high compared to ethanol most of the revenues (>70.4%) comes from this source (see Table 8.6). The total revenues account for M\$ 7.025. The NPV in \$ and the MSP in \$/kg of ethanol was then calculated for different interest rates (IR) and ethanol selling prices. Results are shown in Fig. 8.5. The NPV is above zero for interest rate values lower than 8.5% for an ethanol selling price of 1 \$/kg. The IRR is higher than 10% for an ethanol selling price higher than 1300 \$/t.

8.4 Discussion and Conclusions

Currently, most of the citrus industries use their orange peels wastes for cattle feed (Rivas-Cantu et al. 2013) or dispose them as wastes without any recovery of value-added products, while very few of them extract the essential oils (Anagnostopoulou 2005). The option of using citrus wastes as animal feed provides low profits as the production process reduces significantly the overall profit due to intensive drying and the transportation cost. Apart from economic issues for not extracting the essential oils, there are also environmental concerns due to the fact that volatile compounds are emitted to the atmosphere during the drying process of the citrus waste when it is used as animal feed.

Rivas-Cantu et al. (2013) stressed the necessity to cover the technological gaps for successful hydrolysis of citrus waste by optimizing process conditions and equipment as this material differs from lignocellulosic biomass. The authors have emphasized on the improvement in the hydrolysis process regarding the enhancement of sugar production yield and the reduction of processing time by reducing the size of citrus peel particles (Rivas-Cantu et al. 2013).

Table 8.2 Equipment cost of the citrus waste biorefinery

UNIT	Description	f.o.b. cost (M$)	Source	CEPCI	FM	C_{BM} (M $@2012)
C101	CS, 0.7 m width, 100 m length	0.180	PTW, $@2002	396	1.7	0.450
M101	CS, 12.5 t/h	0.271	JBEI, $@2008	576	2.38	0.654
R101	SS316, 1.67 m^3, 4 units	4 × 0.487	JBEI, $@2008	576	2.0	3.943
V101	SS316, 1.15 m diameter, 3.45 m height	0.030	PTW, $@2002	396	2.0	0.088
V102	SS316, 1.285 m diameter, 3.85 m height	0.030	PTW, $@2002	396	2.0	0.088
V103	SS304, 25.2 m^3	0.065	NREL, $@2009	522	1.8	0.130
V104	SS304, 1000 m^3, includes agitator	0.453	NREL, $@2009	522	2.0	1.013
E101	SS304/CS, 1.53 m^2	0.008	PTW, $@2002	396	2.2	0.027
E102	SS316/CS, 35.8 m^2	0.021	PTW, $@2002	396	2.2	0.066
E103	SS316/CS, 50.6 m^2	0.025	PTW, $@2002	396	2.2	0.079
E104	CS/CS, 13.2 m^2	0.005	PTW, $@2002	396	2.2	0.015
E105	CS/CS, 74.8 m^2	0.010	PTW, $@2002	396	2.2	0.033
Total installed equipment cost of area 100 (M$)						**6.586**
R201	SS304, 1000 m^3, includes agitator & coil	–	NREL, $@2009	522	–	1.645
	5th seed bioreactor 100 m^3, SS304	–	NREL, $@2009	522	–	0.328
	4th seed ferm. 10 m^3, SS304, skid complete	0.081	NREL, $@2009	522	1.8	0.162
	3rd seed ferm. 1 m^3, SS304, skid complete	0.061	NREL, $@2009	522	1.8	0.122
	2nd seed ferm. 0.1 m^3, SS304, skid complete	0.040	NREL, $@2009	522	1.8	0.080
	1st seed ferm. 0.01 m^3, SS304, skid complete	0.023	NREL, $@2009	522	1.8	0.046
T201	SS316, D = 0.4 m, H = 7 m	0.036	PTW, $@2002	396	2.0	0.106
	15 sieve trays	0.013	PTW, $@2002	396	1.0	0.019
T202	SS316, D = 0.84 m, H = 38.3 m	0.360	PTW, $@2002	396	2.0	1.065
	57 sieve trays	0.048	PTW, $@2002	396	1.0	0.070
V201	SS316, D = 0.7 m, H = 2.1 m, horizontal	0.006	PTW, $@2002	396	2.0	0.018
V202	CS gr. C, V = 144 m^3, floating roof	0.083	NREL, $@2009	522	1.7	0.158
E201	SS304/CS, 130 m^2	0.025	PTW, $@2002	396	2.2	0.079

(continued)

Table 8.2 (continued)

UNIT	Description	f.o.b. cost (M$)	Source	CEPCI	FM	C_{BM} (M $@2012)
E202	SS304/CS, 8 m^2	0.004	PTW, $@2002	396	2.2	0.021
E203	SS304/CS, 43 m^2	0.008	PTW, $@2002	396	2.2	0.041
F201	Centrifuge, 2 kg/s solids, SS316	0.200	PTW, $@2002	396	1.7	0.500
BH201	Boiler, 5 t/h of HPS	1.513	JBEI, $@2008	576	1.8	2.758
Total installed equipment cost of area 200 (M$)						**7.218**
Total installed equipment cost (M$)						**13.8**
Fixed capital investment (M$)						**16.6**

Table 8.3 Energy requirements and calculation of the utilities cost

UNIT	Electricity (kW)	HPS (t/h)	CW (t/h)
C101	1		
M101	70		
A101	170		
F201	14		
E101			5.6
E102			65.6
E203			51.5
E104		1.2	
TOTAL	255	1.2	122.7
Cost (M$/y)	**0.061**	**0.080**	**0.008**
Total utilities cost			**0.150**

Grohmann (2007) evaluated the viability of an ethanol production plant from citrus peel by implementing experimental trials in pilot scale bioreactors (0.38 and 3.78 m^3) and in an industrial scale bioreactor (37.9 m^3). Grohmann (2007) evaluated two options for bioconverting citrus wastes into ethanol. The first one was by enzymatic hydrolysis followed by fermentation, while the second approach involved steam pretreatment of citrus wastes followed by D-limonene removal and finally simultaneous saccharification and fermentation. The advantages of the second approach were numerous regarding both economic and technical issues. D-limonene provides an essential income on plant's revenues, microbial inhibition is decreased as D-limonene is a toxic compound, steam pretreatment pasteurize citrus wastes and hence contamination issues are reduced. Grohmann (2007) also compared the cost of ethanol production from citrus waste with the one obtained via corn processing. Ethanol production from a citrus waste processing plant lead to a higher total income per liter (0.576 $/L contrary to 0.544 $/L from corn processing).

Zhou et al. (2007) have also carried out an economic analysis of ethanol production from citrus peel waste. The authors compared the production cost of

Table 8.4 Number of workers required for each unit operation

Type of equipment	Number of units multiplied by required workers	Number of workers
Towers or vessels	8 × 0.25	2.0
Heat exchangers	8 × 0.1	0.8
Bioreactors	1 × 0.5	0.5
Boiler	1 × 0.5	0.5
Filter	1 × 0.5	0.5
Cutter	1 × 0.5	0.5
Conveyor	1 × 0.5	0.5
	Total number of workers	5.3

Table 8.5 The cost of raw materials

Material	kg/h	t/y	Unit cost $/t	Total $/y
Orange peel waste	12,500.0	50,000.0	0.0	0
H_2SO_4 98%	39.0	156.0	100.0	15,600
Process water	2650.0	10,450.0	0.5	5225
NaOH	31.2	125.0	400.0	50,000
Corn steep liquor	33.0	132.0	60.0	7920
diammonium hydrogen phosphate	4.4	17.4	1000.0	17,400
TOTAL C_{RM}	96,145			

Table 8.6 Calculation of the annual revenues of the orange peel waste biorefinery

Product	t/y	Unit price ($/t)	Revenues in M$
Ethanol	2075	1000	2,075,000
Limonene	495	10,000	4,950,000
Total revenues			**7,025,000**

ethanol produced from three different raw materials: starch, cellulose and citrus peels. It seems that the ethanol production cost from citrus waste (0.325 $/L) is considerably lower than the production cost of ethanol from cellulose (0.430 $/L) and only slightly higher than the production cost of ethanol from starch (0.264 $/L). The main contributor to the ethanol production cost was the cost of chemicals, waste disposals and utilities, while the plant producing ethanol from citrus waste was benefited by the high revenues obtained due to D-limonene recovery (Zhou et al. 2007).

Pourbafrani et al. (2010) carried out a laboratory study for the valorization of citrus wastes for the production of ethanol, biogas, D-limonene, pectin and animal feed. The authors first implemented a diluted acid hydrolysis in a 10 L high pressure reactor with the addition of steam. Optimum conditions were examined by the authors by conducting a central composite design. Limonene was recovered by flashing the content of the reactor after hydrolysis. The hydrolysates were then processed in a centrifugation. The solid fraction was used in the anaerobic digester for biogas production, while the liquid stream was used for pectin extraction and

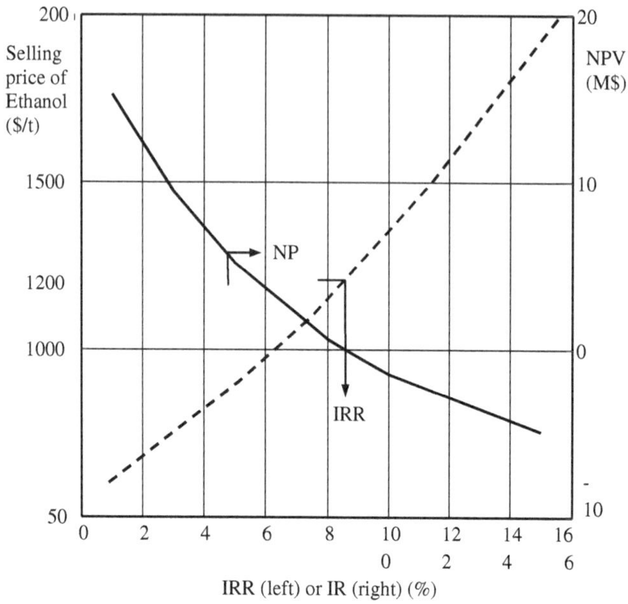

Fig. 8.5 NPV and selling price of ethanol for different IR and IRR, respectively

ethanol production. From 100 kg of citrus waste with a moisture content of 80%, 0.89 L of D-limonene and 3.88 kg of pectin were extracted and 3.96 L of ethanol and 45 m^3 of methane were produced (Pourbafrani et al. 2010). The previous experimental study was developed in a process design and economic analysis without the extraction of pectin in Aspen Plus (Lohrasbi et al. 2010). Increasing the plant capacity results in decreasing ethanol production cost from around 2.5 $/L at 25,000 t/y to approximately 0.5 $/L at 400,000 t/y. Apart from the credit from D-limonene, there is also a significant income from biogas produced during the anaerobic digestion. The authors have considered a cost of raw material equal to 10 $/t due to transportation. If the facilities of the CW plant are integrated in an existing juice production plant, this cost can be reduced to zero resulting in an ethanol production cost of around 0.3 $/L for a plant capacity of 400,000 t/y (Lohrasbi et al. 2010).

All the studies presented above focusing on techno-economic evaluation of food waste valorisation, including the studies on the valorization of citrus waste, underline the necessity of using as many as possible, if not all, fractions of food waste for the production of various chemicals together with biofuels and energy. Economically viable biorefineries can be realized only if preliminary techno-economic studies illustrate key factors that affect the profitability criteria of these new investments. This chapter examines the economic sustainability of a base case scenario utilizing citrus waste for the production of D-limonene and ethanol. Also, high pressure steam is generated for the needs of the facility from the citrus residues after the fermentation process. In the citrus waste biorefinery, compounds with high added value such as D-limonene, pectin and flavonoids contained in citrus waste should be extracted first leaving the lignocellulosic fraction to be used as feedstock in the fermentation process for the production of biofuels or chemicals. The results presented in this chapter illustrate that more than 70% of the revenues are coming from D-limonene. If only bioethanol was produced, profits could be only reached for very large production capacities (i.e. higher than 200,000 t/y). Furthermore, the application of heat integration is essential in order to minimize the cost of utilities. In the proposed process, the fractions that are not used for the production of bioethanol are burnt in a boiler generating around 80% of the steam requirements of the plant. Profitability indicators are also expected to be improved if bioethanol production is replaced by chemical production via fermentation.

References

Anagnostopoulou M (2005) Analysis, structure identification and study of the antioxidant activity of flavonoids compounds of the peel of *Citrus sinensis*. PhD Thesis, Aristotle University of Thessaloniki, Department of Chemical Engineering

Anonymous (1956) Chemistry and technology of citrus, Citrus products and byproducts. Agricultural Research Service, United States Department of Agriculture. Washington, D.C. Agriculture Handbook No. 98

Aravantinos G, Oreopoulou V, Tzia C, Thomopoulos CD (1994) Fibre fraction from orange peel residues after pectin extraction. LWT-Food Sci Technol 27(5):468–471

Dimou C, Kopsahelis N, Papadaki A, Papanikolaou S, Kookos IK, Mandala I, Koutinas AA (2015) Wine lees valorization: biorefinery development including production of a generic fermentation feedstock employed for poly(3-hydroxybutyrate) synthesis. Food Res Int 73:81–87

FAOSTAT, Food and Agricultural commodities production

Food Wastage Footprint, Impacts on natural resources, *Technical Report* http://www.fao.org/docrep/018/ar429e/ar429e.pdf

Galanakis C (2012) Recovery of high added-value components from food wastes: conventional, emerging technologies and commercialized applications. Trends Food Sci Tech 26:68–87

Grohmann K (2007) Final Technical Report, Project Title: *Citrus Waste Biomass Program*. USDA Agricultural Research Service CRADA 58-3K95–4–1053. Available at: www.osti.gov/bridge/servlets/purl/898345–9P4Slw/898345.pdf. Accessed 3 Feb 2016

Han W, Fang J, Liu Z, Tang J (2016) Techno-economic evaluation of a combined bioprocess for fermentative hydrogen production from food waste. Bioresource Technol 202:107–112

FAOSTAT, http://faostat.fao.org/site/339/default.aspx (visited December 2015)

Hull WQ, Lindsay CW, Baier WE (1953) Chemicals from oranges. Ind Eng Chem 45(5):876–890

Humbird D, Davis R, Tao L, Kinchin C, Hsu D, Aden A, Schoen P, Lukas J, Olthof B, Worley M, Sexton D, Dudgeon D (2011) Process design and economics for biochemical conversion of lignocellulosic biomass to ethanol dilute-acid pretreatment and enzymatic hydrolysis of corn stover. National Renewable Energy Laboratory (NREL), Cole Boulevard Golden, Colorado

Koutinas AA, Chatzifragkou A, Kopsahelis N, Papanikolaou S, Kookos IK (2014a) Design and techno-economic evaluation of microbial oil production as a renewable resource for biodiesel and oleochemical production. Fuel 116:566–577

Koutinas AA, Vlysidis A, Pleissner D, Kopsahelis N, Lopez Garcia I, Kookos IK, Papanikolaou S, Kwan TH, Lin CSK (2014b) Valorization of industrial waste and by-product streams via fermentation for the production of chemicals and biopolymers. Chem Soc Rev 43(8):2587–2627

Koutinas AA, Yepez B, Kopsahelis N, Freire D, Machado A, Papanikolaou S, Kookos IK (2016) Techno-economic evaluation of a complete bioprocess for 2,3-butanediol production from renewable resources. Bioresource Technol 204:55–64

Kwan TH, Pleissner D, Lau KY, Venus J, Pommeret A, Lin CSK (2015) Techno-economic analysis of a food waste valorization process via microalgae cultivation and co-production of plasticizer, lactic acid and animal feed from algal biomass and food waste. Bioresource Technol 198:292–299

Lin CSK, Pfaltzgraff LA, Herrero-Davila L, Mubofu EB, Abderrahim S, Clark JH, Koutinas AA, Kopsahelis N, Stamatelatou K, Dickson F, Thankappan S, Mohamed Z, Brocklesby R, Luque R (2013) Food waste as a valuable resource for the production of chemicals, materials and fuels. Current situation and global perspective. Energy Environ Sci 6(2):426–464

Lohrasbi M, Pourbafrani M, Niklasson C, Taherzadeh MJ (2010) Process design and economic analysis of a citrus waste biorefinery with biofuels and limonene as products. Bioresource Technol 101:7382–7388

Lopez JAS, Li Q, Thompson IP (2010) Biorefinery of waste orange peel. Crit Rev Biotechnol 30(1):63–69

Mirabella N, Castellani V, Sala S (2014) Current options for the valorization of food manufacturing waste: A review. J Clean Prod 65:28–41

Monier V, Mudgal S, Escalon V, O'Connor C, Gibon T, Anderson G, Montoux H, Reisinger H, Dolley P, Ogilvie S, Morton G (2010) Preparatory study on food waste across EU 27, Technical Report-2010-054, European Commission

Parfitt J, Barthel M, Macnaughton S (2010) Food waste within food supply chains: quantification and potential for change to 2050. Philos T R Soc B 365:3065–3081

Peters MS, Timmerhaus K, West RE (2003) Plant design and economics for chemical engineers, 5th edn. McGraw-Hill, Singapore

Pfaltzgraff LA (2014) The study and development of an integrated and additive-free waste orange peel biorefinery. PhD Thesis, University of York

Pourbafrani M, Forgács G, Horváth IS, Niklasson C, Taherzadeh MJ (2010) Production of biofuels, limonene and pectin from citrus wastes. Bioresource Technol 101:4246–4250

Rivas-Cantu RC, Jones KD, Mills PL (2013) A citrus waste-based biorefinery as a source of renewable energy: technical advances and analysis of engineering challenges. Waste Manag Res 31:413–420

Summers H, Ledbetter R, McCurdy A, Morgan M, Seefeldt L, Jena U, Hoekman S, Quinn J (2015) Techno-economic feasibility and life cycle assessment of dairy effluent to renewable diesel via hydrothermal liquefaction. Bioresource Technol 196:431–440

Turton R, Bailie RC, Whiting WB, Shaeiwitz JA (2009) Analysis, synthesis and design of chemical processes, 3rd edn. Prentice Hall International Series, Boston, MA

Ulrich GD (1984) A guide to chemical engineering process design and economics. John Wiley & Sons, USA

Van Antwerpen FJ (1941) Utilization of citrus wastes. Ind Eng Chem 33:1422–1426

Vlysidis A, Binns M, Webb C, Theodoropoulos C (2011) A techno-economic analysis of biodiesel biorefineries: assessment of integrated designs for the coproductionof fuels and chemicals. Energy 36:4671–4683

Zhou W, Widmer W, Grohmann K (2007) Economic analysis of ethanol production from citrus waste. Proc Fla State Hort Soc 120:310–315

Part II
Regulation and Policy Analysis

Chapter 9
Bio-Based Economy: Policy Framework and Foresight Thinking

Luana Ladu and Rainer Quitzow

A goal without a plan is just a wish
—Antoine de Saint-Exupéry

Abstract The bioeconomy, understood as the production of renewable biological resources and their conversion into food, feed, bio-based products and bioenergy via innovative and efficient technologies, has the potential to tackle current grand challenges like natural resource scarcity, climate change, food supply and energy. Improved and systematic foresight investigations with a focus on regulations, policies and technologies are needed for better decision-making in the future and for enabling the bio-based economy to timely tackle those challenges. A common understanding of the challenges and of the capacities available is a basis for conducting foresight. This chapter, after providing an overview of the drivers and challenges of the bioeconomy and of the European policy framework governing it, explains the concept of foresight thinking and its potential contribution to the achievement of the targets of the bio-based strategy. It explains the potential role that regulatory foresight can play in establishing a sustainable circular bio-based economy and provides an overview of existing foresight studies directed to improve understanding of the future in the following dimensions:

- Biomass availability and trends
- Technology development and horizon scanning of emerging technology
- Market acceptance of bio-based products
- Regulatory and policy-framework.

Keywords Bioeconomy · Biomass · Bio-based products · Foresight · Regulatory foresight

L. Ladu (✉) · R. Quitzow
Department of Innovation Economics, Technische Universität Berlin,
Secr. MAR 2–5, Room: MAR 2.066, Marchstraße 23 - D,
10587 Berlin, Germany
e-mail: luana.ladu@tu-berlin.de

P. Morone et al. (eds.), *Food Waste Reduction and Valorisation*,
DOI 10.1007/978-3-319-50088-1_9

9.1 Introduction

The production system adapted to current consumption patterns implies the use of large amounts of raw materials and energy, as well as the generation of an enormous amount of waste, which produces externalities that negatively affect both the environment and quality of life (European Commission 2009). Existing societal challenges, such as climate change, rapid resource depletion, food security and increases in the global population, demand smart, sustainable and green economic growth models that reconcile the goals of economic development with the sustainable use of resources. The promotion of a bioeconomy, in which materials and production processes are based on sustainable renewable biological resources, represents an important step in this direction. It is expected that the bioeconomy has the potential to promote environmental sustainability by reducing greenhouse gas (GHG) emissions and dependence on fossil fuels, as well as stimulate the economy by encouraging the use of locally produced bio-based materials and products that reduce dependency on imports and create local jobs (Scarlat et al. 2015). The sustainable production and exploitation of biological resources will allow for the production of 'more with less' by utilizing waste products and increasing resource efficiency in production (State of Green 2015).

To date, more than forty countries are actively promoting the concept of a bioeconomy with the intention of meeting the grand societal challenges of sustainable development (Communiqué Global Bioeconomy Summit 2015). Policies and regulations have an important role in promoting a bioeconomy to realize its potential as a source of economic growth and as an avenue towards more sustainable economic development. In particular, a supportive regulatory framework, which promotes the use, re-use and recycling of bio-based products and establishes a sustainable management of bio-waste plays a crucial role in achieving a resource-efficient society and a bioeconomy in Europe.

However, this potential, already recognized in the European and national bioeconomy strategies, has yet to be deployed on a large scale. The bioeconomy sector is growing, but at a slower rate than expected with important uncertainties and technological, political and commercial challenges remaining. Accelerating the transition toward a sustainable bioeconomy will depend on:

- A sustainable supply of biomass
- Further advancements in technology and innovations
- The social acceptance and commercialization of bio-based products
- The existence of a supportive and coherent policy and regulatory framework, including sustainability assessment schemes.

Reducing the uncertainties in these four areas is critical for stimulating investment in biorefineries and growth in the bioeconomy.

Measures to enhance our understanding of possible future trajectories of the bioeconomy and their potential implications in these domains represents an important avenue for increasing the effectiveness of bioeconomy strategies today,

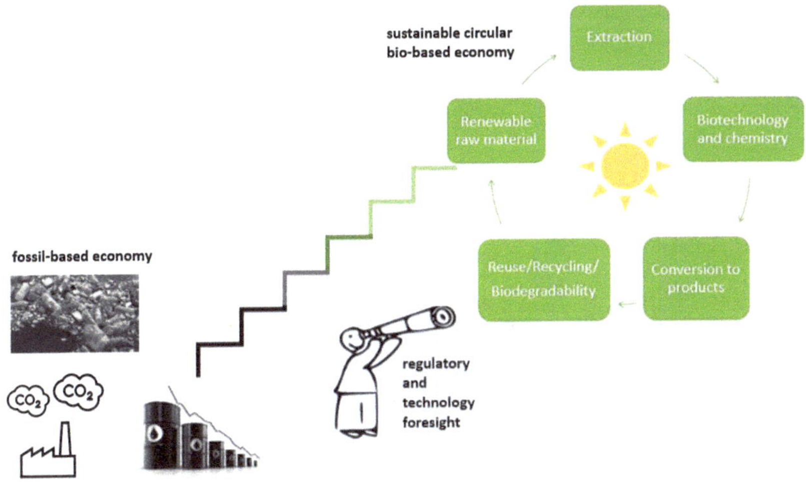

Fig. 9.1 Foresight as a tool for facilitating the transition to the post-petroleum society

while preparing for needed future actions. Foresight analysis represents an important tool for identifying and anticipating needs and challenges that might prevent an efficient and full deployment of a sustainable bioeconomy (see Fig. 9.1). It can provide a more systematic assessment of possible futures and, in turn, reduce uncertainty by proactively anticipating possible challenges.

In particular, an ex-ante, forward-looking regulatory foresight is essential for the identification of possible improvements needed to increase the efficiency along the bio-based value-chain and among the various sectors (e.g. improving waste management efficiency), as well as identify legal barriers hindering the uptake of bio-based products. These include the lack of a level playing field and supportive regulatory framework for the market uptake of bio-based products, the existence of only incipient initiatives and tools for conducting sustainably assessments of bio-based products, and the absence of new value chains based on the development of sustainable biomass collection and supply systems, while valorising waste biomass.

This chapter provides an overview of different foresight methodologies and their potential for addressing various future challenges related to the development of a bioeconomy. It explains the potential of foresight as a vehicle for moving from a fossil-based economy to a sustainable circular bio-based economy (see Fig. 9.1). It then provides an overview of the drivers and challenges of the bioeconomy and the relevant European policy framework. Following this, the concept of foresight and its potential contribution to fostering the development of a bioeconomy will be discussed. Finally, an overview of existing foresight studies is provided to support a better understanding of the future in the following areas:

- Biomass availability and sustainability
- Trends in innovation and technological change
- Market trends and social acceptance of bio-based products
- Regulatory and policy-frameworks for the bioeconomy.

9.2 Bioeconomy

9.2.1 Definition of Bioeconomy and Biomass

The European Commission defines the bioeconomy as the production of renewable biological resources and the conversion of these resources and waste streams into value added products, such as food, feed, bio-based products[1] and bioenergy via innovative and efficient technologies provided by industrial biotechnology. It is an economy-wide concept in the sense that it includes the sectors of agriculture, forestry, fisheries, food and pulp and paper production, as well as parts of chemical, biotechnological and energy industries (European Commission 2012). The bioeconomy encompasses a comprehensive range of activities, situated along a multitude of different value chains, each including producers, suppliers, distributors and purchasers (Golbiewski 2013).

The bio-economy is intrinsically dependent on biomass, which is defined by the CEN[2] European Standard EN 16575:2014 as material of biological origin excluding geological formations and/or fossilized (e.g. whole of parts of plants, trees, algae, marine organisms, micro-organisms, animals, etc.). Biomass is a mixture of organic molecules that always contains hydrogen and is always carbon based. It can be derived from both plant and animal materials and represents the basis for all bio-based products and processes. As biomass is the basis for all bio-based products and processes, and, as indicated in Fig. 9.2, it represents the starting point of all bioeconomy-related value chains (Hasenheit et al. 2016).

Depending on its origin, biomass can be classified as:

- Biomass from agriculture: energy and agricultural crops and agricultural primary residues (e.g. agricultural by-products)
- Biomass from forestry: forestry biomass, dedicated lignocellulosic crops, primary forestry residues and secondary forestry residues
- Biomass from marine environment: fresh water plants, algal and aquatic biomass

[1]The term bio-based product refers to products wholly or partly derived from biomass, such as plants, trees or animals (the biomass can have undergone physical, chemical or biological treatment). CEN—Report on Mandate M/249. A standard defining general terms to be used in the field of bio-based products, EN 16575, was published by CEN in August 2014.

[2]European Committee for Standardization.

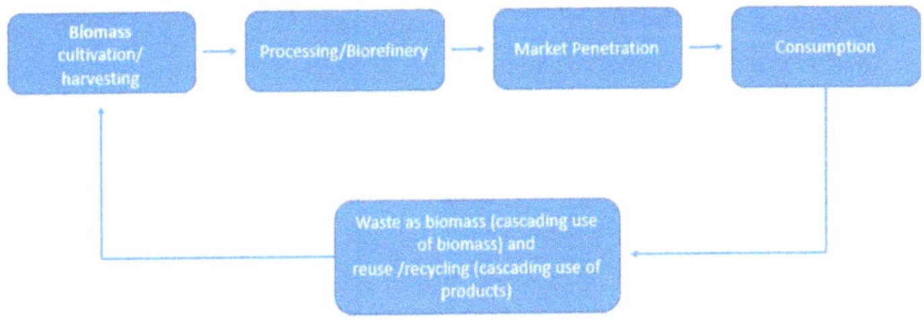

Fig. 9.2 Bioeconomy value chains

- Biomass from waste[3]: primary, secondary and tertiary residues and waste (e.g. municipality solid waste and non-hazardous commercial and industrial waste).

Biomass feedstocks are primarily utilised in three main economic sectors: food & feed, energy & fuels and industry (material production). Biomass as food for humans and as feed for animal husbandry still represents its main use. Additionally, it is increasingly being used for the production of biofuel, such as ethanol and biodiesel, in the transport sector. Biomass is also used as biomaterial, for example, as a raw material in various pharmaceutical products or in bioplastics. Biomaterials are still in the early stage of product development and therefore still constitute relatively small niche markets.

9.2.2 *Main Drivers of the Bioeconomy*

9.2.2.1 Environmental Drivers

The bioeconomy promotes the production and use of sustainable products obtained from local biological resources and raw materials, effectively reducing heavy dependency on fossil resources, while decoupling economic growth and the emission of green gashouses (GHG) Philippidis et al. (2016). The climate change mitigation potential of the bioeconomy is estimated at 2.5 billion tonnes of CO_2 equivalent per year by 2030 (Mattana 2015). The use of biomass feedstock for the production of bio-based products by European industry, has the potential to support Europe's industrial and economic growth while significantly reducing environmental burdens and resource dependency, through the displacement of fossil-based products with bio-based alternatives (Nattrass et al. 2016).

The use of biomass and bioeconomy-related innovations provides opportunity for new production processes and potentially solutions to the current resource

[3]Food Waste: The EU defines 'food waste' as food lost from the food supply chain, not including food diverted to material uses such as bio-based products, animal feed, or sent for redistribution.

efficiency challenges that the EU is facing (European Bioplastics 2016). According to the EU, in the coming decades, there will be increased competition for limited and finite natural resources worldwide, especially petroleum, and an estimated 70% increase in the world food supply is needed to feed the global population of 9 billion by 2050. The introduction of innovative bio-based solutions across the entire biological value chain are intended to support the efficient use of biological resources in a sustainable and integrated manner and pave the way toward a production of 'more with less' (Horizon 2020, Work Program 2016–2017).

9.2.2.2 Economic Growth, Regional Development and Job Creation

In the EU, the bioeconomy is one of the biggest segments of the economy with an estimated turnover of 2.2 euros trillion. It includes agriculture, food and beverage, agro-industrial products, fisheries and aquaculture, forestry, wood-based industries, biochemicals, enzymes, biopharmaceuticals, biofuels and bioenergy. It is estimated that the European bioeconomy already employs almost 22 million people, representing 9% of total employment in the EU (Scarlat et al. 2015; Piotrowsky et al. 2015). Many of these employees live in coastal and rural areas, which are dependent on these sectors for economic sustainability in a world where most employment opportunities are found in urban areas. The most relevant sectors employing individuals in the bioeconomy are the primary sectors, including: agriculture (53%), the manufacturing of food products (21.3%) and fisheries (10.6%). The bioeconomy can play a crucial role in promoting economic growth, regional development, job creation and preventing rural exodus. The promotion of local growth and the integration with local agronomical value chain is crucial in this regard.

9.2.2.3 Technological Innovations

Innovative materials are the solutions to many present problems. Therefore, there is considerable economic and political pressure to ensure that novel technologies deliver innovations in line with societal priorities. Europe currently has a global leadership position in science and technology, including biotechnology; therefore, kick-starting a globally competitive bioeconomy represents an important opportunity (BBI JU 2012).

In the European Union, six types of Key Enabling Technologies (KET) have been identified. They include:

1. Micro and nanoelectronics;
2. Nanotechnology;

3. Industrial biotechnology[4];
4. Advanced materials;
5. Photonics;
6. Advanced manufacturing technologies.

These technologies have a multitude of applications in various sectors and are believed to have the potential of strengthening the EU's industrial and innovation capacity and be essential components of the solutions to modern and future societal challenges, especially those related to dependence on fossil fuels. Moreover, they are said to aid in the transition to the bioeconomy as it is expected that innovative KET will result in divergence from a fossil fuel based economy towards an innovative bioeconomy. Such a transition is expected toenable European industry to deliver high-value products which satisfy evolving consumer needs, create new commercial opportunities and reduce possible risks to human health, as well as the environment (Horizon 2020). Industrial biotechnology is considered a KET that enhances the global economy and promises dynamic growth opportunities (Festel 2010). It is commonly accepted as a promising approach to minimize the impacts of climate change and diminishing of fossil resources and bridge the divide toward a post-petroleum bio-based circular economy.

To maintain this position, R&D funding agencies need to continue their investments in the sector, and policy developments need to be conducive to increasing private R&D expenditures.[5] In addition, a Public-Private Partnership of 3.7 billion euros between the EU and 30 leading bio-based industries, called Bio-based Industries Joint Undertaking (BBI JU), has been created. It aims at increasing investment in the development of a sustainable bio-based industry in Europe. The payoff for current investments in innovation is expected to be great: the EU bioeconomy strategy estimates that each Euro invested in EU-funded bioeconomy research and innovation today could generate ten euros of value added in the bioeconomy sector by 2025 (European Commission 2012). Thus, Horizon 2020, the EU's core research and innovation program launched in 2014 includes a fit-for-purpose bioeconomy research and innovation agenda.

9.2.2.4 Policies and Regulations

Strategies and policies at a European and national level are important to promote and maximize the economic impacts and value added from the bioeconomy and

[4]Industrial biotechnology (IB)—the use of biological substances (e.g. plants, algae, marine life, fungi, micro-organism), systems and processes to produce materials, chemicals and energy. IB uses biotechnological knowledge to develop new processes for making products, such as industrial enzymes or chemical building blocks. These are used, in turn, in the production of chemicals, detergents, textiles, paper, and much more.

[5]Initiatives like the EU Bioeconomy Strategy, the National Research Strategy BioEconomy 2030 in Germany or the National Bioeconomy Blueprint of the US White House in 2012 emphasize the development towards a bioeconomy in the near future.

represent important potential drivers to establish a path toward a resource-efficient, competitive and sustainable economy (Staffas et al. 2013). *Innovating for Sustainable Growth: A Bioeconomy for Europe* is the primary policy framework of the European bioeconomy as defined by the 2012 communication of the European Commission.[6] The strategy proposes a comprehensive approach to address *five inter-connected societal challenges*[7] and promotes advancements in bioeconomy research and innovation to improve the management of European renewable biological resources and open new and diversified markets for food and bio-based products.

Additional relevant strategies include *Growth Within: a circular economy vision for a competitive Europe* (2015). Which confirms the need for sustainable approaches to economic growth; the *Resource Efficiency Europe* (EC 2011) flagship with the objective to support the shift toward a resource-efficient and low carbon economy; and the Lead Market Initiative, which supports a number of demand-side innovation policies for promoting the market uptake of bio-based products *Taking bio-based from promise to market*.

Many European countries (see Table 9.1) have already adopted or are developing national bioeconomy strategies linked to long-term industrial policies with the objective of promoting the development of innovative technologies linked to efficient resource use and the creation of sustainable, low impact value-chains with a holistic multilateral approach that involves all areas and stakeholders, from primary production to final consumption.

9.2.3 *Turning Challenges into Opportunities*

9.2.3.1 Food Security

Food security is seen as one of the most important challenges of the bio-based economy. Indeed, if not properly managed, the demand for biomass for the production of bio-based products could further stress land availability and create concerns about negative environmental impacts associated with crop cultivation. Changes to the current policy framework governing the different sectors of the bioeconomy have the potential to manage those negative impacts and turn those challenges into strengths. For example, policies that promote the establishment of an effective after-use economy, one that uses waste and residues as biomass (see Box 9.1), reduces the need for virgin feedstock, and thus offers a potential route to overcome those concerns (Scarlat et al. 2015). In addition, negative impacts of

---

[6](COM (2012)60), adopted on February and 13th 2012.

[7](1) ensuring food security, (2) managing natural resource sustainability, (3) reducing dependency on non-renewable resources, (4) mitigating and adapting to climate change and (5) creating jobs and maintaining European competitiveness.

Table 9.1 European countries have already adopted or are in the process of developing a bioeconomy strategy

EU Countries with bio-based economy strategies	EU Countries developing bio-based economy strategies	EU Countries using alternative methods
• Finland • Germany • Netherlands • Sweden • United Kingdom	• Austria • Denmark • Ireland • Italy • Spain	• Estonia • France • Poland

specific bio-based products, such as the production of palm oil, could be managed with the promotion of standards that ensure, for example, the sustainable production of imported biomass.

BOX 9.1 Potential of waste as biomass

Sources of biomass should not compete with food crops. Consequently, waste biomass, especially agricultural and food waste (van der Hoeven 2014), represents an improvement of the food security challenge and an attractive and viable option as a potential substitute feedstock for fossil fuels. The introduction of new chemical technologies (e.g. green chemistry) will allow for the commercial use of industrial waste.

The unused potential is about 100 million tonnes of bio-waste–a valuable bio-based resource and secondary raw material (European Bioplastics 2016). In 2014, the EU announced targets for the circular economy (European Commission 2014a) recognizing agricultural and food waste as 'secondary raw materials' with the objective of increasing the security of supply. This is in line with the promotion of the cascading use of biomass and cascading use of bio-based materials/products with several reuse and recycling cycles, which can be facilitated by the innovation potential of the bioeconomy. However, realising this potential depends on the level of investments in integrating bio-refineries capable of processing biomass and bio-waste for different end-uses. At the moment, the EU market of secondary raw materials is still small due to technical and non-technical barriers, such uncertainty of the quality of the materials, fragmented waste management regulation at national and regional level and the absence of EU-wide waste management standards.

The identification of solutions for existing and upcoming challenges that might hamper the development of the bio-based economy represents an important step in the shift towards a sustainable bioeconomy. In this vein, a critical debate on the bioeconomy and its potential future development trajectories is essential. This means taking existing criticism seriously and engaging in a public discussion with the relevant stakeholders. Such a societal process helps shape political decisions by identifying priority areas for action and consequently allocating public and private resources to the development of solutions.

The *Food versus Fuel* debate provides an important example of this phenomenon: Initial evidence linked first-generation biofuels to a food price spike, while other studies indicated that it was not clear as to how and to what extent biofuels would affect global food prices and hence global food security (dos Santos et al. 2009). Nonetheless, the *Food versus Fuel* debate led to the emergence of second generation biofuels and the incorporation of organic waste sources (cellulosic agricultural and forestry waste, municipality waste and industrial waste) as feedstocks in future bio refineries (European Biofuels Technology Platform 2014).

Following this view, in the JRC Foresight on Global Food Security 2030, the Joint Research Centre (JRC) states that food security is not only a systematic challenge, but also an opportunity for the EU to play a role in innovation, trade, health, genomics and geopolitics. For some of these challenges (food security issues), genomics (without GMO) could potentially provide solutions, e.g. for increasing the efficiency of food and crop production (more efficient plants that use less water). At this stage, it is quite clear that there are significant advances in the efficiency of food production to be made, even without genetic modification (Jimenez-Sanchez 2016). It is clear that food and feed production have to be given priority when considering biomass use. However, the efficiency of food chains can be increased by improvements in agricultural productivity, land management, logistics and storage (Star COLIBRI 2011).

9.2.3.2 Food Waste as an Opportunity for a Bio-Based Economy

The EU estimated that around 88 tonnes of food are wasted annually in the EU (173 kg of food waste per capita annually), with an associate cost estimated at 143 billion euros (Stenmarck et al. 2016). Food is wasted throughout the entire supply chain from agricultural primary production to food preparation and consumption (Vanham et al. 2015) and the social, economic and environmental impacts of this wastage are enormous.

Municipal solid wastes (MSW), including food waste, are usually incinerated for energy recovery or placed in landfills, which can generate problems, such as liberation of harmful compounds like dioxin and furan. When excess food waste is disposed of in a landfill, it decomposes and is a significant source of methane gas, a particularly harmful GHG. In addition, the heating value of food waste is low, therefore it is difficult to recover energy from the waste incineration processes. Another possible treatment consists in anaerobic digestion, a process that breaks down waste into digestate (which can be used as fertilizer) or biogas (which can be used as energy source).

These conventional food waste processes, including composting and anaerobic digestion, miss the large opportunity to exploit the molecular complexity that exists in bio waste for added-value products (Royal Society of Chemistry 2014). Waste feedstock includes a huge diversity of functionalized chemicals components, including sugars, lignin and oils (many of which cannot be found in traditional petrochemicals feedstocks), that could potentially be valorised as important biomass sources for the production of sustainable bio-based material. Sugars, oils and other compounds in bio-waste could be converted into building blocks that have a high transformation potential into new families of useful molecules such as lubricants, flavours, nutriceuticals, solvents, polymers, and pharmaceuticals (Royal Society of Chemistry 2014).

There is an important opportunity to use these components for higher value applications, such as polymers for packaging, solvents for printing, resins for inks and as surfactants. For instance, Rodenburg Company in Oosterhout, Netherlands, produces starch-based plastics from wastewater from the production of French fries. These plastics are 100% locally sourced, 100% bio-based and 100% biodegradable. Typical items made from these bioplastics are biodegradable plant pots. They have the important advantage that the pot does not have to be removed from the soil when the plant has grown, saving labour costs. Another relevant example is the use of food waste from Starbucks for the production of fertilizers (Eshelman 2012). In addition, food waste has the potential to provide sources of economically useful food ingredients, including: flavouring and colouring.

9.2.3.3 Challenges Related to the Existing Regulatory Framework

Value chains valorizing food waste within the bioeconomy offer a high potential for regional innovation and new productive investments. They can contribute to azero-waste circular economy at a local level, reduce resource depletion and create local job opportunities (European Commission 2014b). Waste biomass does not compete with food and therefore it is an attractive and viable option as a potential substitute feedstock to fossil fuels. To realize this potential, an integrated, supportive regulatory framework will be needed to encourage the use of bio waste (including food waste) for value added products including chemicals, materials and fuels, and not solely for energy generation via incineration.

However, observers have identified a number of shortcomings within existing regulatory frameworks for advancing an innovative bioeconomy. Critics identify the lack of a level playing field vis-à-vis existing fossil-based products and the lack of a supportive regulatory framework for the market uptake of bio-based products (Peuckert and Quitzow, forthcoming). Moreover, observers have called for more advanced tools for conducting sustainability assessments of bio-based products to better demonstrate both the potential of bio-based products to solve important sustainability challenges and reduce concerns regarding possible negative implications. Furthermore, it remains an important challenge to put in place a regulatory framework that can adapt to the pace of change of new technologies and innovations. Finally, the cross-sectorial nature of the bioeconomy makes the development and adaptation of relevant policies a particular challenge.

Overcoming these challenges requires a forward-looking approach to regulation, which anticipates future developments and their inter-dependencies with regulatory regimes in areas like waste management, natural resource protection and raw material production. Relevant examples of regulatory efforts include 7th Environment Action Programme, the Resource Efficiency Roadmap and the Raw Materials Initiative (European Commission 2016d).

9.3 Foresight for the Bioeconomy

Foresight analysis has an important role to play in the establishment of a cutting-edge, sustainable, bio-based economy for Europe by helping to reduce uncertainty in the four critical areas already referred to above:

- Biomass availability and sustainability
- Trends in innovation and technological change
- Market trends and social acceptance
- Regulatory and policy frameworks for the bioeconomy.

These areas can be attributed to four corresponding areas of foresight (see Fig. 9.3). After providing a brief introduction to foresight research, this section provides an overview of existing methodological approaches to foresight in each of the four areas and the related studies, which have been conducted. The review reveals that past exercises have mainly focused on technological trends, biomass availability, and to a lesser extent market development. An important gap remains in the sphere of regulatory foresight. The section concludes with a discussion of the importance of regulatory foresight for a more effective promotion of a sustainable bioeconomy.

9.3.1 *Foresight and Foresight Methods*

Although it is not possible to precisely predict the future, upcoming trends can be explored, analysed and estimated by conducting foresight analyses.

Foresight can be defined as the implementation of a forward-looking analysis, which provides different methods for enhancing future thinking and gaining insights about future developments via the systematic gathering of anticipatory intelligence from a wide range of knowledge sources. It is said to be an 'opening to the future with every means at our disposal, developing views of future options, and then choosing between them' (Slaughter 1995 p. 1). It provides techniques designed to extract information and produce conclusions from data sets, while considering and anticipating important future trends and change-inducing variables (e.g. existing and upcoming policies, technologies development and markets development). Foresight can be described as a structured dialogue about possible future developments among

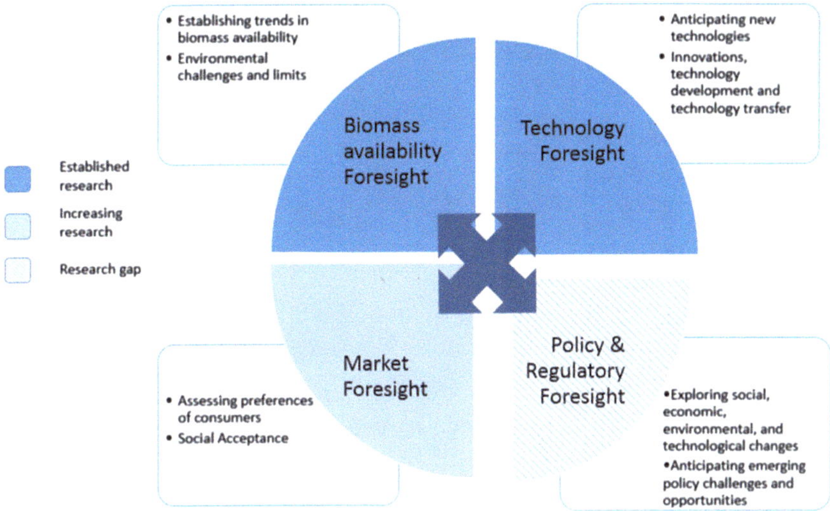

Fig. 9.3 Areas of foresight analysis

relevant stakeholders (Giaoutzi 2013). It is a process that combines strategic intelligence, sense-making activities and their links to the policy cycle (European Forum on Forward Looking Activities 2014) to allow thoughtful debate to be used for shaping the future.

Foresight represents a relevant decision-making tool and policy instrument based upon five distinguishing features (Vecchiato and Roveda 2014):

(1) Anticipation
(2) Participation
(3) Networking
(4) Vision
(5) Action.

Moreover, the aforementioned features allow for the anticipation of the implications of present-day actions to identify the interacting dynamics that are shaping future developments. This is done through the implementation of multidisciplinary and multifactor analyses of the views of multiple stakeholders, representative of different sectors and value chains. This information can then be used in policy development and planning processes.

Foresight analysis is comprised of three main tasks (see Fig. 9.4):

- The identification of new events and drivers of change
- The investigation of their possible evolution and potential impacts/consequences for a specific industry/sector (e.g. the bio-based economy)
- The provision of recommendations for strategic decisions.

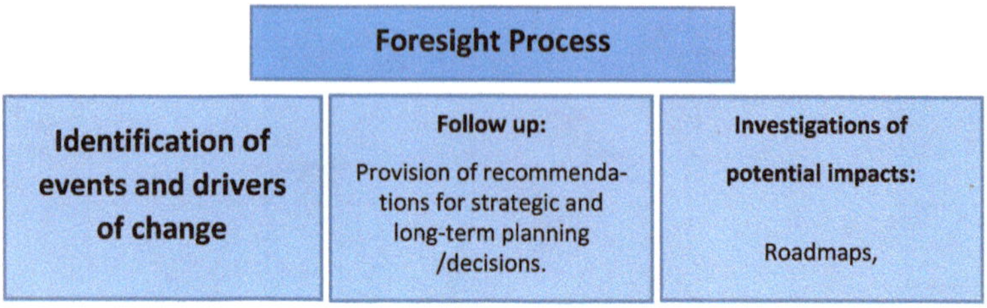

Fig. 9.4 Foresight process

Different methods for conducting foresight exercises are available and can be classified as:

- *Qualitative methods*: which provide meaning and subjective interpretations to events and perceptions
- *Quantitative methods*: statistical analysis and modelling of measureable variables
- *Semi-quantitative methods*: analyses for quantifying subjectivity and views of experts.

As shown in Table 9.2, the above-mentioned methods can be used in conducting the following types of foresight exercises (Popper 2008):

- *Interaction-based exercise*: usually bottom-up participatory methods, such as citizen panels and stakeholder workshops, which bring together experts and non-experts to think about the future and potential impacts of actions.
- *Evidence-based exercise*: attempt to explain and/or forecast a particular phenomenon with the support of reliable documentation and means of analysis. These activities are particularly helpful for understanding the actual state of development of the research issue. For this reason, quantitative methods (e.g. benchmarking, bibliometric, data mining and indicator-based work) have become popular given that they are supported by statistical data or other types of indicators. They are fundamental tools for technology and impact assessment and scanning activities (Porter et al. 1980).
- *Expertise-based exercise*: like road mapping, relevance trees or morphological analysis that are usually reserved for experts. They require skill and knowledge on the topic. These methods are frequently used to directly support decision-making, and provide advice to policy-makers.
- *Creativity-based exercise*, such as backcasting or essays that can be used by creative thinkers, often also those that have a background in technology or even science fiction.

There is not a one-size-fits-all method for conducting foresight analyses, and the reliability of the results is greatly improved by the development and adoption of a solid foresight methodology and the selection of the most suitable combination of

Table 9.2 Foresight methods

Methods	Qualitative methods	Quantitative methods	Semi-quantitative methods
Participatory exercise	Conference, workshop Survey, multi perspective analysis, morphological analysis, citizen panel	Benchmarking, meta-modelling indicators, quantitative scenarios	Polling/voting Stakeholder analysis Web-based crowdsourcing
Explanatory and evidence based exercise	Logic chart, interview Scanning, weak signal Literature reviewing	Social network analysis (SNA), extrapolation, patent analysis, regression analysis, bibliometric, indicator/index System, dynamics/simulation	Data/text mining
Expertise advisory based exercise	Expert panel	Rule based forecast Impact analysis	Road mapping, Delphi Prediction market, multi-criteria analysis, key technologies, cross impact/structural analysis
Exploratory exercises	SF, wild card, scenario vignette, Gebius/expert forecast, backcasting, role play/gaming, Teepese analysis, Swot, brainstorming Scenario workshop Relevance trees		

Source Popper (2008)

methods (foresight process). The selection of the best methodology is dependent upon the goals, motivations and objectives of the exercise. One way to select suitable methods is to consider the level of uncertainty involved and choose more sophisticated tools, e.g. scenario development, forecasting, modelling and simulation, when complexity abounds and the time horizon is long term Phaal et al. (2004). When there is less uncertainty and the time horizon is short term, approaches like trend impact analysis/extrapolation and Delphi methods may be suitable (Jackson 2013). For priority setting where wider consultations are important, Delphi surveys, focus group exercises and interviews are useful methods. For the analysis of large, unstructured datasets of textual data generated by heterogeneous sources,[8] text mining allows for a comprehensive analysis of these datasets and for the identification of relevant information (Cuhls et al. 2015; Ortner et al. 2014a, b).

[8]Data sources include among others: standards, twitter, patents, scientific publications, newspapers, blogs, Rss-feeds.

9.3.2 Foresight in the Bioeconomy

Long-term horizon scanning and foresight in strategic bio-economy decision making could represent an important tool for reducing uncertainty about trends in the bioeconomy and increasing the responsiveness of policy and investment decisions to expected future developments. (Bio-based economy for Europe 2011). In doing so, regular foresight exercises could promote a more innovation- and investor friendly bioeconomy. In the following section, four key areas for foresight analysis on the bioeconomy are presented.

9.3.2.1 Foresight on Biomass Availability and Sustainability

Considering that biomass is a renewable, but limited resource whose production requires land and supplemental resources, it is important to analyse the demand for biomass in relation to the existing supply potential, land availability, expected technological trends, societal challenges and the fulfilment of the United National (UN) Sustainable Development Goals[9] (SDGs). For achieving this, forward-looking policy decisions have to be made about the most appropriate use of available natural resources in a given location and situation. Foresight analysis plays a crucial role in the identification of the most suitable species considering local conditions (e.g. dedicated low-impact crops) and biomass availability (including waste) to be used efficiently in local bio-refinery processes. This in turn has implications for the reduction of GHG emissions and other negative effects associated with the use of land. Furthermore, by involving different stakeholders from relevant bioeconomy value-chains, foresight exercises have the potential to create new supply chains within primary sectors, subsequently ensuring competitive conditions within biomass production processes.

A brief literature review shows that numerous foresight studies on biomass availability and supply have been implemented in Europe in the last decades. Most of those studies estimate future biomass demand for food, feed and energy utilization in the EU. Relevant data on the production and availability of biomass in Europe are collected by the Bioeconomy Observatory (BISO) at JRC-IPTS, which aims at compiling qualitative and quantitative data of relevance for the monitoring of the bioeconomy. Existing studies differentiate themselves by: geographical coverage (all EU countries vs. national level studies); type of biomass feedstock considered (full range of biomass types vs. one type); methodology applied for conducting the analysis including definition of time frame; and final use of biomass (food & feed, energy or production of materials). For example, the Biomass Futures Project estimated that the biomass sustainable potential of EU-27 for energy might reach 15.675 PJ in 2020, of which 2.075 PJ is derived from waste, 7.000 OJ from forestry and 6.600 PJ from agriculture (Elbersen et al. 2012).

[9]http://www.un.org/sustainabledevelopment/sustainable-development-goals/.

A high level of variation in the results of foresight studies focused on biomass availability indicates that the real potential of biomass supply is difficult to define and depends on assumptions made about important variables included in the applied model, such as land area availability, population growth, technology improvements and political-decision making frameworks, which may never take place. Therefore, it is important that foresight models are based on variables grounded in assumptions that will lead to the sustainable production and consumption of biomass, in order to cope with existing societal challenges. Policy makers and governments can play a key role in ensuring a sustainable and effective use of biomass by establishing framework conditions needed for a more coherent, harmonized and complete approach to assess biomass availability. They can promote the design of forward-looking European, national and regional policies initiatives resulting from participatory foresight exercises that consider societal challenges within a sustainable vision, which considers the different dimension of biomass use (food and feed, fuel and energy and bio-based materials). This can help harmonize the results of future trajectories for sustainable biomass availability by promoting the use of common terminologies, a collective understanding of sustainability and the most appropriate applications for sustainable biomass.

The 4th SCAR Foresight Exercise on biomass availability in the agricultural sector has taken such an approach (European Commission 2015a). It explores the interactions between the primary sector and the broader bioeconomy. The study describes the state of play in the bioeconomy and presents possible scenarios related to the developing paradigm of the bioeconomy with the fundamental constraint of sustainability to identify the principles that would enable primary production to address the complexity of the challenges. Three scenarios for the period 2011–2050 (called BIO-MODESTY, BIO-BOOM and BIO-SCARCITY) are defined by alternative futures of the two most important uncertainties, which are:

- Biomass demand for materials and energy
- Biomass supply growth.

The study analyses dilemmas, possible conflicts, opportunities and threats and identifies guiding principles for future actions. The report concludes that, in order for the bioeconomy to achieve its goals regarding food security, environmental care, energy security, climate change adaptation and mitigation and employment creation, it needs to be implemented according to the set of principles: food first, sustainable yields, cascading approach, circularity and diversity (European Commission 2015a).

9.3.2.2 Foresight and Innovation and Technological Change

The world is rapidly changing and technology plays a pivotal role in this evolution Georghiou et al. (2008). It is expected that the convergence of new technologies from the field of ICT, biotechnology, nanotechnology and molecular biology will enable new systems and services to establish a sustainable bioeconomy that respects the environment. In particular, the further investments in industrial biotechnology, a KET for the bioeconomy, will shape its future development trajectory. In order to promote these advancements in a sustainable and discerning manner, technology foresight can help identify possible future technological trends and developments. It can offer an assessment of the potential impacts of emerging and breakthrough technologies. A systematic monitoring of expected technological trends is necessary for flexible and timely strategic decision making and to allow for the exploitation of emerging and critical technologies in a manner that is most beneficial to society in terms of both economic and social benefits via the development and implementation of flexible and strategic policy framework (Mikova et al. 2014).

Bioeconomy-technology foresight focuses on the identification of key technologies and alternative paths of technological futures that will drive the competitiveness and sustainable growth of the European bioeconomy sectors and value-chain over a set time horizon (usually 10–30 years). The aim should be the identification of possible technology trajectories and the related areas of technology for which private and public investments are the most promising. Furthermore, in addition to anticipating future trends of strictly technological or process innovations, it aims at identifying non-technological innovation potential, such as product or functional innovations, that even if not central to the innovative solutions, help enable them (European Commission 2010). Indeed, investment in enabling new technologies will play a decisive role in promoting sustainably intensified production, employment and exports, while minimizing and reducing damage to the environment. Therefore, foresight should help identify strategic investments in ground-breaking research that could potentially enhance the performance of the bioeconomy.

Existing technological foresight studies

Significant EU financial resources has been allocated to key KET biotechnology foresight and financing approaches (including modelling and simulation) with the objective of providing comprehensive and dependable information about the future industrial biotechnology scenario.

Pertinent examples of technology foresight include:

- **The Science and Technology Option Assessment**—STOA Studies commissioned by the European Parliament in 2013, *Technology option for feeding 10 billion people*, examines and reviews bio-refinery technology options that exist to convert biomass in the form of agricultural crop and forestry residues and waste from the whole wood chain into biomaterials and bioenergy. The study shows that advanced biofuels and innovative bio-based pathways based on waste and residues have considerable potential and should be further developed, especially as Europe already has a lead in relevant technologies. However, the study highlights that there are also considerable uncertainties for investors and indeed all market participants and thus a major task is to ensure transparency and better information concerning the availabilities of the waste and residue streams, the opportunities for processing, and the benefits to consumers. Foresight analysis could play a crucial role in establishing a solid framework for the use of food waste as biomass.

- **The Forest Fibre Industry: 2050 Roadmap to a low-carbon bioeconomy**—is an example of a successfully developed roadmap designed to 'model various pathways to the future' for the forest fibre industry. The study identified a variety of potential technologies, presented a means for bringing the technologies into use and made a call for related policy action. The roadmap included a projection of both consumer and societal demands in 2050. It then examined the expected internal and external constraints and proposed proactive solutions as to how to meet the identified demands, maintain competitiveness and achieve the goal of an 80% reduction of industrial CO_2 emissions by 2050 (The Confederation of European Paper Industries CEPI 2011).
- **Bioeconomy to 2030: Designing a Policy Agenda (2009)** presents the published results of interdisciplinary foresight efforts on biotechnology that provides an analysis of expected future developments in agriculture, health and industry. According to this study, the use of advanced knowledge of genes and complex cell processes will be incorporated in new production processes and products. The potentially disruptive power of biotechnology will be turned into economic advantage, if barriers to biotechnology innovations can be reduced and the integration of biotechnology research across commercial applications is promoted.
- **BIO-TIC (2012)** aimed to identify, examine and comprehend innovation hurdles in biotechnology across Europe with the intention of formulating action plans to overcome the identified obstacles.

Foresight on market trends and social acceptance

The market for bio-based products includes both innovations with important new functionalities, such as bio-based packaging materials with advances functionalities, as well as products, which offer similar functionalities based on new materials and production processes. For both cases, foresight can help assess relevant market trends and growth potentials and identify key factors for realizing identified market potentials.

In the former case, a critical task of foresight analyses is an assessment of potential market demand in different market sectors and regions. Assessments will typically build on existing analyses of relevant societal trends and develop assumptions on how these trends may influence the development of market demand. Such studies should provide estimations of expected growth in turnover and employment, based on various scenarios and their underlying assumptions. This type of foresight study is critical for market actors to make informed investment decisions and is frequently supplied by industry-based consulting firms. In early stages of market development, there may be a case for public support for such studies to illustrate market potential to policy makers and firms.

In the second case, the development of market demand for the relevant applications–whether based on renewable or non-renewable resources—may be fairly certain or may be the subject of existing foresight studies, as they represent existing market sectors. In this case, bioeconomy foresight may analyse market trends in existing segments of the economy and make an assessment of how these might affect the development of bio-based products and materials. It can help provide a better understanding as to how bio-based materials and building blocks fit different applications and value chains (including recycling) and develop scenarios for their integration in these existing industrial sectors.

In this context, bio-based products face the challenge of entering existing markets and competing with products for which production and end-use has frequently been optimized for decades and are well-known in the entire supply chain (BIOCHEM 2010). Uncertainty is thus also related to the ability of bio-based alternatives to capture relevant market shares and displace existing products. Understanding the attributes that may motivate potential buyers to purchase bio-based alternatives represents a crucial challenge in this context. Foresight studies can help assess how preferences of different potential buyers are expected to evolve in relationship to relevant product attributes and how market acceptance may be expected to develop on the basis of different scenarios and related assumptions regarding technological change, production costs, etc. Conversely, foresight studies may explore potential risks to social acceptability of bio-based products, assuming a scenario of widespread diffusion and industrial-scale production.

Finally, market penetration of bio-based products is frequently restricted by weak market transparency and an absence of tools for assessing and clearly communicating to consumers the environmental competitive advantages of bio-based products, implying subsequent social acceptance of bio-based products. Hence, questions of market acceptance also depend on optimization and improvement of current standards and regulations governing the bioeconomy. Forward-looking, market research can play an important role in identifying entry-points for opening existing and new markets for bio-based products by anticipating and identifying tools and initiatives needed for fostering their demand. Indeed, market foresight has the potential to anticipate and analyse consumers' environmental needs. On this basis, studies may identify key criteria that bio-based products will have to fulfil to extract a green premium from different types of potential buyers. Similarly, anticipating sustainability preferences to be inserted in sustainability standards may represent an important market push for bio-based products. In this way, foresight on market trends and social acceptance may represent key inputs to the development of more focused exercises on specific regulatory and standardization issues, a topic that is explored in more detail in the following section.

Existing studies on bio-based markets focus on the market availability and potential of specific bio-based products related to industries that have being growing rapidly, including biopolymers and plastics. In addition, other studies related to the identification of existing barriers that prevent the market uptake of bio-based products. Pertinent studies and projects on foresight exercise for bioeconomy market trends and social acceptance include:

- **Global Visions for the Bioeconomy—an International Delphi Study (2015)**, which was an international meeting of experts with the purpose of recommending the most important fields for innovation and policy to the German Bioeconomy Council. In the framework of this study, three roadmaps focusing on the market potential, Research & Development (R&D) priorities and non-technological hurdles of industrial biotechnology innovation have been developed and made publicly available. In particular: (i) the market roadmap that gives an overview of the current markets for a selection of five industrial biotechnology business cases for Europe and market projections extending to 2030; (ii) the technological roadmap that aims to gain insight into the R&D-related hurdles that are impeding the full realization of Europe's industrial biotechnology market potential in 2030; (iii) the non-technological roadmap aims to identify regulatory and non-technological hurdles that may inhibit industrial biotechnology innovation towards identified market opportunities in the market roadmap.
- **The Open Bio** research project explored the most important market barriers and drivers of bio-based products. Two Delphi surveys were conducted among experts from the business community and public procurement. This foresight method drew on the knowledge of this pool of experts to identify drivers and barriers to the future development of the business-to-business market for bio-based products as well as drivers and barriers to their uptake in public procurement. The drivers included a positive public image, independence from fossil fuels, savings in CO_2 emissions and compliance with environmental regulation. Thus the B2B market is driven by the positive image and environmental benefits of bio-based products. Most important market barriers were the higher cost of production, uncertainty about future regulation, volatility of feedstock prices and unsupportive regulatory environment (Peuckert and Quitzow, forthcoming).
- **The BIOCHEM** project conducted a study called "Assessment of the Bio-based Products Market Potential for Innovation" (BIOCHEM 2010) Chap. 2 of this report provides future growth estimates for bio-based products. The total volume growth of major bio-based chemical groups between 2008 and 2020 is estimated at 2.1 Mt (5.3% pa). Assuming similar market value growth, the market is estimated to grow from 21 billion EUR in 2008 to 40 billion EUR in 2020. This will increase the market share of bio-based products from 4% in 2008 to 6% in 2020, providing 43,600 new jobs within the biochemical industry only. Future growth will be affected by the cost of biomass feedstocks but also by fossil fuel prices and by the level of public support. The volume of the European bio-plastics market totalled 0.13 Mt in 2008 and is estimated to grow to 0.9 Mt in 2020 (growth rate 16% pa). At the current state of technology, 5–10% of the plastics market could theoretically be bio-plastics and the long-term potential (2030 onwards) is significantly higher (70–100%). In the initial phase of market introduction, products are often used in niche markets, but some bio-based polymer applications have already gained an established position in the market.

- **The STAR COLIBRI** research project estimated that by 2030, the bio-based economy is expected to have grown significantly in Europe. A pillar of this, both now and in the future, is bio refining, the sustainable processing of biomass into a spectrum of marketable products and energy. The European sector of 2030 is estimated to evolve from established biorefinery operations for products like food, biofuels, paper and board, to a broader, more mature sector. In 2030, biorefineries are expected to use a wider range of feedstocks and produce a greater variety of end-products than today. Achieving the European Biorefinery 2030 vision will require future biorefineries to be better integrated, more flexible and operating more sustainably.

Regulatory foresight for the bioeconomy[10]
Current policies and regulations shape future trajectories of the bioeconomy. Early development of the previously described emerging and enabling technologies can be supported by technology push measures, such as public R&D funding and other types of R&D incentives. Following this early phase, policy and regulation can play a central role in creating niche markets, in which technologies can further develop and be scaled-up and important learning takes place. Tools may include eco-labels or green public procurement. Finally, further market development and widespread diffusion will depend on adapting broader regulatory regimes. This may imply the removal of regulatory barriers, new regulations favouring the new technology, as well as regulations to avoid undesired side-effects, representing potential barriers to market acceptance.

A number of studies have identified regulatory uncertainty and unsupportive regulation as two key barriers for the development of markets for bio-based products (Peuckert and Quitzow, forthcoming). This includes, among other things, a frequently changing and incoherent regulatory framework governing the use of biomass in various application sectors. A forward-looking analysis of policy and regulations would help to anticipate needed changes to be implemented by considering the policy landscape that directly influences biomass availability, including indirect land use change (ILUC) directives, Waste to Energy (WtE) initiatives, the EU's Circular Economy Package and the 2030 C&E Framework, as well as new policies related to biomass.

Regulatory foresight can help anticipate regulatory needs and facilitate the timely adaptation of regulatory environments. Data originating from regulatory foresight can be used to produce recommendations that can guide policy makers in their efforts to adapt regulations to support needed changes and stimulate innovation in the bioeconomy. This is especially the case for waste management. A supportive regulatory framework that promotes the development of a dynamic waste management sector with additional waste management solutions (e.g. organic

---

[10] Regulatory foresight: strategic activities carried out by policy makers to identify future requirements for regulation or reregulation (including formal standards) in existing and emerging technologies in order to shape pro-active innovation-promoting regulatory framework conditions crucial to the competitiveness of innovation systems (Blind 2008).

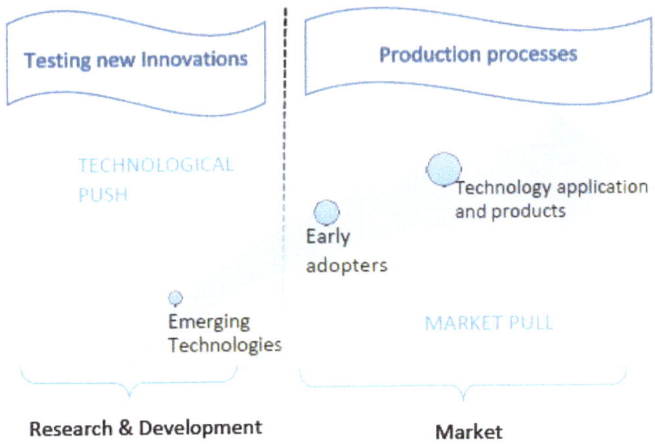

Fig. 9.5 Regulatory foresight for fostering the leap from R&D to market application

recycling) and the market for secondary raw materials represents a crucial factor for deploying the full potential of waste in the bioeconomy. Moreover, it can help anticipate regulatory challenges related to the scaling-up of emerging technologies, and fostering the leap from research and development to market formation (Fig. 9.5). Finally, within the EU, a particular concern is the harmonization of regulations to enable innovators to benefit from the scale and scope of the single market (Better regulations for innovation-driven investments at EU level). At the same time, diversity in regulation may offer opportunities for experimentation and innovation. Hence, careful assessment of alternative paths of regulatory development can offer important insights on their implications for shaping development trajectories in the bioeconomy.

Figure 9.5 illustrates the importance of regulatory foresight as a tool for anticipating and promoting measures from technology push to market pull for innovations. It helps identify the most important fields for innovation support and where policies should be elaborated upon. The selection of available, emerging and breakthrough technologies that are likely to have the biggest future influence and the greatest potential to transform the different sectors of the bioeconomy represents the starting point for conducting technological foresight. In conjunction with investment patterns, they represent two important variables for analysing technology trends.

Systematic and transparent discussion in the context of foresight exercises on various regulatory options between policy makers, industry and other stakeholders can help define future regulatory paths and identify appropriate triggers for regulatory action, thereby increasing the predictability of regulatory frameworks. A key to regulatory foresight is, therefore, the participation of relevant stakeholders in the exercise. In this way, regulatory foresight is not only a vehicle for signalling future regulatory action to stakeholders, but also an instrument for increasing acceptance among affected stakeholders before the regulatory measure is formally introduced,

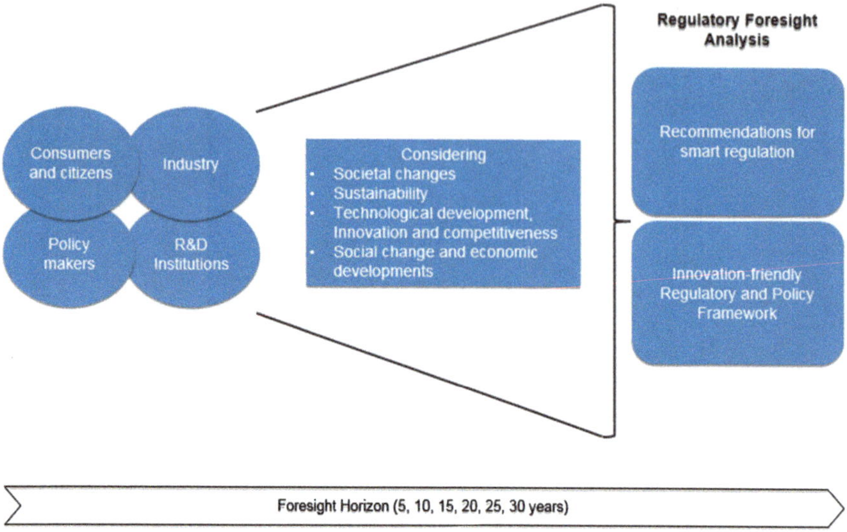

Fig. 9.6 Foresight as a new approach to regulation and policy development

thereby further increasing predictability. Such an anticipatory approach can also provide the basis for engaging in pro-active experimentation and collaborative approaches to developing solutions. Blind (2008) classifies regulatory foresight methodologies in three main categories:

- Indicator-based approaches
- Surveys
- Delphi studies

For those methodologies is possible to provide original empirical evidence.

As shown in Fig. 9.6, regulatory foresight can be considered a new approach to regulation and policy development, by involving relevant bioeconomy stakeholders and considering important external variables such as:

- Societal challenges and sustainability
- Technological development, innovation and competitiveness
- Social change and economic development.

Furthermore, foresight processes can support the identification of the need for technical standard[11] development to keep pace with rapidly developing

[11]Standard—Voluntary documents that define technical or quality requirements with which current or future products, production processes, services or method may comply. Standards results from voluntary cooperation between industry, public authorities and other interested parties collaborating within a system founded on openness, transparency and consensus (A strategic Vision for European Standards, EU 2011).

technologies and business practices and the evolution of social and environmental concerns. As alluded to above, regulatory foresight may also anticipate the needs for standardization and certification in order to provide evidence of the competitive advantages of bio-based products. To establish a tenable bioeconomy, standards for environmental sustainability are essential to support consumers in selecting products, policy makers in dealing with sustainability issues and to guide the industry toward sustainable practices and environmental friendliness. Anticipating the needs for standards in a more systematic manner through the use of foresight is considered an important vehicle for accelerating the standardization processes underpinning sectoral development (Scapolo et al. 2014). Europe's growth strategy, Europe 2020, reiterates the importance of standardization as a vital element to stimulate and enable innovation and competitiveness in Europe. A pro-active approach to standardization can facilitate harmonization across supply chains, as well as create the necessary enabling environment for investment by lowering risks and increasing potential returns (i.e. translating into an improved risk-return ratio) (European Commission 2016a).

Finally, regulatory foresight may be used to identify rules that may then be subject to so-called 'Innovation Deals'. These deals represent a new option to address regulatory obstacles to innovation in a pragmatic, open and transparent manner (European Commission 2016b). This takes the form of voluntary cooperation between the EU, innovators and local, regional and EU regulating authorities with the objective of overcoming legislative obstacles in order to allow for faster market uptake of innovative products through the development of more modern and responsive administration (European Commission 2016c).

In conclusion, regulatory foresight can provide a systematic assessment of future regulatory and standardization needs and how different regulatory options may influence development paths within the bioeconomy. On this basis, policy makers can develop an anticipatory policy agenda that is able to tackle regulatory challenges in a manner that further enhances market developments. Such a future-oriented approach to regulation and standardization can be a critical factor in accelerating market uptake of bio-based products and thus translates into an important competitive advantage for the given regulatory environment. Despite its potential importance, focused regulatory foresight exercises are relatively scarce. This represents a gap in the current foresight literature.

9.4 Conclusions

The bioeconomy is a highly dynamic and in many areas still emerging field of technology and industry. Hence, its future development remains highly uncertain. As outlined in this article, key areas of uncertainty relate to the sustainable supply of biomass, technological developments, market and societal trends and the evolution of the relevant regulatory framework. Foresight exercises thus have an important role to play in reducing uncertainty regarding developments in these key areas, while also highlighting key inter-dependencies between the various fields. Indeed, none of the fields highlighted in this article can be considered in isolation and developments are highly inter-related. Nonetheless, the distinction of the four focus areas offers a useful framework for distinguishing different focus areas for analysis.

The role of foresight in reducing uncertainty for market actors in the bioeconomy can take two basic forms. Firstly, foresight exercises can provide stakeholders with a systematic and structured overview of available information. This may involve the analysis of existing information and the collection of new data from previously untapped sources. Secondly, foresight can aid stakeholders by bundling the existing expertise of market actors and by supporting the development of shared visions on the future development of the bioeconomy.

In the first case, foresight is critical for providing stakeholders with a systematic overview of developments and trends that may in large part be exogenous to the relevant sub-sector of the bioeconomy. By reducing uncertainty in regard to these exogenous trends, it can reduce investment risks for firms in the sector as well as policy risks for political decision makers, thus acting as a stimulus for action by both types of actors.

In the second case, foresight exercises address factors that may be influenced by stakeholders themselves. Hence, in addition to its function of providing and systematizing information to stakeholders, foresight has the potential to enhance coordination and exchange of information between stakeholders. It offers a tool for stakeholders to agree on desired development paths and coordinate activities to realize an agreed vision.

In practice, most foresight exercises have elements of both, albeit with differing emphasis on one or the other. In particular, technology foresight frequently includes the aim of enhancing the degree of coordination of research and development activities in both the private and public sector. Regulatory foresight represents an equally promising yet underutilized instrument for realizing such a coordination function among stakeholders. Regulatory foresight offers not only the potential to identify key entry-points for enabling regulatory action, but for optimizing policy strategies by anticipating future regulatory challenges and agreeing on integrated, forward-looking solutions. In doing so, it has been argued that regulatory foresight can be a critical instrument for the further development of the European bioeconomy. It allows for the design of innovation-friendly policies and regulations that can accelerate developments in the sector. Indeed, such an anticipatory policy framework may represent a key competitive advantage for the development of bio-based industries in Europe.

To reduce regulatory and administrative burners, simplify policy and level the playing field between sectors, as needed for promoting innovation and competition in the bioeconomy. Current agricultural, energy and waste policies need to be re-designed based on the results of accurate regulatory foresights in order to make the transition to a vibrant bioeconomy.

References

Blind K (2008) Regulatory foresight: methodologies and selected applications. Technol Forecast Soc Chan 75(4):496–516

BIOCHEM (2010) Report on the assessment of the bio-based products market potential for innovation. http://www.biochem-project.eu/download/innova/reports/Assessment%20of%20the%20Bio-Based%20Products%20Market%20Potential%20for%20Innovation.pdf

Bio based Industry Consortium BBI JU. (2012) Accelerating innovation and market uptake of biobased products. http://biconsortium.eu/sites/biconsortium.eu/files/publications/BIC_Vision_web_0.pdf

Cuhls K, Kayser V, Grandt S, Hamm U, Reisch L, Daniel H, El-Chichakli, B, Dieckhoff P (2015) Global visions for the bioeconomy: an international Delphi-study. Accessed 04 May 2016. http://biooekonomierat.de/fileadmin/Publikationen/berichte/Delphi-Study.pdf

dos Santos GJ, Luana Ladu, daCosta Mata HT, Garcia de Oliveira G, Germani GI, deAthayde Couto V (2009) Agrocombustiveis, segurançia e sobernaria alimentar: element do debate internationale e analise do caso brasileiro. Bahia Analise & Dados (Jornal Superintendência de Estudos Econômicos e Sociais da Bahia) 18(4):549–556

Elbersen B, Boywer C, Kretschmer B (2012) Atlas of EU biomass potentials: summary for policy makers: policy briefing under D6.4. Unpublished manuscript, last modified 13 July 2016. http://www.biomassfutures.eu/public_docs/final_deliverables/WP6/D6.4%20Atlas%20of%20EU%20biomass%20potentials.pdf

Eshelman RS (2012) Food waste from starbucks turned into useful products. Accessed 04 May 2016. http://www.scientificamerican.com/article/food-waster-from-starbucks-turned-into-useful-products/

European Biofuels Technology Platform (2014) Food versus fuel. Accessed 13 July 2016. http://biofuelstp.eu/food-vs-fuel.html

European Commission (2009) Sustainable food consumption and production in a resource-constrained world: 3rd SCAR Foresight Exercise. https://ec.europa.eu/research/scar/pdf/scar_3rd-foresight_2011.pdf.

European Bioplastics (2016) Relevant EU Policies. Accessed September 18, 2016. http://www.european-bioplastics.org/policy/

European Commission (2010) Europe 2020 flagship initiative innovation union. SEC(2010) 1161

European Commission (2011) A strategic vision for European standards: moving forward to enhance and accelerate the sustainable growth of the European economy by 2020. COM(2011) 311 final. Accessed 04 May 2016. http://eur-lex.europa.eu/LexUriServ/LexUriServ.do?uri=COM:2011:0311:FIN:en:PDF

European Commission (2012) Bioeconomy strategy: innovating for sustainable growth: a bioeconomy for Europe. Accessed 04 May2016. http://ec.europa.eu/research/bioeconomy/pdf/bioeconomycommunicationstrategy_b5_brochure_web.pdf

European Commission (2014a) Closing the loop—an EU action plan for the circular economy. COM 2015 614 final. Accessed 13 July 2016. http://ec.europa.eu/environment/circular-economy/index_en.htm

European Commission (2014b) Towards a circular economy: a zero waste programme for Europe. COM/2014/0398. http://eur-lex.europa.eu/legal-content/EN/TXT/?uri=CELEX:52014DC0398R%2801%29

European Commission (2015a) Sustainable agriculture, forestry and fisheries in the bioeconomy: a challenge for Europe: 4th SCAR foresight exercise. Unpublished manuscript, last modified 13 July 2016. https://ec.europa.eu/research/scar/pdf/ki-01-15-295-enn.pdf#view=fit&pagemode=none

European Commission (2015b) The junction of health, environment and the bioeconomy: foresight and implications for european research & innovation policies. Accessed 04 May 2016. http://www.kowi.de/en/Portaldata/2/Resources/horizon2020/coop/Foresight-Junction-Health-ENV-Bioeconomy-.pdf

European Commission (2016a) Better regulations for innovation-driven investment at EU level. Unpublished manuscript, last modified 13 July 2016. https://ec.europa.eu/research/innovation-union/pdf/innovrefit_staff_working_document.pdf

European Commission (2016b) Innovation deals for a circular economy. https://ec.europa.eu/research/innovation-deals/index.cfm?pg=home

European Commission (2016c) Open innovation, open science, open to the world: a vision for Europe. Unpublished manuscript, last modified 13 July 2016. https://ec.europa.eu/digital-single-market/en/news/open-innovation-open-science-open-world-vision-europe

European Commission (2016d) Environment action programme to 2020. http://ec.europa.eu/environment/action-programme/

European Forum on Forward Looking Activities EFFLA (2014) Towards standards in forward looking activities for the EC. EFFLA POLICY BRIEF 14. Unpublished manuscript, last modified 13 July 2016. https://ec.europa.eu/research/innovation-union/pdf/expert-groups/effla-reports/effla_pb_14_-_towards_a_foresight_standard.pdf

Festel G (2010) Industrial biotechnology: market size, company types, business models, and growth strategies. Ind Biotechnol 6(2):88–94

Georghiou L, Cassingena J, Keenan M, Miles I, Popper R (eds) (2008) The handbook of technology foresight: concepts and practice. Edward Elgar Publishing Ltd, Cheltenham, UK

Giaoutzi M (2013) Recent developments in foresight methodologies. Springer, Berlin

Global Bioeconomy Summit edition (2015) Communiqué global bioeconomy summit 2015: making bioeconomy work for sustainable development

Gołębiewski J (2013) Value chains within bioeconomy. International Scientific Electronic Journal Earth Bioresources and Life Quality Founded by National University of Life and Environmental Sciences of Ukraine (NUBiP of Ukraine) and Global Consortium of Higher Education and Research for Agriculture (GCHERA) (4)

Hasenheit M, Gerdes H, Kiresiewa Z, Beekman V (2016) Summary report on the social, economic and environmental impacts of the bioeconomy

Jackson M (2013) Practical foresight guide. Accessed 04 May 2016. https://www.shapingtomorrow.com/media-centre/pf-ch03.pdf

Jiménez-Sánchez G, Philp J (2016) Genomics and the bioeconomy. In: Kumar D, Chadwick R (eds) Genomics and society: ethical, legal, cultural and socioeconomic implications. Academic Press an imprint of Elsevier, Amsterdam, pp 207–238

Mattana P (2015) The role of bioeconomy in climate change mitigation. News release. September 29. Accessed 04 May 2016. http://ebcd.org/wp-content/uploads/2015/07/Press-Release-The-Role-of-Bioeconomy-in-Climate-Change-Mitigation1.pdf

Mikova N, Sokolova A (2014) Selection of information sources for identifying technology trends: a comparative analysis. Higher School of Economics Research Paper No. WP BRP 25

Nattrass L, Biggs C, Bauen A, Parisi C, Rodríguez-Cerezo E, Gómez-Barbero M (2016) The EU bio-based industry: results from a survey

OECD (2009) The bioeconomy to 2030: designing a policy agenda. 1. Aufl. s.l. OECD. http://gbv.eblib.com/patron/FullRecord.aspx?p=437750

Ortner H, Pfurtscheller D, Rizzolli M, Wiesinger A (eds) (2014a) Datenflut und Informationskanäle, 1st edn. Innsbruck University Press, Innsbruck

Ortner H, Pfurtscheller D, Rizzolli M, Wiesinger A (2014b) Zur Einführung–Datenflut und Informationskanäle. In: Ortner H, Pfurtscheller D, Rizzolli M, Wiesinger A (eds) Datenflut und Informationskanäle, 1st edn. Innsbruck University Press, Innsbruck

Peuckert J, Quitzow R (forthcoming) Acceptance of bio-based products in the business-to-business market and public procurement: expert survey results biofuels, bioproducts & biorefining

Phaal R, Farrukh CJP, Probert DR (2004) Technology roadmapping—a planning framework for evolution and revolution. Technol Forecast Soc Chang 71(1):5–26

Philippidis G, M'barek R, Ferrari E (2016) Drivers of the European bioeconomy in transition: bioeconomy 2030. Accessed 04 May 2016. http://publications.jrc.ec.europa.eu/repository/bitstream/JRC98160/lfna27563enn.pdf

Piotrowski S, Carus M, Essel R (2015) Global bioeconomy in the conflict between biomass supply and demand. Accessed 04 May 2016. http://agricultura.gencat.cat/web/.content/de_departament/de02_estadistiques_observatoris/27_butlletins/02_butlletins_nd/documents_nd/fitxers_estatics_nd/2015/0165_2015_ERenovables_Biomassa_biomassa.pdf

Porter A, Rossini A, Carpenter S, Roper A, Larson R, and Tiller J (1980) Guidebook for technology assessment and impact analysis. North Holland

Popper R (2008) Foresight methodology. In: Georghiou L, Cassingena J, Keenan M, Miles I, Popper R (eds) The handbook of technology foresight: concepts and practice. Edward Elgar Publishing Ltd, Cheltenham, UK, pp 44–88

Royal Society of Chemistry (2014) Waste opportunities: stimulating a bioeconomy. Accessed 04 May 2016. http://www.rsc.org/globalassets/04-campaigning-outreach/realising-potential-of-scientists/research-policy/global-challenges/waste-opportunities-response-november-2013.pdf

Scapolo F, Churchill P, Viaud V, Antal M, Córdova H, de Smedt P (2014) How will standards facilitate new production systems in the context of EU innovation and competitiveness in 2025? Accessed 13 July 2014. https://ec.europa.eu/jrc/sites/default/files/jrc-foresight-study-web_en.pdf

Scarlat N, Dallemand J-F, Monforti-Ferrario F, Nita V (2015) The role of biomass and bioenergy in a future bioeconomy: policies and facts. Environ Dev 15:3–34

Scapolo F, Churchill P, Viaud V, Antal M, Córdova H, de Smedt P (2014) How will standards facilitate new production systems in the context of EU innovation and competitiveness in 2025? Accessed 13 July 2014. https://ec.europa.eu/jrc/sites/default/files/jrc-foresight-study-web_en.pdf

Slaughter R (1995) The foresight principle: cultural recovery in the 21st century Praeger Westport. Praeger Westport, Conneticut

Staffas L, McCormick K, Gustavsson M (2013) A global overview of bioeconomy strategies and visions. Accessed 04 May 2016. http://lup.lub.lu.se/luur/download?func=downloadFile&recordOId=3814329&fileOId=3814737

Star COLIBRI (2011) Joint European biorefinery vision for 2030: Strategic targets for 2030-collaboration initiative on biorefineries. http://www.forestplatform.org/files/Star_COLIBRI/Vision_document_FINAL.pdf

State of Green (2015) Producing more with less: Danish strongholds in bioeconomy & resource-efficient production. https://stateofgreen.com/files/download/7415

Stenmarck Å, Jensen C, Quested T (2016) Estimates of European food waste levels. http://www.eu-fusions.org/phocadownload/Publications/Estimates%20of%20European%20food%20waste%20levels.pdf

Unfold the Future. A Project of The Confederation of European Paper Industries CEPI. (2011) The forest fibre industry: 2050 roadmap to a low-carbon bio-economy. Accessed 04 May 2016. http://www.unfoldthefuture.eu/uploads/CEPI-2050-Roadmap-to-a-low-carbon-bio-economy.pdf

van der Hoeven D (2014) Circular economy: recycling agricultural and food waste. Accessed 04 May 2016. http://www.biobasedpress.eu/2014/04/circular-economy-recycling-agricultural-food-waste/

Vanham D, Bouraoui F, Leip A, Grizzetti B, Bidoglio G (2015) Lost water and nitrogen resources due to EU consumer food waste. Environ Res Lett 10(8): 84008

Vecchiato R, Roveda C (2014) Foresight for public procurement and regional innovation policy: the case of Lombardy. Res Policy 43(2):438–450

Chapter 10
Bringing a Sharing Economy Approach into the Food Sector: The Potential of Food Sharing for Reducing Food Waste

Pasquale Marcello Falcone and Enrica Imbert

Abstract According to the UN estimates, world population will increase to over 8 billion by 2030. Increasing demand for food and raw materials will place additional pressure on limited natural resources. In this context, the current levels of food waste in advanced economies are no longer economically, socially and environmentally sustainable over the long term. Structural changes will be needed along the whole supply chain as well as in consumers' attitudes and behaviours. The sharing economy is actually playing an important role in trying to achieve more sustainable patterns, also within the food sector. In particular, several initiatives and start-ups are being developed in the US and Europe, involving the collection and use of the excess of food from consumers and retailers and the promotion of collaborative consumption models. However, the correlation between food sharing practices and reduced food waste cannot be taken for granted. This chapter identified the literacy contours of this relationship, highlighting how food sharing is frequently undermined by social factors and that to make it effective specific skills are needed. Moreover, a major effort towards general routines and practices, which underpin individual-level behavior, is required to tackle food waste in a more effective manner.

Keywords Sharing economy · Sustainable consumption · Food waste · Food sharing

10.1 Introduction

In recent times, with global climate change challenges and its various consequences on ecosystems and on resource depletion, the socio-economic as well as the environmental impacts of the mass-consumption economy have become important

---

P.M. Falcone · E. Imbert (✉)
BiT Bioeconomy in Transition Research Group,
Unitelma-Sapienza University of Rome, Rome, Italy
e-mail: enrica.imbert@uniroma1.it

P. Morone et al. (eds.), *Food Waste Reduction and Valorisation*,
DOI 10.1007/978-3-319-50088-1_10

issues gaining momentum in the international debate. Particularly, the amount of waste (especially in term of food waste) is expected to increase at an alarming rate unless effective policies and alternative production and consumption patterns are implemented to address the problem. According to the US Environmental Protection Agency (EPA), food waste currently represents the largest share of waste entering landfills. In this context, there is a unanimous viewpoint according to which a holistic approach to food waste prevention can bring to a reduction of GHG emissions, i.e. due to the reduction in the methane and carbon dioxide emissions arising from degradation of food in landfills (Hall et al. 2009), as well as the decrease of natural resources depletion used for food production and distribution.

While in developing countries food waste arise largely at early stages of the supply chain and can be due to financial, managerial and technical constraints in harvesting and storing practices, in developed countries food is wasted generally at later stages of the supply chain because of consumers' behaviour (FAO 2011b). Therefore, food waste reduction at the consumption level represents for medium and high income countries a key objective on the policy agenda of national and international institutions (e.g. Monier et al. 2010; Braun 2012).

Recently, several initiatives and practical solutions (e.g. packaging and alternative storage technics) have been proposed to moderate the waste of food by acting also on household behaviours through unconventional consumption models. In this context, attention has been paid to the sharing economy approach as many food sharing initiatives have been launched around Europe and the United States based on collecting and utilizing the excess of food from consumers and retailers and by promoting collaborative consumption models (e.g. Foodsharing, Growington, Feastly, etc.). As suggested by Belk, the sharing economy approach entails 'the act and process of distributing what is ours to others for their use and/or the act and process of receiving or taking something from others for our use' (2007: 126).

Against this background, the sharing economy principles could provide a new way of thinking based, essentially, on environmental effectiveness and economic efficiency by potentially offering a successful way to reduce food waste so as to accelerate the transition toward a more sustainable development. However, the assumption that this type of approach necessarily leads to food waste reduction, with benefits for the environment, local municipal bodies and household savings, is not a straightforward conclusion. In this chapter we aim at identifying the literacy contours of this theoretically beneficial relationship.

Section 10.2 highlights the theoretical background surrounding the sharing economy approach. Section 10.3 explores the link between sharing economy and sustainable development. Section 10.4 provides different definitions to analyses and understands food waste issues at macro (i.e. developed countries general trends) and micro level (i.e. households' food behaviours). Section 10.5 reviews different literature contributions on the impact of food sharing on food waste reduction. Finally, Sect. 10.6 concludes the chapter and provides some final thoughts on the topic under investigation.

10.2 Sharing Economy: A Theorical Background

The term *sharing* usually refers to two different meanings. Typically, it denotes anything that is shared between or among two or more people. Moreover, it can imply that two or more people are characterized by something in common (Zvolska 2015). These two meanings of the word *sharing* actually represent what Tomalty (2014) identified as 'zero-sum' and 'non zero-sum'. While the former represents a situation in which each person is left with less of something when he/she shares it (for instance, a cake), the latter implies a configuration in which the people who share something are left with the same amount of it (for instance, a genetic trait).

Although *sharing* does not represent a new phenomenon-'Sharing has probably been the most basic form of economic distribution in hominid societies for several hundred thousand years' (Price 1975 cited by Belk 2010, p. 715) the *sharing economy* is a new form of sharing that is gathering attention in the last years. The foundation of the sharing economy is a 'zero-sum' type of sharing of people's own assets and their reliance on the use of modern information technology. In this context, as emphasized by Zervas et al. (2015), the success of the sharing economy crucially depends on the existence of network platforms, enabled by the information and communication technologies (ICT), able to link consumers' needs to the sharing economy activities. As a result, different economic initiatives based on new types of consumption models are gaining more and more influence in a large majority of developed countries. In particular, as emphasized by Botsman and Rogers (2010), business models based on *collaborative consumption*, also known as *sharing economy* activities, are growing in many sectors ranging from transportation (e.g. Car2go, Uber), accommodation (e.g. Airbnb) to finance (e.g. Indiego). Therefore, diffusion and uptake of Internet technologies, on the one side, and the growing crisis of traditional models based on consumerist society on the other,[1] have been key factors for the emergence and diffusion of this new economic models. Nowadays, consumer interest is no longer just on ownership. The aim is, rather, access to goods and services (Rifkin 2000). Sharing and collaborative practices are indeed characterized by temporary access to goods and services and dependence on Internet–mainly Web 2.0.

However, given its early developmental stage, there is still no harmonized definition of the sharing economy in the literature. As emphasized by Botsman (2013a), the academic debate over the sharing economy definition is still ongoing as the terms mentioned so far, such as *collaborative consumption*, *sharing economy* and *accessed based consumption* are frequently used synonymously.[2] Indeed, access-based consumption has been defined 'as transactions that may be market

[1]Especially after the global economic crisis of 2008.

[2]Europe Economics defined the sharing economy as 'The use of digital platforms or portals to reduce the scale for viable hiring transactions or viable participation in consumer hiring markets (i.e. 'sharing' in the sense of hiring an asset) and thereby reduce the extent to which assets are under-utilised' (Goudin 2016, p. 11).

mediated in which no transfer of ownership takes place' (Bardhi and Eckhardt 2012: 881). Collaborative consumption encompasses access based-consumption and is 'embedded within the sharing economy which involves access-based consumption of products or services that can be online or offline' (Barnes and Mattsson 2016: 200).

While a commonly agreed definition on the sharing economy seems still not to exist, researchers further disagree whether it is based on monetary or non-monetary exchanges, or both, and whether it includes *Peer-to-Peer* (P2P) models or also *Business-to-consumer* (B2C) and *Business-to-Business* (B2B) models (Zvolska 2015).

Botsman (2013b) in distinguishing between the sharing economy and the peer economy uses an inclusive definition of the former. The former refers to both B2C and P2P models, while the latter involves only the P2P segment. In particular, the author outlines the sharing economy as 'an economic model based on sharing underutilized assets from spaces to skills to stuff for monetary or non-monetary benefits'. However, such demarcation of the sharing economy was not exempt from receiving criticisms. Especially, Belk (2014) stresses as it seems too general and thus, unable to stand out from a mere gift giving or sharing. Conversely, other authors adopt a more narrow definition. According to Frenken et al. (2015) the sharing economy includes only P2P initiatives, suggesting that businesses such as B2C (e.g. car rental), the second-hand economy and on-demand economy, should not be considered as part of the sharing economy. On the same wavelength, Schor provided the following definition of the sharing economy: 'an economic activity that is Peer-to-Peer, or person-to-person, facilitated by digital platforms' (2015: 14). Moreover, the author contends that when assets' owners make profit by sharing, they no longer share but rent. Therefore only, few non-profit platforms purely concern sharing. Finally, B2B sharing models refers to 'the sharing of services, utility, and by-product resources among industries' (Geng et al. 2014: 1). Such models represent certainly 'the next generation of the sharing economy' (Slagen 2014). Currently, some of the main B2B companies (e.g. WeWork, Floow2 etc.) are allowing firms to provide access to everything from shared office space to underutilized machinery in the supply chain. However, very little attention has been paid toward them to the present, resulting sometimes completely ignored from the sharing economy literature.

10.3 Sharing Economy and Sustainable Development

Beyond the theoretical debate over the sharing economy boundaries, a number of relevant and controversial questions in the environmental as well as in the social and economic fields are gathering attention. In particular, this topic is of great interest as it brings new perspectives to today's debate on the effects of the sharing economy approach on sustainable development. It is well recognized that sustainable development, to be thought jointly from an economic, social and environmental point of view, occupies a central role in the global agenda. While, on the one hand, as regards its social and economic dimension, some benefits coming from the sharing economy have been pointed out[3] (see for instance Goudin 2016), on the other hand, there is a growing interest in the sharing economy as a model for sustainable consumption practices. On this ground, Heinrichs (2013) sees the sharing economy as a 'potential new pathway to sustainability', Botsman and Rogers (2010) perceive it as a potential way out from the unsustainable consumption practices on which current developed economies are based. Their central argument is founded on a beneficial transition of culture from a 'consumer's own assets' (i.e. traditional linear economy) towards a 'consumers share access to assets' (i.e. sharing economy) able to connect consumers and allow them to make a more efficient use of underutilized available goods and services. However, although several successful experiences (i.e. Airbnb, Uber, etc.) positively contributed to the debate surrounding the sharing economy approach, the stakeholders' discourses and opinions concerning the link between sharing economy and sustainability are, oftentimes, framed in contrasting and contradictory way. In particular, by examining the actors sharing discourses Martin (2016) found that, although the sharing economy can be seen as a socio-technical niche able to foster more sustainable consumption and production practices, at same time it could paradoxically strengthen the current unsustainable economic paradigm making unlike the transition towards a more sustainable consumption and production practices.

In this context, Goal n.12 of the 2030 Agenda for Sustainable Development aims to ensure sustainable consumption and production patterns. The sustainable consumption and production (SCP) concept, introduced in the early Nineties and in 2002 during the World Summit on Sustainable Development (WSSD), is recognized as one of the three essential requirements for sustainable development. 'SCP aims at doing more and better with less, by reducing resource use, environmental degradation, waste and pollution along the whole life cycle of goods and services, while at the same time increasing quality of life for all' (UNEP 2011: 10). Therefore a key issue is the extend to which an increase in environmental awareness and related effects can be attained through:

- a resource efficiency in the production processes (i.e. less resource inputs to achieve the same or improved output)
- a transition towards greener consumption and production patterns (e.g. more efficient and less polluting goods and services)
- a transition towards more efficient consumption models (i.e. consumers share access to their own goods and services).

Specifically, looking at the consumption side, 'Sustainable consumption is not just about buying the more sustainable products. Refusing to consume when not necessary and engaging in alternative means of satisfying needs are also important. Sustainable product design, switching from products to services and collaborative

[3]Likewise, concerns have been raised. As reported by Martin et al. (2015) recovering Morozov's view of the sharing economy as a 'neo liberism on steroids', a strong criticism has been put forward by the literature with a focus on the sharing economy ability to bypass environmental and social laws.

consumption are examples of approaches to sustainable lifestyles' (UNEP 2015: 118). In this context, Geels et al. (2015) provide a critical review about the SCP literature and propose a new view focusing on transitions in socio-technical systems and daily life practices. In particular, beside the well-known *reformist SCP-position* (see Lebel and Lorek 2008) and *revolutionary SCP-position* (see among the others: Belk 2010; Schor 2014), the authors offer a new position (*reconfiguration SCP-position*), focusing either on macro-contexts or on individuals' attitudes and behaviors. Specifically, transition towards new SCP systems and practices (i.e. transport, electricity, heat, food, etc.) is viewed as a multilevel process in which heterogeneous actors engage by going beyond individual consumers and firms behaviors to involve social movements, media, public opinion, advisory bodies, researchers, and special-interest groups as well. Such *reconfiguration SCP-position* concentrates mainly on the adjustments in existing sociotechnical systems and related (re)alignments among different new and old elements rather than the development of a technological niche. A particular example in the transport domain concerns a transition towards a reconfigured system in which vehicle utilizes alternative fuels, the cities are characterized by a more developed and environmentally friendly public transport, consumers share vehicles, and so on.

In considering the role played by the sharing economy in accelerating sustainable consumption and production patterns in cities worldwide Cohen and Munoz (2015) provide an integrated framework for theorizing five ideal sharing categories: energy, food, goods, mobility and transport and space sharing. Particularly, the food sector is indeed recognized as a strategical area for sustainable consumption and production implementation (Tukker et al. 2008). In principle and both from a macro and micro perspective, food sharing may have a positive impact on all three dimensions of sustainable development by boosting savings, helping to create and/or consolidate existing social relations and by reducing waste generation. Currently, numerous initiatives and start-ups are springing up in the US and Europe, concerning the collection and use of the excessing food from consumers and retailers and the development of collaborative consumption models (e.g. Foodsharing, Growington, Feastly, etc.). On these types of new models we will be focusing in Sect. 10.5.

10.4 Food Waste in Higher Income Countries: Conceptualization and General Trends

Roughly one third of the food produced in the world for human consumption annually gets lost or wasted of which about 30% are cereals, 40–50% are root crops, fruits and vegetables, 20% are oilseeds, meat and dairy and 35% are fish.[4] In the EU-28 alone it has been estimated that, on average, 173 kg of food waste is

[4]See http://www.fao.org/save, http://www.fao.org/save-food/resources/keyfindings/en/-food/resources/keyfindings/en/.

produced per person per year (FUSIONS 2016). An enormous loss of resources with long-lasting detrimental effects on global nutrition, environment and savings.

The food sector, in general, contributes significantly to environmental degradation, accounting for about 30% of the world's total energy consumption and for more than 20% of total Greenhouse Gas emissions (FAO 2011a). Specifically, regarding its waste food wastage ranks as the third top GHG emitter after USA and China (FAO 2013). Such an environmental impact is owing to the waste of resources employed for its production (e.g. land and water), transport and for its final disposal.

Also considering food waste economic impact, costs are very high. For example, Lipinski et al. (2013) stress that in China alone US$32 billion worth are lost due to food waste. Further, food waste has a significant social impact since it reduces food security in developing countries (Kummu et al. 2012; Foley et al. 2011; Godfray et al. 2010).

Accordingly, international institutions as well as national governments treat food waste as a serious problem to be tackled with multiple types of interventions at different levels of the food supply chain. However, these institutions themselves are faced with a pressing issue related to food waste measurement. Definitions of food waste are indeed still not harmonized (Lebersorger and Schneider 2011) as the terms *food waste* and *food loss* are frequently used synonymously and/or defined in different ways. This obviously affects food waste quantification, hampering international comparisons (Monier et al. 2010) and effective sustainable development strategies.

HLPE report (2014) identified three main approaches to food waste definition in the literature. The first definition, to which this contribution refers to, is based on the stage in which losses and waste materially occur. Although food is lost and wasted along all levels of the food chain, affecting all countries, a distinction has been made since food losses concern mostly developing countries occurring at early stages of the supply chain[5] while food waste is more specific to developed countries, taking place at downstream phases[6] (Parfitt et al. 2010, FAO 2013).

On the other hand, the second definition relates to the origin of loss or waste, distinguishing from behavioral/deliberate waste and involuntary losses while the third utilizes food waste (or wastage) as an all-encompassing term.

In Table 10.1 we report a summary of the most referenced definitions of food waste pinpointing for each definition quantified and not-quantified elements.

Looking at Table 10.1, we can observe that definitions developed by the UK Waste and Resources Action Programme (WRAP) and the European-funded food waste prevention project (FUSIONS) refer only to the term food waste while the Food and Agriculture Organization of the United Nations (FAO) definition includes three notions, i.e. food loss, food waste and food wastage. Additionally, FAO's report in 2013 included both edible and nonedible parts of food while the estimations of the 2011 report (Gustavsson et al. 2011) were based only on edible parts of

[5]Mainly due to underdeveloped infrastructures, premature harvesting and poor storage.

[6]At retail and consumption stages.

Table 10.1 Most referenced definitions of food waste

	Quantified elements	Not-quantified elements
Food waste is any food that had the potential to be eaten, together with any unavoidable waste, which is lost from the human food supply chain, at any point along that chain (WRAP 2015)	– Food produced for human consumption – All food and drink types, all disposal routes, and all sectors of the supply chain – Avoidable (edible) and unavoidable (inedible) food waste	– Food or food surplus used as animal feed
Food waste is any food, and inedible parts of food, removed from the food supply chain to be recovered or disposed (including composted, crops ploughed in/not harvested, anaerobic digestion, bio-energy production, co-generation, incineration, disposal to sewer, landfill or discarded to sea (Fusions 2014)	– Edible and inedible food and drink waste – Fish discarded to sea	– Edible and inedible food used as animal feed – Edible and inedible food for bio-material processing or other industrial uses
Food loss refers to a decrease in mass (dry matter) or nutritional value (quality) of food that was originally intended for human consumption. These losses are mainly caused by inefficiencies in the food supply chains, such as poor infrastructure and logistics, lack of technology, insufficient skills, knowledge and management capacity of supply chain actors, and lack of access to markets. In addition, natural disasters play a role Food waste refers to food appropriate for human consumption being discarded, whether or not after it is kept beyond its expiry date or left to spoil. Often this is because food has spoiled but it can be for other reasons such as oversupply due to markets, or individual consumer shopping/eating habits Food wastage refers to any food lost by deterioration or waste. Thus, the term wastage encompasses both food loss and food waste. (FAO 2013) Food losses refer to the decrease in edible food mass throughout the part of the supply chain that specifically leads to edible food for human consumption. Drawing on Parfitt et al. (2010) food waste relates to food losses occurring at the end of the food chain, i.e. retail and final consumption (FAO 2011b)	– Edible and inedible food waste – Food that was originally meant to human consumption which fortuity gets out even if it is then directed to a non-food use (feed, bioenergy…)	– Feed and parts of products which are not edible

food.[7] Moreover, the FUSIONS food waste definition logic is driven by the final destination of food. For example, when food is reused for bio-material processing it is considered to be exploited in a productive way so it is not accounted as waste, but instead, as 'valorisation and conversion' (FUSIONS 2014). Garcia-Garcia et al. (2015) highlighted, however, that the unplanned use of food originally thought for human consumption but subsequently devoted to other uses such as for bio-material processing is still accounted as waste by FAO. Once again, it should be noted that a major challenge is to ensure common definitions and measurement methods to generate more accurate and comparable data. Currently, FAO, FUSIONS, WRAP, together with the World Resources Institute (WRI), Consumer Goods Forum (CGF), UNEP, World Business Council for Sustainable Development (WBCSD) are jointly working toward reaching this ambitious goal.[8]

At the regional level, America, Oceania and the European Union have been identified as the greatest wasters (Gustavsson et al. 2011) and despite the above mentioned limitations, country and regional level studies (e.g. Mason et al. 2011; Quested et al. 2011; Ventour 2008; FUSIONS 2016) have helped to identify most critical issues. Thus, prevention and mitigation measures have been put in place in recent years[9] and specific laws at country level, such as in France and Italy, have been enacted. The United States, for which food waste has been estimated at 40% of the entire food supply (Hall et al. 2009), launched in 2013 the U.S. Food Waste Challenge which has been adopted at all levels of the food supply chain.[10]

Overall, Kummu et al. (2012) estimated that, roughly, food supply losses could be reduced by half and that, potentially, the best results could be achieved in agricultural losses and at consumption waste which will be the focus of the following paragraph.

10.4.1 Food Waste at Consumption Level: Households' Food Behavior

Although in wealthy countries there is empirical evidence of consistent quantities of food waste at upstream stages (e.g. Fine et al. 2015), it is widely agreed that in high and medium income countries food is largely wasted at retail and, especially, at the consumer level due to households' attitudes and behaviors (Parfitt et al. 2010; Gustavsson et al. 2011). This latter tendency seems to be more prevalent in western countries when comparing data with countries such as China where food waste has

---

[7]The 2013 report outlined that since its main objective was to calculate food waste enviromental impact, caused by both edible and non edible parts of food, the two components were quantified.

[8]http//www.wri.org/sites/default/files/uploads/FLW_Standard_Executive_Summary_PrePublication Version_2016_April.pdf.

[9]However, these measures themselves may entail economic and environmental costs. 'Obviously, from the environmental point of view, the negative impacts of measures to reduce food loss and waste should be lower than the benefits' http://www.fao.org/3/a-i4068e.pdf.

[10]This initiative is jointly coordinated by USDA and EPA.

been found mostly at restaurant and catering sector rather than in domestic settings (Liu 2014).

Households account for 53% of EU-28 food waste in 2012 (FUSIONS 2016). In the UK, food waste costs the average household £470 a year (WRAP 2013). The Waste and Resources Action Programme (WRAP) and the Global Commission on the Economy and Climate report (2015) estimated that a decrease in consumer food waste by 20–50% could save the global economy between US$120 and 300 billion per year by 2030. However, in considering reasons for food waste at household level, Segrè et al. (2014: 31) outlined that consumers are 'influenced by a number of cultural, psychological and social aspects that do not always follow criteria related to economic rationality'.

The literature identified several reasons among which poor knowledge and understanding of food date labels (Halloran et al. 2014), inadequate purchase planning and insufficient home economic skills and food knowledge (Moomaw et al. 2012) as well as the influence of promotional offers and packaging (Williams et al. 2012) and the carless attitude of wealthy consumers (Gustavsson et al. 2011). As stated by Woolley et al. (2016: 374) 'the management of food inventory is difficult in domestic environments'.

Aschemann-Witzel et al. (2015) emphasized the impact on the amount of food waste of both psychographic factors and socio-demographics. In fact, food waste at individual level could also be influenced by a wide range of factors including household composition and size, income, age, level of education and type of employment.

Most studies found, as was expected, that the larger households' components, the greater the amount of waste becomes; even though this result does not hold on a per capita basis as single households waste more in proportion (see for instance Ventour 2008, Koivupuro et al. 2012). Additionally, as outlined by Ventour (2008) it is important to pay attention to household composition, since families with children under 16 years old, for example, tend to waste more.

On the other hand, establishing correlation with households' age is more challenging (Jorissen et al. 2015). Some studies suggested that old people waste less (especially for their austerity experiences during the Second World War) while other empirical evidence (e.g. Ventour 2008) found this correlation not so strong.

Also the correlation between income and food waste is rather problematic. Although some studies have found that low income households produce less food waste compared to the wealthiest (e.g. Monier et al. 2010; Secondi et al. 2015) this is not always valid (see for example Porpino et al. 2015). Several studies found little or no correlation (Parfitt et al. 2010). For example, Koivupuro et al. (2012) outlined that no correlation was found even though they observed more waste in consumers with little sensitivity to promotional offers, indicating a tendency to waste by wealthy households. Lastly, as regards gender correlation women have been found to be, in general, more sensible to sustainable consumption initiatives (Oecd 2008).[11] The overall variety of the methods used for the analyses and the

[11]For further readings on the relationship between women and food waste, see among others.

heterogeneity of the samples, both in regards to the size and the composition, still makes it difficult, however, to make comparisons so as to establish assured and statistically significant correlations (Katajajuuri 2013).

One of the targets of the above mentioned UN Sustainable Development Goal 12 aims at halving per capita global food waste at the retail and consumer level. According to the United States Environmental Protection Agency (EPA), food waste can be tackled on several levels in decreasing order of effectiveness to reach best environmental results. Firstly, by reducing it at the source through prevention activities. Secondly, by reusing it within the human food chain through food recovery activities (e.g. donations) and finally, when all these options may not be possible, by recycling it (EPA 2015). The so-called hierarchy of the three R's: Reduce-Reuse-Recycle. In suggesting to consider the waste hierarchy as 'a flexible guideline for formulating waste policies', Rasmussen et al. (2005) stress that this approach, in general, is too much focused on environmental aspects (neglecting the socio-economic aspects), yet it has the benefit of highlighting the importance of prevention activities in achieving food waste minimization.

Over recent years, targeted initiatives have been put in place at the household level and the results have been encouraging especially for countries that have benefited from well-structured awareness campaigns. For example, in the UK, avoidable food waste was reduced by 21% between 2007 and 2012, saving around £13 billion (WRAP 2013).

Among the most concrete initiatives developed at national level, the EU reported about the German Food sharing movement.[12] Surplus food management by introducing a sharing economy approach deserve, therefore, special attention since food sharing initiatives may be important pathways for more sustainable food waste behaviors.

10.5 Food Sharing and Food Waste Reduction

The concepts of food recovery and food redistribution have gained, in recent times, increasing influence within food waste reduction strategies of industrialized countries[13] (see Gram-Hanssen et al. 2016). At the same time, the food sector in general has drawn growing attention also from consumers, leading to the development of new food movements among which food sharing (Rombach and Bitsch 2015).

Beyond being recognized as an everyday inter-family practice, the sharing of food among different households has been firstly described by anthropological studies on primitive and contemporary hunter gatherer societies (e.g. Hunt 2000; Jaeggi and Gurven 2013; Ziker and Schnegg 2005). Over the last few years, there

[12]http://www.europarl.europa.eu/RegData/bibliotheque/briefing/2014/130678/LDM_BRI(2014)130678_REV1_EN.pdf.

[13]Food is redistributed for example through national food banks and local charity organizations.

has been a growing body of literature examining this practice from another perspective, since food sharing initiatives have also been rising in most developed societies through a variety of forms such as web food networks, underground restaurants, public refrigerators or simply private initiatives within specific households consisting of nonrelated people like students (e.g. Kera and Sulaiman 2014; Morone et al. 2016). Therefore, food sharing can take the form of selling as well as donating and bartering initiatives.

Most often, they are start-ups aimed at exchanging leftovers. Examples of these include Foodsharing[14]; LeftoverSwap[15]; S-Cambia Cibo,[16] even though there are also many initiatives whose main goal is to cook and eat together (e.g. Cookening; Feastly). As a result, food sharing practices have been investigated also from a social perspective with a particular focus on the relationships that may be built in urban settings where more and more citizens live in alienating conditions (e.g Kera and Sulaiman 2014). It should be noted, indeed, that many of these initiatives are developing in big cities. This is an important point since the level of urbanization is positively associated with food waste production (Secondi et al. 2015).

As emphasized in the above paragraphs on the sharing economy, new internet communication tools have been key drivers as, in most cases, it is the online platform enabling consumers to reach each other (Kera and Sulaiman 2014). In parallel, internet technologies are also playing an important role through a rising number of mobile apps specifically designed to reduce domestic waste by improving households' food management efficiency (Farr-Wharton et al. 2014).

Saving of money is among the main objectives of people participating in food sharing initiatives. However, as stated by Ganglbauer et al. (2014) in their qualitative analysis of the German community platform foodsharing.de, few members acknowledged their economic motivation. Yet, this finding is consistent with a number of studies which found that respondents, both from low and middle income classes, are generally ashamed of telling about their concerns over their economic needs (e.g. Cappellini 2009). Overall, despite a growing public awareness over environmental issues and the focus by many scholars on the environmental impact, empirical evidence showed that, in general, consumers choosing sharing economies initiatives are mostly driven by economic rather than environmental reasons (e.g. Barnes and Mattson 2016). Also concerning the specific context of food waste reduction efforts from households, the influence of ecological motivations appears to be less relevant (Quested et al. 2013; Graham Rowe et al. 2014).

Despite food sharing initiatives have attracted in recent years enthusiastic media and public attention, there have been also many criticisms. A number of weaknesses have also been identified by the literature. In testing a food sharing practice as a preventive way to reduce food waste in a University setting, Lazell (2016) found that the socio-cultural context in which food is perceived is a critical factor.

[14]See https://foodsharing.de/.

[15]See http://leftoverswap.com/.

[16]See http://www.scambiacibo.it/.

A finding in line with several studies (e.g. Evans 2012; Papargyropoulou et al. 2014). Specifically, the lack of social relationships and, consequently, of trust have shown to have the most negative impacts on food sharing practices (Lazell 2016; Farr-Wharton et al. 2014). For example, many consumers stated they do not know how the food was stored and, therefore, whether was safe. In general, leftovers 'are perceived as food that has lost its original qualities and aura' (Cappellini 2009: 370). There is, indeed, a conflicting relationship between food waste and food safety (Watson and Meah 2013; Kera and Sulaiman 2014).

Moreover, as Evans (2012) outlines, recirculation of food surplus is particularly complicated when sharing takes place outside the intimacy of the domestic setting since culinary performances and habits of people offering food are open to criticism.

Another important issue concerns participants' willingness to undertake an initiative involving great organizational efforts (Ganglbauer et al. 2014), since the practice of reusing leftovers is, already in itself, extremely challenging at the household level (Cappellini 2009).

Also the assumption that the adoption of food sharing practices automatically leads to food waste reduction is not a foregone conclusion. Although several initiatives and start-ups are being developed in Europe and US, little attention has been paid to test the effectiveness of the possible sharing of consumer-side food surplus. In this context, a first attempt to assess the existence (or not) of a casual relation between food sharing and waste reduction has been provided by Morone et al. (2016). Through a framed field experiment on twenty students sharing private accommodations, the authors assessed the impact of food sharing on waste production, controlling for several other variables influencing subject's behaviors. Specifically, preliminary results showed that sharing practices associated with food purchase and consumption could give rise to a reduction in the amount of the organic food waste for those households showing a certain degree of environmental and economic awareness (e.g. previous engagement in separate waste collection, acquaintance of food shopping expenses, etc.), adequate domestic skills (e-g. appropriate food storage, etc.) and collaborative behaviors.

Finally, Lazell (2016, p. 7) showed that the sharing of food is undermined by the fact that 'food consumption behavior is interlinked with other behaviors in the form of sets of action that determine routines and habits'. There is, indeed, a growing body of the literature which illustrates the importance of acting on collective routines and habits rather than focusing on measures based on behavior change at individual-level (e.g. Shove 2010; Moloney and Strengers 2014).

10.6 Conclusion

With global population nowadays consuming more food than ever before, the food sector has assumed even greater importance (Moomaw et al. 2012) and has been recently recognized as a strategical area for SCP implementation (Tukker et al. 2008).

In particular, there is a growing middle class from emerging countries that will increase demand of most resource intensive food, such as meat (Dobermann and Nelson 2013). This is coupled with an extremely high population growth rate in the least developed countries, which add pressure onto the food supply chain. In this context, the solution lies not only in countries' ability to increase food production. Also the excessive production of food waste needs to be addressed urgently (Garcia-Garcia et al. 2015).

Food waste at the consumer stage has become a plague to developed countries. As described above, greater consistency in its measurement still remains a major challenge since it plays a crucial role in laying down the foundations for more effective preventive and minimization actions (Garrone et al. 2014). Nevertheless, several measures have been put in place in the last years and some positive results have been achieved. More recently, a number of food initiatives related to the sharing economy model have attracted public and media interest.

The sharing economy gives us a new way of thinking based, in principle, on environmental effectiveness and economic efficiency. Our research showed that this new economic model could offer some important opportunities in accelerating the transition towards new and more sustainable consumption and production models. In the more specific context of food waste, food sharing can lead, in theory, to more efficient use of resources reducing at the same time the amount of waste produced. However, the literature points to a number of critical elements such as the rebound effect. For instance, Rutten et al. (2013) present a scenario where savings from reduced food waste are spent for other commodities and/or more expensive food such as meat. 'Therefore, the sharing economy may be presented as a tool for ecological transition only if it meets a number of conditions, such as the durability of the goods or a change in habits in relation to consumption' (Goudin 2016: 17).

Routines and habits indeed play the most critical role in food waste production (Lazell 2016). In general, according to a growing body of the literature based on social practices (e.g. Shove 2010), the prevention and mitigation measures based on behavior change at individual-level cannot effectively change consumption practices as they 'ignores how and why we do what we do, and how practices and routines come to be normal' (Moloney and Strengers 2014, p. 97). Therefore, a more ambitious policy is certainly required.

Moreover, the hypothesis themselves on which food sharing is based need to be further investigated since, as it emerged from the study of Morone et al. (2016), the correlation between food sharing practices and reduced food waste cannot be taken for granted. Since food sharing practices entail great efforts, it is also important that consumers better understand the economic benefits, in terms of savings, the use of this practice can lead to. Likewise, a great deal remains to be done to increase environmental awareness. Just as in studies on the sharing economy, also in studies on food waste it has been found that the environmental concern is still not determinant (Graham Rowe et al. 2014). Lastly, food sharing initiatives should be placed into the more general framework of food waste minimization which is extremely complex as food waste must always be perceived, and thereby tackled, all along the food supply chain not focusing only on the consumer level (Papargyropoulou et al. 2014).

References

Aschemann-Witzel J, de Hooge I, Amani P, Bech-Larsen T, Oostindjer M (2015) Consumer-related food waste: causes and potential for action. Sustainability 7:6457–6477

Bardhi F, Eckhardt G (2012) Access based consumption: the case of car sharing. J Consum Res 39:881–898

Barnes SJ, Mattsson J (2016) Understanding current and future issues in collaborative consumption: a four-stage Delphi study. Technol Forecast Soc Change 104(C):200–211

Belk R (2007) “Why Not Share Rather than Own?”. Ann Am Acad Polit Soc Sci 611:126–140

Belk R (2010) Sharing. J Consum Res 36(5):715–734

Belk R (2014) You are what you can access: sharing and collaborative consumption online. J Bus Res 67(8):1595–1600

Botsman R (2013a) The sharing economy lacks a shared definition. Co. Exist, 21 Nov 2013

Botsman R (2013b) The sharing economy lack a shared definition: giving meaning to the terms. An article produced for Fast Company, via fastcoexist.com

Botsman R, Rogers R (2010) What’s mine is yours. How collaborative consumption is changing the way we live. Harper Collins Publishers, London

Braun S (2012) Food waste report on the situation and recent activities in Germany, Stuttgart University. Retrieved online at http://ec.europa.eu/dgs/health_foodsafety/dgs_consultations/docs/ag/summary_ahac_05102012_3_s_usanne_braun_en.pdf

Cappellini B (2009) The sacrifice of re-use: the travels of leftovers and family relations. J Consum Behav 8:365–375

Cohen B, Muñoz P (2015) Sharing cities and sustainable consumption and production: towards an integrated framework. J Clean Prod 134:87–97

Dobermann A, Nelson R (2013) Opportunities and solutions for sustainable food production. background paper for the high-level panel of eminent persons on the post-2015 development agenda. Sustainable development solutions network. Retrieved at: http://www.post2015hlp.org/wp-content/uploads/2013/05/Doberman-Nelson_Solutions-for-Sustainable-Food-Production.pdf

EPA (2015) Los Angeles food recovery, links and resources. Available at: https://www3.epa.gov/region9/waste/solid/pdf/r9-food-waste-recovery-guide-los-angeles.pdf

Evans D (2012) Binning, gifting and recovery: the conduits of disposal in household food consumption. Environ Plann D Soc Space 30(6):1123–1137

FAO (2011a) Energy-smart food for people and climate issue. Paper available at: http://www.fao.org/docrep/014/i2454e/i2454e00.pdf

FAO (2011b) Global food losses and food waste—extent, causes and prevention. Available at: http://www.fao.org/docrep/014/mb060e/mb060e00.pdf

FAO (2013). Food wastage foot print: impacts on natural resources. Technical report. http://www.fao.org/docrep/018/ar429e/ar429e.pdf

Farr-Wharton G, Choi JH-J, Foth M (2014) Food talks back: exploring the role of mobile applications in reducing domestic food wastage. In: Robertson T, O’Hara K, Wadley G, Loke L, Leong T (eds) Proceedings of the 26th Australian computer-human interaction conference on designing futures: the future of design. Association for Computing Machinery (ACM), Sydney, Australia, pp 352–361

Fine F, Lucas J-L, Chardigny J-M, Redlingshöfer B, Renard M (2015) Food losses and waste in the French oilcrops sector. OCL 22:A302. doi:10.1051/ocl/2015012

Foley JAN, Ramankutty KA, Brauman ES, Cassidy JS, Gerber MJ et al (2011) Solutions for a cultivated planet. Nature 478:337–342
Frenken K, Meelen T, Arets M, van de Glind P (2015) Smarter regulation for the sharing economy. An article lead-produced by Koen Frenken, Chair of Innovation Studies at the Copernicus Institute of Sustainable Development at Utrecht University, for The Guardian, accessed on 09/05/2015 via theguardian.com
FUSIONS (2014) Fusions definitional framework for food waste. Retrieved at: http://www.eu-fusions.org/phocadownload/Publications/FUSIONS%20Definitional%20Framework%20for%20Food%20Waste%202014.pdf
FUSIONS (2016) Estimates of European food waste levels. Retrieved at: http://www.eu-fusions.org/phocadownload/Publications/Estimates%20of%20European%20food%20waste%20levels.pdf
Ganglbauer E, Fitzpatrick G, Subasi O, Güldenpfennig F (2014) Think globally, act locally: a case study of a free food sharing community and social networking. In: Proceedings of CSCW'2014. ACM Press, New York, pp 911–921
Garcia-Garcia G, Woolley E, Rahimifard S (2015) A framework for a more efficient approach to food waste management. Int J Food Eng 1(1):65–72
Garrone P, Melacini M, Perego A (2014) Opening the black box of food waste reduction. Food Policy 46:129–139
Geels FW, McMeekin A, Mylan J, Southerton D (2015) A critical appraisal of sustainable consumption and production research: the reformist, revolutionary and reconfiguration positions. Global Environ Change 34:1–12
Geng Y, Liu Z, Xue B, Dong H, Fujita T, Chiu A (2014) Emergy-based assessment on industrial symbiosis: a case of Shenyang Economic and Technological Development Zone. Environ Sci Pollut Res 21(23)
Godfray HCJ et al (2010) Food security: the challenge of feeding 9 billion people. Science 327:812–818
Goudin P (2016) The cost of non-Europe in the sharing economy, European Parliament, EPRS, European added value unit, Brussels. Retrieved at: http://www.europarl.europa.eu/RegData/etudes/STUD/2016/558777/EPRS_STU(2016)558777_EN.pdf
Graham-Rowe E, Donna C, Jessop DC, Sparks P (2014) Identifying motivations and barriers to minimising household food waste. Resour Conserv Recycl 84:15–23
Gram-Hanssen I, Jørgen Hanssen O, Hultén J, Silvennoinen K, Werge M, Stenmarck Å, Aare AK (2016) Food redistribution in the Nordic region phase II: identification of best practice models for enhanced food redistribution. Available at: http://ec.europa.eu/food/safety/food_waste/library/docs/nordic_council_report_food_donations2_201602_en.pdf
Gustavsson J, Cederberg C, Sonesson U, van Otterdijk R, Meybeck A (2011) Global food losses and food waste: extent, causes and prevention. Food and Agriculture Organisation of the United Nations (FAO), Rome
Hall KD, Guo J, Dore M, Chow CC (2009) The progressive increase of food waste in America and its environmental impact. PLoS ONE 4:7940
Halloran A, Clement J, Kornum N, Bucatariu C, Magid J (2014) Addressing food waste reduction in Denmark. Food Policy 49:294–301
Heinrichs H (2013) Sharing economy: a potential new pathway to sustainability. Gaia 22(4):228–231
HLPE (2014) Food losses and waste in the context of sustainable food systems. A report by the high level panel of experts on food security and nutrition of the Committee on World Food Security, Rome
Hunt RC (2000) Forager food sharing economy: transfers and exchanges. Senri Ethnological Stud 53:7–25
Jaeggi A, Gurven M (2013) Reciprocity explains food sharing in humans and other primates independent of kin selection and tolerated scrounging: a phylogenetic meta-analysis. Proc Roy Soc B 280:20131615
Jorissen J, Priefer C, Bräutigam K-R (2015) Food waste generation at household level: results of a survey among employees of two European research centers in Italy and Germany. Sustainability 7:2695–2715

Katajajuuri JM (2013) Food waste and related climate impacts in the Finnish food chain, Waste Along the Supply Chain OECD Food Chain Analysis Network. In: 4th meeting, 20–21 June 2013 OECD Conference Centre, Paris, France
Kera D, Sulaiman N (2014) Fridgematch: design probe into the future of urban food commensality. Futures 62:194–201
Koivupuro H-K, Hartikainen H, Silvennoinen K, Katajajuuri J-M, Heikintalo N, Reinikainen A, Jalkanen L (2012) Influence of socio-demographical, behavioural and attitudinal factors on the amount of avoidable food waste generated in Finnish households. Int J Consum Stud 36:183–191
Kummu M, de Moel H, Porkkaa M, Siebertd S, Varisa O, Wardb PJ (2012) Lost food, wasted resources: global food supply chain losses and their impacts on freshwater, cropland, and fertiliser use. Sci Total Environ 438:477–489
Lazell J (2016) Consumer food waste behaviour in universities: sharing as a means of prevention. J Consum Behav 15:430–439
Lebel L, Lorek S (2008) Enabling sustainable production-consumption systems. Ann Rev Environ Res 33:41–275
Lebersorger S, Schneider F (2011) Discussion on the methodology for determining food waste in household waste composition studies. Waste Manag 31(9–10):1924–1933
Lipinski B et al (2013) Reducing food loss and waste. Working paper, installment 2 of creating a sustainable food future. Washington, DC, World Resources Institute. Available online at: http://www.worldresourcesreport.org
Liu G (2014) Food losses and food waste in China: a first estimate, OECD food, agriculture and fisheries papers. No. 66, OECD Publishing, Paris. doi:
Martin C (2016) The sharing economy: a pathway to sustainability or a nightmarish form of neoliberal capitalism? Ecol Econ 121:149–159
Martin C, Upham P, Budd L (2015) Commercial orientation and grassroots social innovation: insight from the sharing economy. Ecol Econ 118:240–251
Mason L, Boyle T, Fyfe J, Smith T, Cordell D (2011) National food waste data assessment: final report. Prepared for the Department of Sustainability, Environment, Water, Population and Communities, by the Institute for Sustainable Futures, University of Technology, Sydney
Moloney S, Strengers Y (2014) 'Going Green?': The limitations of behaviour change programmes as a policy response to escalating resource consumption. Environ Policy Gov 24(2):94–107
Monier V, Shailendra M, Escalon V, O'Connor C, Gibon T, Anderson G, Hortense M, Reisinger H (2010) Preparatory study on food waste across EU 27. European Commission (DG ENV) Directorate C-Industry. Final report
Moomaw W, Griffin T, Kurczak K, Lomax J (2012) The critical role of global food consumption patterns in achieving sustainable food systems and food for all, A UNEP discussion paper. United Nations Environment Programme, Division of Technology, Industry and Economics, Paris
Morone P, Falcone PM, Imbert E, Morone M, Morone A (2016) Tackling food waste through a sharing economy approach: an experimental analysis. MPRA working papers. Available at: https://mpra.ub.uni-muenchen.de/70626/
OECD (2008) Promoting Sustainable Consumption, good practices in OECD countries. Available at: http://www.oecd.org/greengrowth/40317373.pdf
Papargyropoulou E, Lozano R, Steinberger J et al (2014) The food waste hierarchy as a framework for the management of food surplus and food waste. J Clean Prod 76:106–115
Parfitt J, Barthel M, Macnaughton S (2010) Food waste within food supply chains: quantification and potential for change to 2050. Philos Trans Roy Soc B Biol Sci 365(1554):3065–3081
Porpino G, Parente J, Wansink B (2015) Food waste paradox: antecedents of food disposal in low income households. Int J Consum Stud 39:619–629
Quested TE, Parry AD, Easteal S, Swannell R (2011) Food and drink waste from households in the UK. Nutr Bull 36:460–467
Quested TE, Marsh E, Stunell D, Parry AD (2013) Spaghetti soup: the complex world of food. Resour Conserv Recycl 79:43–51

Rasmussen C, Vigsø D (2005) Rethinking the waste hierarchy? Environmental Assessment Institute (Institut for Miljøvurdering), Copenhagen. Available at: http://www.dors.dk/files/media/graphics/SynkronLibrary/Publikationer/IMV/2005/waste_hierarchy.pdf
Rifkin J (2000) The age of access. Penguin, New York
Rombach M, Bitsch V (2015) Food movements in Germany: slow food, food sharing, and dumpster diving. In: International food and agribusiness management review. Available at: http://ageconsearch.umn.edu/bitstream/208398/2/201500131.pdf
Rutten M, Nowicki P, Bogaardt M-J, Aramyan L (2013) Reducing food waste by households and in retail in the EU. A prioritisation using economic, land use and food security impacts LEI-report 2013-035
Schor JB (2014) Debating the Sharing Economy. Available at: http://greattransition.org/publication/debating-the-sharing-economy
Schor JB (2015) Getting sharing right. Contexts 14–15
Secondi L, Principato L, Laureti T (2015) Household food waste behaviour in EU-27 countries: a multilevel analysis. Food Policy 56:25–40
Segrè A, Falasconi L, Politano A, Vittuari M (2014) Background paper on the economics of food loss and waste (unedited working paper). Rome, FAO
Shove E (2010) Beyond the ABC: climate change policy and theories of social change. Environ Plann A 42:1273–1285
Slagen D (2014) From P2P to B2B: the next phase of the sharing economy. Retrieved from: http://venturebeat.com/2014/09/08/from-p2p-to-b2b-the-next-phase-of-the-sharing-economy/
Tomalty R (2014) Ours is better than yours. Altern J 40(2):18–19
Tukker A, Emmert S, Charter M, Vezzoli C, Sto E, Munch Andersen M, Geerken T, Tischner U, Lahlou S (2008) Fostering change to sustainable consumption and production: an evidence based view. J Clean Prod 16(11):1218–1225
UNEP (2011) Paving the way for sustainable consumption and production the Marrakech process progress report
UNEP (2015) Sustainable consumption and production global edition. A handbook for policymakers. Retrieved at: https://sustainabledevelopment.un.org/content/documents/1951 Sustainable%20Consumption.pdf
Ventour L (2008) The food we waste. Wrap, Banbury UK
Watson M, Meah A (2013) Food, waste and safety: negotiating conflicting social anxieties into the practices of domestic provisioning. Sociol Rev 60(Suppl S2):102–120
Williams H, Wikström F, Otterbring T, Löfgren M, Gustafsson A (2012) Reasons for household food waste with special attention to packaging. J Clean Prod 24:141–148
Woolley E, Garcia-Garcia E, Tseng R, Rahimifard S (2016) Manufacturing resilience via inventory management for domestic food waste. Procedia CIRP 40:372–377
WRAP (2013) Household food and drink waste in the UK 2012. http://www.wrap.org.uk/content/household-food-and-drink-waste-uk-2012
WRAP (2015) Strategies to achieve economic and environmental gains by reducing food waste. Available at: http://newclimateeconomy.report/2015/wp-content/uploads/2016/04/WRAP-NCE_Economic-environmental-gains-food-waste.pdf
Zervas G, Proserpio D, Byers JW (2015) The rise of the sharing economy: estimating the impact of Airbnb on the hotel industry. Boston U. School of Management Research Paper No. 2013-16. Retrieved from http://people.bu.edu/zg/publications/airbnb.pdf
Ziker J, Schnegg M (2005) Food sharing at meals. Human Nat 16(2):178
Zvolska L (2015) Sustainability potentials of the sharing economy: the case of accommodation sharing platforms. Master thesis IIIEE, Lund University, Sweden. http://lup.lub.lu.se/luur/download?func=downloadFile&recordOId=8055286&fileOId=8055287

Chapter 11
Defining the Meaning of Food Waste as a Matter of Urgency

Monica Delsignore, Margherita Ramajoli and Carola Ricci

Abstract The lack of a uniform definition of waste worldwide applies to food waste as well. International organizations (including regional integration organization as the EU) and State governments refer to different definitions. In the International perspective, policy and definition of food waste have been traditionally developed by the Food and Agriculture Organisation (FAO), with the explicit intent to struggle world hunger. Nonetheless, the same initial goal of combatting global food insecurity has been changing recently adding a new perspective to the traditional narrow concept, brought in by sustainability and its broad interpretation including the circular economy target, contained in the 2015 post-Global Millennium Development Goals. The challenge is offering the chance for a clear definition, distinguishing the European framework from the International context only on the basis of the specific scope of such a peculiar Regional Integration Organization. This Article aims to demonstrate that the European definition of food waste has been targeting so far, as for the legislative perspective, the specific goal of environmental impact reduction, which is just one of the numerous aims identified within the International legal framework. The National systems in the European context ask for an unambiguous definition in order to measure and estimate in a credible, practical and consistent manner the extent of loss and waste and to identify where the loss and waste occur. A precise definition will enable countries, companies and other organizations to take sustainable decisions and program their investments. Having a clear and consistent legal framework will certainly assist

M. Delsignore (✉) · M. Ramajoli
School of Law, Università Degli Studi Di Milano-Bicocca,
Piazza Dell'Ateneo Nuovo 1, 20126 Milano, Italy
e-mail: monica.delsignore@unimib.it

M. Ramajoli
e-mail: margherita.ramajoli@unimib.it

C. Ricci
Department of Political and Social Sciences, Università Degli Studi Di Pavia,
Corso Strada Nuova 65, 27100 Pavia, Italy
e-mail: carola.ricci@unipv.it

P. Morone et al. (eds.), *Food Waste Reduction and Valorisation*,
DOI 10.1007/978-3-319-50088-1_11

businesses and regulators to make decisions on a more certain basis. That is a key factor in order to achieve the ambitious target of the circular economy. In the European Union system, there is no specific definition of food waste under the Waste Framework Directive (WFD, No 2008/98/EC). The previous Directive (No 75/442/EEC) containing a definition was amended in 1991 with the addition of "categories of waste" (Annex I) and the omission of any reference to national law. A lack of legal clarity under EU law regarding the distinction between waste and non-waste could hinder the efficient use of by-products. That is why the European Parliament has recently called on the EU Commission to develop guidance on the implementation of Article 5 of the WFD which defines by-products. The distinction between waste, by products and end of waste is a key issue in the Circular Economy Package and would also help in raising awareness among food industries, retailers and consumers.

Keywords Food waste · Food loss · Circular economy

11.1 Introduction

Fighting food waste is the most fashionable policy nowadays, especially in Italy where the Expo exhibition *Feeding the planet, energy for life* has just taken place.

Much has been written and said about the urgent need to reduce food loss and waste, although little has actually been done about it (Gonzalez Vaqué 2015).

In European Union policy makers are trying to draw a complete picture of food waste and set ambitious targets to be achieved.

Furthermore, across the world, there is growing recognition that the prevailing model of economic growth, grounded in ever-increasing resource use and pollutant emissions, cannot be sustained indefinitely.

The increased interest in food waste valorization springs from the actual situation, which is assumed becoming alarming.

Recently in Paris the Climate Change Conference COP21 highlighted the environmental impacts of food waste. Food waste is today the single largest type of waste entering landfills in most high-income countries, with a major impact on the environment (in fact, food in landfill decomposes over long periods of time, creating potent gases–like methane, a gas with the 21 times the global warming potential of carbon dioxide).

In European Union policy makers are trying to draw a complete picture of food waste and set ambitious targets to be achieved. The *Circular Economy Package* asks for food waste reduction and prevention. In order to achieve those targets Member States needs to know which substances and materials are to be considered food waste.

11.2 The Lack of a Harmonized Definition of Food Waste

As the European Environmental Agency confirmed in its annual report for 2014, Europe's food system is part of a global market in which food and animal fodder are increasingly traded across the globe. Imports of food and fodder to the EU are increasing, indicating that a considerable share of life-cycle environmental pressures and impacts related to food consumption in Europe is felt outside its borders. Food is the household consumption category with the highest embedded environmental pressures. Large amounts of food losses and food waste across the whole food chain are responsible for a considerable share of environmental impacts and a waste of resources.

Environmental impacts from food production and food waste in Europe can be mitigated through regulation and market-based instruments, including the removal of environmentally harmful subsidies.

Business and civil society have surely an important role to play through greening of supply chains and changes in consumption behavior, as economics are stressing.

Nevertheless, our point is that regulation has the key role.

The importance of default rules is evident especially when environmental issues are at stage:

> Well-chosen default rules, attentive to the full set of costs and benefits, are likely to emerge as a significant contributor to efforts to protect human health and the environment-a tool in the regulatory repertoire that is potentially more effective, in many cases, than either information and education or substantial economic incentives (Sunstein and Reisch 2013, p. 158)

European Union needs to find good and beneficial use for safe food that is presently thrown away, through different ways, but first of all needs a clear ruled definition of food waste.

Currently the legal status of food waste is still unsettled.

The uncertainty of the term *waste* has not been sufficiently examined, assuming that waste can be inserted un-problematically into existing legal frameworks. Even more, the distinction between waste and not waste is sometime misunderstood although the jurisprudence seeks to address the problem. Definitions of food waste differ a lot because what food waste consist of has a strong impact on its qualification. This contributes to create even more uncertainty as regulatory authority is dispersed across different actors and level of government (Baldwin et al. 2012).

As it will be explained, food waste apparently finds different declinations in International, European and National law. The only explicit definition can be found at the International level, even if it is not a globally valid definition, whether other levels do not define specifically food waste.

In the International perspective, policy and definition of food waste have been traditionally developed by the Food and Agriculture Organisation (FAO), with the explicit intent to struggle world hunger. That peculiar perspective doesn't find an equivalent in the other systems. Nonetheless, the same initial goal of combatting global food insecurity has been changing recently adding a new perspective to the traditional narrow concept, brought in by *sustainability* and its broad interpretation having implications for the same international definition of food waste and losses.

The Article aims to demonstrate European definition of food waste is targeting, as for the legislative perspective, the specific goal of environmental impact reduction, which is just one of the numerous aims identified within the International legal framework.

What is at stake is that a considerable amount of food is being discarded and that food waste give rise to both environmental and ethical problems and economic and social costs, so that measures have to be taken towards halving food waste and preventing generation of bio-waste. This perspective is far away from the initial FAO notion of food waste but not so distant from the more recent one contained in the post-Global Millennium Development Goals and the challenge is offering a clear definition, distinguishing the European framework from the International context only on the basis of the specific scope of such a peculiar Regional Integration Organization.

The National systems in the European context ask for an unambiguous definition in order to measure and estimate in a credible, practical and consistent manner the extent of loss and waste and to identify where the loss and waste occur. A precise definition will enable countries, companies and other organizations to take sustainable decisions and program their investments. Having a clear and consistent legal framework will certainly assist businesses and regulators to make decisions on a more certain basis. That is a key factor in order to achieve the ambitious target of the circular economy.

The first paragraph will examine the evolution of the definition of food waste in International law. The International level is targeting the *Zero Hunger Challenge*, especially after 2012 and Rio+20, recognizing interconnectedness of world's food systems and impact on poverty, hunger, malnutrition, natural resources and climate and lately adding a new perspective of the circular economy, inviting all the States, civil society and other stake-holders not only to prevent and reduce waste but also to reuse and recycle. That perspective became finally in some way convergent with the actual European standpoint of sustainable choices. As it will be explained in paragraph two, the European Community had been considering, for a long time, food just as a good in the market. The Commission, only few years ago, within the Communication of 30 July 2010, establishing the High Level Forum for a Better Functioning Food Supply Chain, pointed out the importance and centrality of Agro Food Industry for the competitiveness and functioning of internal market. Since last decade, the European legislator has mainly spent its efforts towards introducing rules to protect water from waste and for waste management (Bruno 2014) and food waste has not been for long time a specific issue.

Just recently, in the resolution of the 19 January 2012,[1] the European Parliament reasons on how to avoid food wastage. The resolution underlines health and environmental implications since unconsumed food make a major contribution to

[1]P7_TA(2012)0014.

global warming and food waste produces methane, which as a greenhouse gas is 21 times more powerful than carbon dioxide. The Parliament calls 'on the Council, the Commission, the Member States and players in the food supply chain to address as a matter of urgency the problem of food waste along the entire supply and consumption chain'.

Finally in 2014, the Commission has drafted a proposal for amending waste directives, underlining the necessity of providing the 'definitions of municipal waste, food waste, backfilling' in the Directive 2008/98/EC.

Elaborating this new definition, looking at the circular economy targets, represents a central issue for a sustainable food supply chain, as turning food waste into a resource is an essential part of increasing resource efficiency and closing the loop and countries, companies and research laboratories need a clear definition of food waste in order to plan their investments and research.

11.3 Food Waste in International Law: From 'Other' to the '3R' Approach

The definition of food waste being crucial, one would expect to find at least a uniform concept within the global legal framework. Waste is formally defined in different national jurisdictions where 'definitions relate to particular points of arising and are often framed in relation to specific environmental controls'(Parfitt et al. 2010, pp. 3065–3081) even if it is true that it generally remains 'a stigmatized Other, [...] to be kept at bay', which '[...] has long been systematically dissociated, separated, and isolated from production, distribution, and consumption, [...] kept away, not merely from eyes and nostrils, but even from awareness and consciousness'(Corvellec 2015, p. 12). With the exception of such a common feeling though, a general legal definition is missing.

As Fig. 11.1 shows, the closest try can be found within the UN when it established its specialized agency in 1945, the Food and Agriculture Organization (FAO), fixing reduction of food losses within its mandate to tackle the plague of food insecurity. By 1974, the first World Food Conference identified reduction of post-harvest losses as part of the solution in addressing world hunger (FAO 1981). Consequently, it established the *Special Action Programme for the Prevention of Food Losses*. The main focus was initially on reducing losses of durable grain; by the early 1990s, the scope of work had been broadened to cover roots and tubers, and fresh fruits and vegetables (FFVs). Poor adoption rates for interventions led to the recognition that a purely technical focus was inadequate for solving problems within the sector and a more holistic approach was developed (FAO 2002). There has been no account of progress towards the post-harvest loss reduction target, since in 2011, FAO asked the Swedish Institute for Food and Biotechnology (SIK) to update data in order to determine a more recent situation profile (Gustavsson et al. 2011). Finally, within the Swedish study (at p. 2) a specific

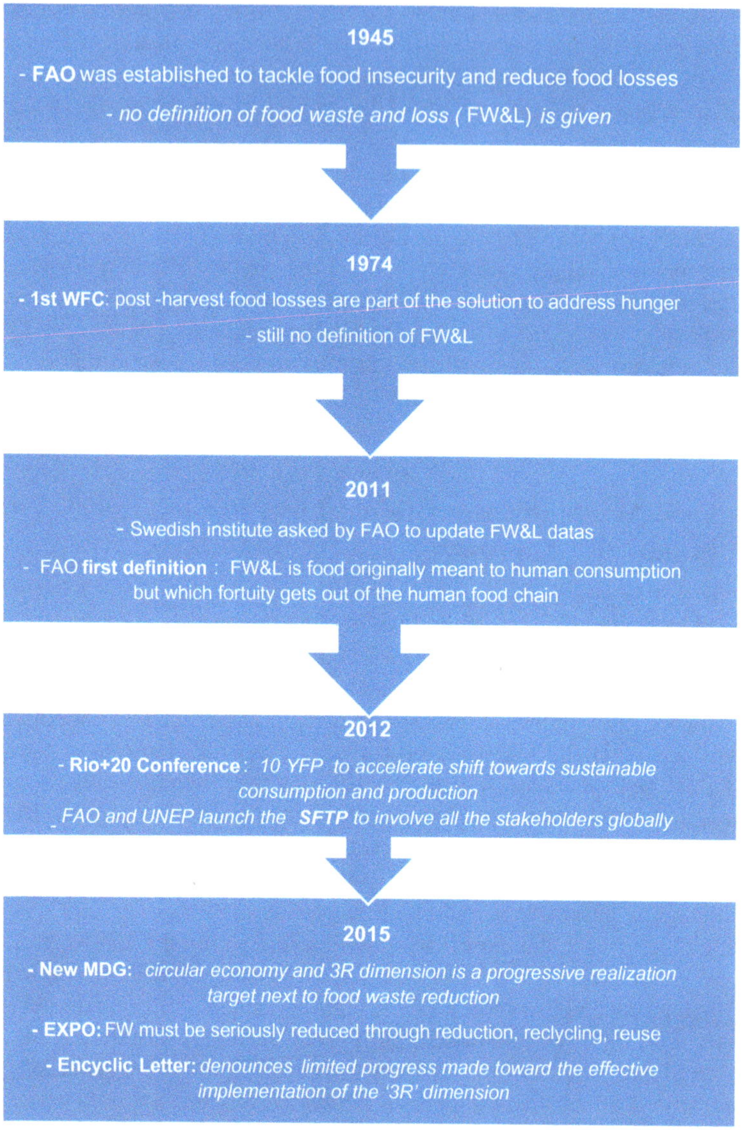

Fig. 11.1 Most relevant phases in the attempt to define food waste in the international context

definition of food waste was given, distinguishing '*planned non-food uses* to *unplanned non-food uses*, which are hereby accounted under losses', implying therefore that 'food that was originally meant to human consumption but which fortuity gets out the human food chain is considered as food loss or waste even if it is then directed to a non-food use (feed, bioenergy…)' (Gustavsson et al. 2011, p. 2). In the Swedish Study there was no specific reference to the challenging opportunity for 'food got out of the human food chain' to be reused and recycled in order to contribute to circular economy efficiency; the analysis was ending in defining the visible dimension of such 'Other-than-edible-food' as waste and loss, being the main focus, as it has been already mentioned, to reduce food insecurity.

Nonetheless, in just few years, approaching the 2015 checkpoint for the *Zero Hunger challenge* set in the Millennium Development Goals (MDG), the same concept of 'the right to *adequate* food' included in the binding 1966 International Covenant on Economic Social and Cultural Rights (ICESCR) has been developed broadly as to include *sustainability*. Article 11(2)(a), in fact, recognizes the fundamental right of every individual to be free from hunger and the duty of the State to adopt, individually and through international cooperation, all the measures deemed to be necessary as to assure this right through the improvement of food production and the conservation and distribution methods, using the *most* advanced technical and scientific know-how and implementing those reforms of the national agricultural systems required to obtain *the best degree of development and employment of resources* (Snyder 2006, Van der Meulen 2010, Ricci 2013).

Such a holistic vision of sustainability to be assured through the entire food chain had been re-elaborated recently after the Rio+20 Conference which reaffirmed that sustainable consumption and production is a cornerstone of sustainable development within the *10-Year Framework of Programmes on Sustainable Consumption and Production Patterns* (10YFP).[2] This is a global framework of action to enhance international cooperation to accelerate the shift towards sustainable consumption and production (SCP) in both developed and developing countries. The framework supports capacity building, and facilitates access to technical and financial assistance for developing countries for this shift. The 10YFP 'aims at developing, replicating and scaling up SCP and resource efficiency initiatives, at national and regional levels, decoupling [...] economic growth from the rising rates of natural resource use and the environmental impacts that occur in both consumption and production stages of product life cycles'. Sustainable food systems are key to ensuring sustainable development; thus UNEP and FAO, within the 10YFP initiative, building on previous work under the so called Agri-food Taskforce, jointly developed the Sustainable Food Systems Programme (SFSP). The two organizations conducted a global survey on SFSP during June-July 2014. This was a public consultation widely disseminated among stakeholders and open to all. It included general questions regarding the proposed goal for the SFSP, challenges, opportunities and key issues for making food systems more sustainable as well as an invitation to express interest to participate in the programme.

---

[2]See the Resolution 67/203 adopted by the General Assembly on 21 December 2012, Implementation of Agenda 21, the Programme for the Further Implementation of Agenda 21 and the outcomes of the World Summit on Sustainable Development and of the United Nations Conference on Sustainable Development; the Resolution 68/210 adopted by the General Assembly on 20 December 2013, Implementation of Agenda 21, the Programme for the Further Implementation of Agenda 21 and the outcomes of the World Summit on Sustainable Development and of the United Nations Conference on Sustainable Development.

Such a globally shared vision has been confirmed recently during the EXPO2015, on the occasion of the adoption of the post-2015 development agenda. On 25th September 2015, the UN announced the seventeen *Sustainable Development Goals* (and 169 targets) that should 'build on the *Millennium Development Goals* and complete what they did not achieve'; the General Assembly states clearly that 'they are integrated and indivisible and balance the three dimensions of sustainable development: the economic, social and environmental' (see the third indent of the Preamble). The challenges and commitments identified at preceding major conferences and summits, held since the 1992 *Rio Declaration on Environment and Development*, are recognized as being interrelated requiring a new approach and integrated solutions, given that

> Sustainable development recognizes that eradicating poverty in all its forms and dimensions, combating inequality within and among countries, preserving the planet, creating sustained, inclusive and sustainable economic growth and fostering social inclusion are linked to each other and are interdependent' (para. 13) (UN 2015).

Within the Declaration the '3R' dimension of the circular economy firstly appears among the progressive realization targets, next to food waste reduction. In fact, under goal twelfth, entitled 'Ensure sustainable consumption and production patterns', all countries and all stakeholders, acting in collaborative partnership, commit themselves not only to

> 'halve *per capita* global food waste at the retail and consumer levels and reduce food losses along production and supply chains, including post-harvest losses' (target no. 12.3), but also to 'substantially reduce waste generation through *prevention*, *reduction*, *recycling* and *reuse* (target no. 12.5) (UN 2015)

To this extent, on the private perspective, companies, especially large and multinationals, are encouraged to adopt sustainable practices and to integrate sustainability information into their reporting cycle (target 12.6), while on the public perspective, public procurement practices that are sustainable, in accordance with national policies and priorities, should be promoted (target 12.7). The same approach is visible in the *Milan Charter* and its annexes, presented to Kofi Annan at the very end of October as the legacy of EXPO2015,[3] affirming that whenever food is not re-usable for human consumption it is important to think to food as a new source not only for bio-energy production but also for diversified markets, recommending a 'use-not-waste' strategy for future food policies. The serious urgency in 'counteracting' the currently prevailing 'throwaway culture' is specifically addressed also in the 2015 *Laudato Si'* Encyclical Letter, where the limited progress made toward the effective implementation of the '3R' dimension are expressly denounced since globally

[3]See the Report elaborated by a specific working group, the no. 15, focused on 'Vietato sprecare' (i.e., 'waste prohibited') visible at http://carta.milano.it/wp-content/uploads/2015/04/42.pdf).

> [...] we have not yet managed to adopt a circular model of production, capable of preserving resources for present and future generations, while limiting as much as possible the use of non-renewable resources, moderating their consumption, maximizing their efficient use, reusing and recycling them (Pope Francis 2015 points 20–22).

11.4 Food Policy and Waste Proliferation in EU

Examining European food law and policy, it can be found they surely had a not irrelevant role in the increasing production of food waste.

EU food policy developed over three major phases, that can be labelled as the internal market phase, the social phase and, nowadays, the sustainable development phase, moving from a technocratic approach to a multi-levelled often participatory policy space.

For a long time, European Law considered food just as a good: free movement was the target to be achieved. After the famous decision *Cassis de Dijon* (Case n. 120/1978), the principle of mutual recognition effectively opened national markets, obtaining through equivalence the result that previously could not be obtained through legislative harmonization. It was applied to objective characteristics of food products and progressively extended to the use of names. Even the jurisprudence of the Court of Justice shows this approach, aiming at privileging the circulations of food products through the European Market. Unfortunately though, measures for harmonization in free movement of goods and services are likely to restrict the Member States' regulatory powers to protect environment. As recently shown (de Sadeleer 2014), there is an awkward relationship between environmental and market issues, and the principle of integration of environment protection across EU policies (Allena and Fracchia 2011) implies a high level of protection and improvement of the existing regulatory framework.

By the end of the '80s with the *Smanor* judgement on yogurt (Case n. 298/1987), a new approach stresses the importance of valorising quality of food and peculiarity of different products, even if environmental protection is not at stage yet.

Technological innovation has played a decisive role in bringing about a radically different relation with food (Albisinni 2014). Change has come about both in the quantity of food provisioning and in the quality of the processing, conservation and logistics of food stuffs.

As well, crisis-such as BSE, dioxin, horse meat or, even recently, cucumber played an important role in developing a different sensibility for health protection.

We come, then, to what I called the social approach, as food is not merely a good in the market. It is therefore necessary to adopt measures aimed at guaranteeing that unsafe food is not placed on the market and at ensuring that systems exist to identify and respond to food safety problems in order, not only, to guarantee the proper functioning of the internal market, but also, to protect human health. This new demand calls for a proper risk regulation which comes with Regulation No. 178/2000, the *General European Food Law*, establishing EFSA (European

Food Safety Authority) (Abels et al. 2014) and covering a wide range of provisions with a direct or indirect effect on the safety of food and feed, including provisions on materials and articles in contact with food, animal feed and other agricultural inputs at the level of primary production.

Nevertheless, food security and safety policies played and play an important role in waste proliferation. Aesthetic defects and quality standards affect the level of food waste, as out-graded vegetables and fruits cannot be sold.

Furthermore the discard rate is seriously increasing as consumers become more and more exigent on features and aspect of food. As well, provisions on food safety and security often require proper storage facilities either safe transportation conditions or particular packing proceedings, which are not always available or too expensive for small producers.

Even quantitative restrictions in production and commerce of agricultural products played and still do play a not minor role. The regulatory framework of the common agricultural policy (CAP) contains instruments and measures which surely influence the waste production, although agricultural issues and waste issues are usually discussed in very different contexts.

Procedural rules concern and control, in example, sugar, milk and potato starch quotas, or limit production potential in the wine, oranges or crops sector. Overproduction has significant consequence in food waste: When production exceeds demand, that means surplus crops, oranges and milk will be sold to processors or as animal feed or just thrown away.

Environmental issues and concerns have been progressively addressed within the boundaries of the internal market legal basis, finally acknowledging environmental protection as one of the Community's essential objectives and Article 3(3) of the Treaty on European Union specifically mentions 'the sustainable development of Europe' and a 'high level of protection and improvement of the quality of the environment'. So we come to what we called the *Sustainable development phase*.

The recent Communication from the European Commission, *Towards a circular economy: A zero waste programme for Europe*, goes even further. It proposes a non-binding target for a reduction in food waste of at least 30% by 2025, in addition to the development, inter alia, of national food waste prevention strategies (EC 2014). We turn, now, to the new approach and to the phase, which I call the sustainable development phase.

Loss and wastage occur at all stages of the food supply chain and value chain. A more economically, socially and environmentally sustainable food system in Europe would imply healthier diets, less food waste and the production and consumption-including from imports-of higher–quality food with lower impacts on climate change and biodiversity in particular.

In addition globalization, deriving from the decline of state authority, the increased role of civil society, and the promotion of self-regulatory mechanisms in a decentralized and multipolar society, complicates the legislative framework, assuming a different perspective, as we will see in detail in next paragraphs.

11.5 The New European Approach

In the *Roadmap to a Resource Efficient Europe* (COM (2011) 571 fin.) the necessity of turning waste into a resource is considered as a priority. Without clear definitions, the prospects for making the waste policy operational in any meaningful way could be bleak. The qualification of a substance or a material as a waste implies for the holder the acquisition of a legal status, which leads to a series of obligations (Pocklington 2011). Each involved company/entity/person in the management of the waste must act in a binding and subject to control check manner, under rules of conduct laid down in National and Community legislation, which expose to possible consequences in case of their violation or breach of the principles mentioned pursuant to Art. 191 of TFUE.

The definitional problems get bigger with the implementation of the waste directives as different national perspectives add, to the variety of legislative levels, variety of jurisdictions and variety of contexts for law implementation and law enforcement.

Waste directives apply to the operators of food supply chain, concerning dissimilar industrial activities.

Under the *Waste Framework Directive* (WFD, No 2008/98/EC), though, there is no specific definition of food waste: waste is generally defined as 'any substance or object which the older discards or intends or is required to discard' at Article 3(1). As well known, the previous Directive (No 75/442/EEC) was amended in 1991 with the addition of 'categories of waste' (Annex I) and the omission of any reference to national law. This was not really helpful in defining waste and therefore the new Directive changed the approach.

Article 2 excludes a number of categories from the scope of the directive, such as the gaseous effluents emitted into the atmosphere. Moreover Art. 2(2) contain an exception from categories insofar as these are covered by other EU legislation. That means, as concerning food waste, the directive does not apply to Biomass as Art. 2, lett. b) of dir. 2001/77/EC defines *biomass* as the biodegradable fraction of products, waste and residues from agriculture (including vegetal and animal substances), forestry and related industries, as well as the biodegradable fraction of industrial and municipal waste. Regarding food waste and considering the whole food chain, Article 2(1)(f) WFD excludes, as well,

> faecal matter, if not covered by paragraph 2(b), straw and other natural non-hazardous agricultural or forestry material used in farming, forestry or for the production of energy from such biomass through processes or methods which do not harm the environment or endanger human health.

The Court of Justice has addressed the point on the exact meaning of waste on numerous occasions.

The *ARCO Chemie Nederland/Epon* (Cases C-418/97 and C-419/97) underlined that the concept of waste is to be interpreted broadly in the light of the Framework Directive's objective of ensuring a high level of environmental protection. Defining the term *discard* has been proven particularly difficult. In *Vessoso and Zanetti*, at first, confirmed in *Tombesi*, the Court ruled that the system of supervision and control established by the directive is intended to cover all objects and substances discarded by their owners, even if they have a commercial value and are collected on a commercial basis for recycling, reclamation or re-use. Only the CJ's judgements can help in establishing the boundaries of the concept of waste even if the CJ does not establish general principles, but simply decide single cases referring to similar precedents.

In the end, one major problem remains: it is still not clear what is the correct interpretation of *to discard* (Jans and Vedder 2012). In order to restrict the field, the WFD introduced the definition of by-product, which is not a waste as stated in a number of previous ruling of CJEU (*Palin Granit*, Case C-9/00 and *Saetti,* Case C-235/02). According to Article 5, a substance resulting from a production process the primary aim of which is not the production of that substance may not be waste, but a by-product if the following cumulative conditions are met:

- Further use of the substance or object is certain
- The substance or object can be used directly without any further processing other than normal industrial practice
- The substance or object is produced as an integral part of a production process
- Further use is lawful, i.e. the substance or object fulfils all relevant product, environmental and health-protection requirements for the specific use and will not lead to overall adverse environmental or human health impacts.

This definition equally refers to general terms and concepts which again called for intervention of the CJ's jurisprudence and of the European Commission, adopting a Communication on waste and by-products.

The interpretative Communication (COM(2007)59 fin.) tried to establish a clear distinction between materials that can be considered as non-waste by-products and those that should be treated as wastes. 'In reality, there is not a black and white distinction, but rather a wide variety of technical situations with widely differing environmental risks and impacts and a number of grey zones' (Keele 2001).

The Community approach on waste management is founded on the fact that a balance must be struck between the need to ensure the functioning of the internal market, on the one hand, and to attain a high level of environmental protection, on the other. If a successfully balance is reached, the Community's waste management can contribute significantly to the pursuit of sustainable development, the Commission noting that 'waste production is one of the best indicators of our progress' towards this goal (EU Focus on Waste Management 1999).

The Commission talks in terms of hierarchy of aims in waste management with prevention being the priority, followed by recovery and then lastly by safe disposal. Specifically, the Community's waste management policy is underpinned by 3 keys objectives. First of all, prevention: the most effective manner in which to eradicate or reduce the impact of waste on environment and human health is to prevent waste being generated in the first place. This can be achieved inter alia by using clean technologies, prohibiting or limiting dangerous substances in products, improving consumer information and providing education. Waste prevention as a priority clearly underlines the need to integrate environmental concerns into the production process. Secondly, the recovery (which contains preparing for re-use; recycling and other recovery). In the event the waste is unavoidably generated, it should be recovered. In this way material can be re-used, or alternatively, waste can either be recycled (material recovery) or utilised as a source of energy (energy recovery). Finally, safe disposal of waste is admitted only when unavoidable.

The choice is not however to be determined exclusively on environmental grounds-whilst the best environmental solution should be considered, account should also be taken of economic and social costs.

If a material is not a waste, this does not mean that it falls completely out of the system of environmental protection set down in Community law. Product based regulation, and other legislation such as the *REACH Regulation* aim at protecting human health and the environment from the potential environmental impacts of products and other materials that are not wastes.

In the waste policy, as synthetically resumed, there is not though a specific consideration for food waste, neither the definition of waste finds a peculiar declination for materials coming from the food chain.

Nevertheless, the European Commission seems taking the issue of tackling food waste very seriously, even without a clear definition of food waste. And that, in our opinion, is a weak point that can prejudice all the given efforts and good purposes.

The Commission is analyzing in close cooperation with industry, consumer and other NGOs, food sector experts and Member State policy experts how to reduce food waste without compromising food safety and it is discussing options for possible EU actions.

The Commission is itself contributing to awareness raising on food waste prevention through production of communication materials. If all these initiatives can move social opinion or consumers behaviors, EU will not achieve the ambitious target of reducing food waste without a clear regulation and a definition of food waste valid for all the Member States.

In 2010, the European Commission set up the High Level Forum for a Better Functioning Food Supply Chain with the aim of assisting the Commission with the development of industrial policy in the agri-food sector by: following the recommendations of the High Level Group on the competitiveness of the Agro-Food Industry (HLG); and implementing the initiatives set out in the 2009 Communication *A better functioning food supply chain in Europe*.

In particular, the HLG recognised the importance of a holistic approach to ensuring the competitive position of the EU's agri-food sector. It acknowledged the need for consistency between all policy areas affecting the EU food chain: agriculture, food safety, nutrition and health, environment, trade, financial markets, research and innovation, and industrial policy more generally. In the meantime, the major factors determining the competitiveness of the whole food supply chain have been analysed extensively.

In 2011, the Commission's *Roadmap to a resource-efficient Europe*,[4] identified food as a key sector where resource efficiency should be improved and called for ambitious action to tackle food waste.

In 2014, the Commission's Communication *Towards a circular economy: a zero waste programme for Europe*, and the related legislative proposal to review recycling and other waste targets put forward objectives for food waste reduction in the EU. It included a proposal for Member States to develop national food waste prevention strategies with the aim of reducing food waste by at least 30% by 2025. Sectors concerned included: manufacturing, retail/distribution, food service/ hospitality and households.

In its 2015 work programme, the Commission has announced that it would withdraw its legislative proposal on waste targets and replace it by end of year with a new, more ambitious proposal to promote circular economy. Withdrawal of the proposal was formalised following consultation of the European Parliament and Council.

The Commission reflected on the scope of the new proposal to promote circular economy including actions to prevent food waste. In order to help inform its work, the Commission has launched a public consultation inviting contributions from citizens, organizations and public authorities. The consultation closed on 20 August 2015.

In December 2015, with the Communication COM (2015) 614 final *Closing the loop*, the Commission formulated the proposal for amending the *Waste Framework Directive*, but yet the proposal does not contain a proper definition of food waste.

The concept of food waste can be found at Art. 9 which is dedicated to *the Prevention of waste,* saying that:

> 1. Member States shall take measures to prevent waste generation. These measures shall: [...] reduce the generation of food waste in primary production, in processing and manufacturing, in retail and other distribution of food, in restaurants and food services as well as in households

Still no definition is offered by the legislator, even if food waste is supposed to occur across the entire food supply chain, from the agricultural production stage to the storage, processing, distribution, management and consumption stages. The Commission states it 'will elaborate a common EU methodology to measure food waste in close cooperation with Member States and stakeholders'. The EU is committed to meeting the target of halving per capita food waste, but it is difficult to measure food waste missing the definition of what is considered food waste.

[4]COM (2011) 571 fin., 20 September 2011.

11.6 Food Waste in National Law

If we now turn to the national perspective, defining food waste becomes a central issue.

In this Multilevel System, National Law cannot be considered in isolation, but must be read and understood in interaction with the broader policy.

Furthermore, especially in environmental provisions, hard law gives way to built-in flexibility; horizontal regulation is preferred to hierarchical systems. It is axiomatic that the effectiveness of legal norms will depend in part at least on extent to which they are effectively enforced (Craig and de Burca 2007) and only clear provisions can be effectively enforced.

As a result of an EU policy in favour of subsidiarity, EU environmental law consists more of directives, and more specifically framework directives, than regulations. The provisions of these framework directives are mostly worded in very general terms, whilst regulations may be extremely precise. Therefore Member State authorities have broad discretion in the choice of form and appropriate means for implementing EU law.

> In tolerating–let alone encouraging administrative diversity, these directives keep uniformity at bay. Clearly, the extent of such a discretion compounds the difficulties faced by the European Commission in verifying the compliance of EU environmental law in 28 Member States (de Sadeleer 2014, p. 195).

That is what we observed going through the waste legislation and lastly with the Waste Framework directive (Tromans 2001).

In the previous paragraph we saw the ECJ, instead of establishing an exhaustive definition, has placed the onus on national courts to determine on a case-by-case basis whether or not a substance should be regarded as waste, and accordingly made subject to relevant regulatory standards. In this sense, the ECJ has favoured the need for environmental protection at the risk of distorting the functioning of the internal market.

Whilst is evident that a broad definition of waste is envisaged, a lack of clarity as to its precise meaning continues to provide difficulties in national implementation of the directive.

The definitions of disposal and recovery operations contained in the Waste Framework directive are general and leave a relatively substantial leeway for interpretation which can be used to undermine obligations for waste destined for disposal by sending this waste to operations which may or may not be recovery operations, as the Commission itself states (Communication from the Commission: Towards a thematic strategy on the prevention and recycling of waste. COM (2003) 301 final, p. 20).

Furthermore, a lack of legal clarity under EU law regarding the distinction between waste and non-waste could hinder the efficient use of by-products. Even the Guidelines on the interpretation of key provisions of Directive 2008/98/EC–guidelines which are not legally binding, but adopted by the DG Environment in July 2012 in order to clarify key concepts-testify the numerous questions raising from the interpretation and implementation by national authorities and private economic operators. Nevertheless they do not solve all the present problems.

These implementation problems and criticism by various stake holders have materialised in litigation both at European and national level as subsidiarity and proportionality leaves spaces to national Authorities in implementing directives.

We do not have to forget that failure to fulfil obligations under the Directives of waste involves financial penalties, as happened in Italy for the waste management in the region of Campania, penalties recently confirmed by the Court of Justice, case C-653/13. Furthermore national provisions, considering, in example, Italian legal system, establish administrative and criminal penalties for treating substances and materials without the authorization of waste management.

Those consequences show how central the definition of waste is for the responsibility and liability in waste treatment and management and how this uncertainty influence even business choices introducing risks and unsolvable doubts.

The inability of the Community Institutions to provide a clear and unambiguous definition of waste has forced operators to a frantic search of the requirements that allow to evade the general administrative regime for treating the substance they are dealing with. Public authorities and judges have generally discouraged such initiatives, convinced that the policy of *all waste* was the most appropriate to ensure effective environmental protection and to prevent circumvention of the discipline. They did not realize that-on the other hand-the increasing amount of the mass of the materials to be disposed of as waste makes it much more difficult to ensure an efficient supervisory activity for the real and proper waste (Dell'Anno 2013).

The debate about the definition of waste is likely to continue, but it must be kept in mind the definition of waste is a legal construction which it may well be possible to improve and that clarification especially for food waste is a fundamental step to let food chain operators be capable of planning the best practices and methods to treat their residues in the production.

Waste legislation is often transposed in a highly decentralised manner in Member States, including on the regional or local level and in multiple legal acts, depending on the administrative structures of the Member States.

> The risk of incorrect transposition and implementation of the Directive, complicate the Commission's task of monitoring the application of EU law. Clear information with respect to the transposition of the revised waste Directives is instrumental in ensuring the conformity of national legislation with their provisions (COM (2014) 397 final).

Lacking a European definition of food waste Member States have started adopted their own.

In example in United Kingdom the Waste and Resources Action Programme (WRAP) has defined food waste as 'all food and drink discarded throughout the entire food chain'. This definition confirms countries, in the European, area do not care about food waste with the scope of reducing food hunger and malnutrition, but aim at reducing the environmental impacts of food waste applying the circular economy to the food sector.

Recently, in France as well supermarkets will be banned from throwing away or destroying unsold food and must instead donate it to charities or for animal feed, under a law set to crack down on food waste.

The French law goes even further than the UK, where the government has a voluntary agreement with the grocery and retail sector to cut both food and packaging waste in the supply chain, but does not believe in mandatory targets.

11.7 Conclusion

The distinction between waste, by products and end of waste is a key issue in the Circular Economy Package, if we want that food waste, or even generally, waste reduction is not a political slogan but a legal rule (de Sadeleer 2014).

There is an inherent and evident difference in the definition of food waste: food waste can be the all food discarded in the entire food chain, as provided by WRAP, or food wasted just at the point of consumption, as in the FAO provisions. Food waste must be food still perfectly edible, which, in the absence of any alternative use, is eliminated or disposed of or European policy have the target to reduce food wasted and loosen trough the all food supply chain?

In the idea of circular economy distinguishing what is waste and what is not waste is, of course, a prerequisite for adopting legislative measures. Food waste in this circular economy approach refers not only to discarded or unused edible food, but fits to the all food supply chain, considering all its steps. Turning food waste into a resource is an essential part of increasing resource efficiency and closing the loop in a circular economy: food can become waste through the all the phases of the supply chain.

The establishment of specific food waste targets for Member States, as part of the waste prevention targets to be reached by Member States, call for a clear definition of waste, with a particular attention for biofuels and bio waste.

The Commission aims to facilitate ‘the clustering of activities to prevent by-products from becoming wastes (industrial symbiosis)’. Therefore a clear and univocal definition of waste, distinguishing by products, is a key issue in order to let the policy be effectively.

Only through collaboration on scientific activities, through greater dialogue and effective communication, through shared activities and shared best practices by the scientific organisations and institutes across Europe, the common necessity of a clear definition of food waste could be satisfied.

The definition must keep in mind the different ingredients of food waste, preparing a general directive for all food waste so that all the different food chain operators have a clear idea of what they can do with the different discards they have to manage with.

References

Abels G, Kobusch A, Trasch J (2014) Scientific regulatory cooperation within the EU: on the relationship between EFSA and national food authorities. In: Alemanno A, Gabbi S (eds) Foundations of EU food law and policy. Ten years of the European food safety authority. Ashgate Publishing Limited

Aginam O (2006) Food safety, South-North asymmetries and the clash of regulatory regimes. (2006/07) VJTL 40.4:1099–1111

Albisinni F (2014) Food law and innovation in European Law. In: Studi in onore di Luigi Costato, vol II. Napoli, p 3

Allena M, Fracchia F (2011) Globalization, environment and sustainable development, in global, European and Italian perspectives. Riv It Dir Pubbl Com 781

Baldwin R, Cave M, Lodge M (2012) Understanding regulation, theory, strategy and practice. Oxford, p 373

Bell S (1999) Refining the definition of waste. 1 ELR 2, 283

Bruno F (2010) Environmental issues in food law. In: Studi in onore di Luigi Costato, vol II. Napoli 2014, p 115

Cheyne I (2002) The definition of waste in EC law. 14 JEL 1 61

Chyau J (2009) Casting a global safety net—a framework for food safety in the age of globalization. FDLJ 64, 313 et seq

Craig P, de Burca G (2007) General editors' preface. In: Wenneras P (ed) The enforcement of EC environmental law, Oxford

Davies PGG (2004) European Union Environmental Law. Ashgate, An Introduction to Key Selected Issues 217

de Sadeleer N (2014) EU Environmental law and the internal market. Oxford

Dell'Anno P (2013) Disciplina della gestione dei rifiuti. In: Trattato di diritto dell'ambiente vol II. Padova, pp 162, 179–180

FAO Statistics Division (1981) Food loss prevention in perishable crops. FAO Agric Serv Bull no. 43

FAO (2002) Post-harvest losses: discovering the full story. Overview of the phenomenon of losses during the post-harvest system. FAO, Agro Industries and Post-Harvest Management Service, Rome

Fracchia F (2010) Lo sviluppo sostenibile, La voce flebile dell'altro tra protezione dell'ambiente e tutela della specie umana. Napoli

Gonzalez Vaqué L (2015) Food loss and waste in the European Union: a new challenge for the food law? Eur Food Feed Law Rev 20

Gustavsson J, Cederberg C, Sonesson U, van Otterdijk R, Meybeck A (2011) Global food losses and food waste. Extent, causes, prevention. Rome

Corvellec H (2015) New directions for management and organization studies on waste, GRI-rapport 2015: 3, Managing Big Cities

Jans JH, Vedder HHB (2012) European environmental law. European Law Publishing, p 476

Keele H (2001) When does waste cease to be waste. 12 ELR 1, 3

Kramer L (2003) The distinction between product and waste in community law. 2 Env Liab 1:3

Lundqvist J, de Fraiture C, Molden D (2008) Saving water: from field to fork—curbing losses and wastage in the food chain, in SIWI Policy Brief. Stockholm

Mariani, MT (2007) The intersection of international law, agricultural biotechnology, and infectious disease. Leiden-Boston

Negri S (2009) Food safety and global health: an international law perspective. Glob Health J III.1, 1 et seq

Parfitt J, Barthel M, Macnaughton S (2010) Food waste within food supply chains: quantification and potential for change to 2050. Philos Trans R Soc B 365:3065–3081

Pocklington D (2011) The law of waste management. Sweet and Maxell, London

Pope Francis (2015) Encyclical letter 'Laudato Si' of the Holy Father Francis on Care of Our Common Home', 24 May 2015, points 20–22
Poto M (2008) Regulations on food safety and the role of African actors in the global arena. J Afr Int Law I.1:107–130
Ricci C (2013) International law as a meta-framework to protect the right to food. In: Lupone A, Ricci C, Santini A (eds) The right to food toward a global governance. Torino, p 9 et seq
Sunstein CR, Reisch LA (2014) Automatically green: behavioral economics and environmental protection (September 19, 2013). Harvard Environ Law Rev 38(1)
Snyder F (2006) Toward an international law for adequate food. In: Mahiou A, Snyder F (eds) La sécurité alimentaire/food security and food safety. Leiden-Boston, p 79 et seq
Tromans (2001) EC waste law—a complete mess? 13(2) JEL 133
van der Meulen B (2010) The global arena of food law: emerging contours of a meta-framework. Erasmus L Rev 3.4, 217 et seq
UN (United Nations) (2015) Transforming our world: the 2030 agenda for sustainable development. Resolution adopted by the general assembly on 25 A/RES/70/1. United Nations, Geneva

Chapter 12
Waste Reduction in Fresh Food Supply Chains: An Operations Research Approach

Gabriella Dellino, Renato Mari and Carlo Meloni

Abstract Sustainability has a high priority for all actors in modern supply chains, and food waste issues attract significant political, market and media consideration. Many retailers have setup programs aimed at tackling it, while the food industry has also launched programs including waste reduction among their main goals. Indeed, food waste is already a crucial theme, and its importance is growing in these years. In the last decade, retailers have achieved relevant progresses in reducing the amount of food wasted in their stores as well as along distribution networks. Nevertheless, there is still room for further improvements: better forecasting, more careful assortment and order decisions, suitable policies promoting products' freshness, and shelf life management can yield significant waste reductions. Besides, retailers can help to reduce waste along the supply chain through closer collaboration with other upstream actors. This chapter considers methods and models devoted to waste reduction in fresh food supply chain operations to be included in a Decision Support System, and presents a case study on a real supply chain dedicated to fresh and perishable packaged products, involving a set of retailers with both small and medium sized stores located in the Apulia region (Italy). Optimization is a crucial issue in such a context and the main criticalities are related to the uncertainty on future sales. This study proposes an integrated and flexible approach that accounts for the following issues: demand forecasting, order planning and delivery optimization. The aim is to support the decision maker to determine operations plans with respect to waste reduction and other different

G. Dellino (✉) · R. Mari · C. Meloni
Istituto per le Applicazioni del Calcolo "Mauro Picone", CNR,
Via G. Amendola 122, 70126 Bari, Italy
e-mail: g.dellino@ba.iac.cnr.it

R. Mari
e-mail: r.mari@ba.iac.cnr.it

C. Meloni
e-mail: meloni@poliba.it

C. Meloni
Dipartimento di Ingegneria Elettrica e dell'Informazione,
Politecnico di Bari, Via E. Orabona 4, 70125 Bari, Italy

P. Morone et al. (eds.), *Food Waste Reduction and Valorisation*,
DOI 10.1007/978-3-319-50088-1_12

criteria, such as shortage, freshness and residual stock of products. Results are reported and discussed enlightening both quality of forecasting and its effects on the order planning activity. The results show the potential benefits of the proposed approach to pursue the waste reduction in the distribution and the retailer supply chain and the possible extensions to contribute to the recovery of fresh food surplus.

Keywords Fresh food supply chain · Waste reduction · Mathematical modeling · Optimization · Operations research

12.1 Introduction

Modeling, forecasting and optimization are key methodologies in the field of complex logistics (Dotoli et al. 2003, 2005, 2006). From an operations research perspective, the fresh food supply chain represents a very interesting applicative context, characterized by several interrelated variables and constraints, and possible sources of uncertainties. Indeed, fresh food is a strategic sector where, in order to increase margins on the overall business activity, an effective management of the logistics operations is needed, and perishability is a particularly critical issue (Alvarez and Johnson 2011; Jacobs et al. 2014; Nahmias et al. 2011).

Unfortunately, the retail models often view waste as a part of doing business. Nevertheless, food waste reduction and an adequate shelf life management are getting a pivotal role in the fresh food supply chain operations management (Garrone et al. 2014).

In this context, the short shelf-life of fresh food products is an important limiting factor which is relevant for different aspects: (i) fresh products cannot be stored for a long time; (ii) the production frequency of fresh products is relatively high compared to other, non-perishable products; (iii) when the product exceeds its nominal shelf-life, it cannot be sold or it can be sold with sensible discounts (in the first case, the cost of wastage is added to lost revenues); (iv) the consumer's requirements for fresh products call for frequent deliveries to the retail stores; (v) stock-out rates present a challenging problem for manufacturers, distributors, and retailers involved in the supply chain.

High stock levels in the stores contribute to contrast the stock-out rates. However, when setting the inventory levels for fresh products, a risk component must be considered to account for stock obsolescence and to avoid (or limit) product outdating and wastage (Jacobs et al. 2014; Nahmias et al. 2011). Usually, consumers evaluate the freshness of a product on the basis of its remaining shelf-life (for packaged products, in particular). If there are not technological methods to extend the shelf-life, the manufacturer and the distributor can only try to produce and supply as closely as possible to the time when demand occurs, in order to guarantee a longer shelf-life to the consumer. Indeed, this effort produces advantages for both the manufacturer, the retailer and the consumer. Firstly, product wastage can be decreased in the retail store as the shelf-life deadlines are less stressful. Secondly, a longer product's shelf-life diminishes the risk of stock outdating in the store, and the overall inventory level of the product can be increased helping to decrease the stock-out rates. Thus, shelf-life should be highly considered at all the levels of the supply chain operations management, and delivering products that are as fresh as possible should become one of the main objectives. From the retailers' point of view, the importance of fresh products can be different. A discounter with a limited assortment, high product volumes and inventory turns should be less concerned about shelf life than a smaller, more traditional retailer with a broad product range and low inventory turns. Therefore, a flexible approach is required in order to consider the characteristics of customers, stores, and products.

12.1.1 Literature Review

The literature clearly shows how sales forecasting plays an even more important role in the market of fresh and highly perishable food, since the shelf life of products is very limited and reliable forecasts are fundamental to reduce and manage inefficiencies such as stockout and outdating (Jacobs et al. 2014; Nahmias et al. 2011).

Packaged fresh food production potentially generates considerable waste through poor planning of operations (Brandenburg et al. 2014). This problem is more relevant for products that have a very short shelf life, and a variable demand. In fact, different supply chain phases are based on forecasted volumes and considerable wastes can be created (as indicated in Fig. 12.1). The research reported in this chapter mainly refers to the retailer stage, indicated in red in Fig. 12.1).

In the last years, different mathematical modeling approaches have been proposed to offer quantitative methods to the decision makers. The optimization issues in the supply chain composed of retailers and potential recipients that practice the food recovery are addressed in Aiello et al. (2014), Muriana (2015). Analytical methods devoted to reduce the overproduction wastes in the convenience food production are proposed in Darlington and Rahimifard (2006), Darlington et al. (2009). Wang et al. (2009) develop approaches to integrate traceability initiatives with operations management objectives for perishable food products. While Van Der Vorst et al. (1998) investigate the effects of supply chain management on logistical performances in food supply chains showing the crucial role of the reduction (or elimination) of uncertainties to improve the overall behavior of the chain. In such a context, a robust supply chain operations management can only be obtained by taking uncertainties of future demand into account and for this reason a good and reliable forecasting plays a crucial role (Dellino et al. 2010, 2012; Fleischmann et al. 2002). At this aim, an extended version of the classical newsvendor problem has been proposed by Huang (2013), to account for specific issues related to random demand and item deterioration over time. Van Donsellar

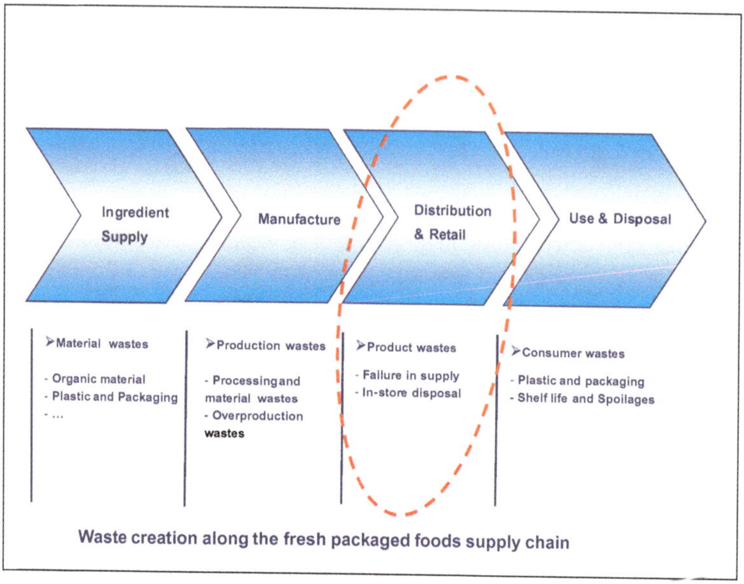

Fig. 12.1 Wastes generation in the supply chain

et al. (2006) investigate inventory control policies for perishable items in supermarkets, providing directions for improving the automated store ordering system currently in use in two Dutch supermarket chains. An important issue related to demand uncertainty across the supply chain is the bullwhip effect (Jacobs et al. 2014). Ma et al. (2013) study the impact of the bullwhip effect on both product orders and inventory, comparing performance of different techniques for demand forecasting.

The close interactions between demand forecasting and inventory management performance are addressed by Babai et al. (2013) focussing on their impact on both inventory costs and service levels. Borade and Sweeney (2015) propose a Decision Support System (DSS) based on an evolutionary algorithm to tackle an inventory routing problem, jointly considering both inventory management and transportation issues for a two-stage logistic system. The study conducted in Rijpkema et al. (2014) shows that in perishable product supply chain design a trade-off should be made between transportation costs, shortage costs, inventory costs, product waste, and expected shelf life, suggesting to adopt a multi-criteria approach. Kaipia et al. (2013) show that the sustainability (in terms of waste reduction) of the perishable food chain can be improved by more efficient information sharing.

A review of the state-of-the-art in the area of planning models for the different components of agri-food supply chains is offered in Ahumada and Villalobos (2009). While the links between sustainability in food supply chains and quantitative methods to support the decision makers are analyzed in details in the surveys (Beske et al. 2014; Soysal et al. 2012).

12.1.2 Contribution

The chapter proposes a fresh food supply chain modeling approach devoted to support a fresh food logistics network and the related operations management. This approach conducts to an integrated and flexible DSS for sales forecasting, order planning and delivery optimization in fresh food supply chain management. The DSS combines a pre-processing module to identify seasonality and noise emerging from historical sales, a forecasting module to derive sales forecasts (for each single item), supported by a module for the automatic tuning of the forecasting model, a multi-objective optimization module with a best alternative selection module, to derive the best order proposal based on a set of Key Performance Indicators (KPIs), and a module implementing an integer mathematical optimization model devoted to the delivery planning.

The order proposal per store is based on the individual item sales forecast (based on a short-term forecast ranging over seven days) for the individual store, possibly taking into account exogenous information (e.g., prices and promotions). The forecasting activity involves a large number of items and stores combinations and is designed as an automatable procedure within the DSS.

The requirements for this forecasting implementation are twofold: (a) an information system to guarantee that (daily) data are available for each item/store pair; (b) effectiveness and speed of the forecasting and optimization modules in the DSS system. Nowadays, most supply chains and even stores already have access to suitable information systems with the possibility to implement and use forecasting and optimization modules. On the other hand, the user or the decision maker in the store should not be exposed to the complexity of these systems (Fleischmann et al. 2002; Meyr 2002). Therefore, an adaptive forecasting system is designed to automatically configure the best performing model. The process is repeated for each individual item in the store and for each forecast session; in this way the model's choice is obtained by setting a relatively small number of simple user's parameters. The proposed approach allows to reduce variability along the supply chain with an accurate demand forecasting and considers an integrated and flexible approach that also accounts for order planning and delivery optimization. The proposed models are able to support the decision maker to determine operations plans with respect to waste minimization and other relevant criteria, such as shortage, freshness and inventory levels.

The chapter is structured as follows. Section 12.2 introduces the proposed DSS. The successive sections describe the modules which compose the DSS. Section 12.3 illustrates the forecasting models considered for the demand prediction. Next, Sect. 12.4 presents the multi-objective approach adopted to determine the order plans, while Sect. 12.5 describes the delivery optimization model. The following Sect. 12.6 reports on some computational results for the considered case study. Conclusions are drawn in Sect. 12.7, also outlining different future research directions.

12.2 A Decision Support System for Fresh Food Supply Chain Operations

In this Section we describe the proposed DSS, whose main building blocks are the following: the first one addresses sales forecasting; the estimated sales forecasts are then used as input for a multi-objective optimization algorithm to define the best order policy, and finally a delivery optimization module is devoted to determine the delivery plan. We analyze real data coming from a set of small and medium sized retailers operating in Apulia region, Italy. Data are pre-processed in order to extract significant information (such as seasonality), identify and remove noise, and apply normalization. The pre-processing phase includes some sophisticated approaches based on independent component analysis (Hyvärinen et al. 2001), that are typically used in signal and image processing, to filter data and remove noise, whenever relevant. Two different forecasting models are applied to the pre-processed data. One of the best known classes of mathematical models for time series forecasting is represented by the Autoregressive Integrated Moving Average (ARIMA) models. They are widely used for several reasons: (a) they are considered as one of the best performing models in terms of forecasting, (b) they are used as benchmark for more sophisticated models, (c) they are easily implementable and have high flexibility due to their multiplicative structure. Nevertheless, they do not take the effect of exogenous variables into account. Indeed, in many cases it may be useful to investigate the impact of external phenomena. In particular, in the case study under investigation, sales of fresh goods are highly influenced by prices and the impact of the latter on the forecasting process should be considered. In the literature several alternative approaches can be found that make forecasting more robust and reliable by including the effect of exogenous variables. The most straightful approach is to adapt an ARIMA model to account for the aforementioned variables, obtaining the so-called ARIMAX model (the interested reader is referred to Box et al. (2008), Makridakis et al. (2008). In this Chapter we adopt another, yet common, technique based on the identification of a transfer function (TF) relating the time series of the variable of interest with the one of the exogenous variable. In this work, we apply ARIMA and TF models on our data set. Specifically, the two models are identified and estimated by varying parameters in predefined intervals. The best parameter setting is selected according to an exhaustive tuning framework that is based on a set of statistical indicators.

The tuned forecasting models provide the input data for the multi-objective optimization method. This is designed to be user-interactive and to provide a Pareto front of optimal order proposals according to some crucial KPIs for fresh and perishable products such as outdatings, stock-out and freshness of goods. We propose to compute the order proposal through a meta-heuristic approach, based on a genetic algorithm, that considers forecasted sales as a proxy of demand. Then, the user (i.e., the manager) specifies the criteria (or even the priority among different criteria) for selecting the best alternative among the pool of non-dominated solutions identified by the optimization module. Therefore, the DSS provides the order

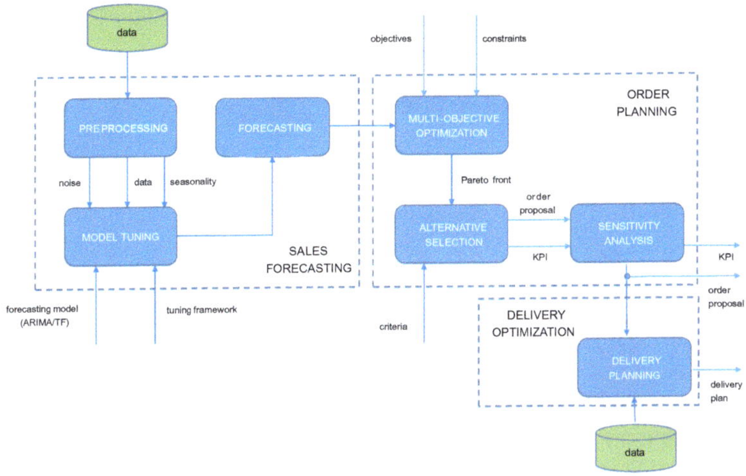

Fig. 12.2 The schematic architecture of the proposed DSS

proposal to be implemented on the basis of the manager preferences; e.g., assigning weights to the KPIs to reflect specific inventory management policies. Demand uncertainty is also taken into account through a simulation analysis to assess the impact of demand perturbations on the optimal order quantity. Another module implementing an integer programming model is dedicated to the determination of the optimal delivery plan taking into account order proposals as well as transportation issues.

A scheme describing the architecture of the proposed DSS is displayed in Fig. 12.2, while each of the next three sections illustrate the three main modules of the DSS.

12.3 The Demand Forecasting Module

In this Section we introduce the two adopted alternative forecasting methods and their main properties. Both methods are available in the literature (Box et al. 2008; Makridakis et al. 2008), nevertheless a unique procedure for tuning their parameters does not exist. To this aim, we design and implement a three-step tuning algorithm that will be described in detail in Sect. 12.3.1. A time series can be considered as the realization of a stochastic process that is observed sequentially over time. Thus, once a time series of data is collected, it is possible to identify a mathematical model to describe the stochastic process.

The first approach we consider is based on the ARIMA models (Box et al. 2008) which represent one of the best known classes of mathematical models for time series forecasting. Let z_t be an observation of time series at time t, and let a_t be a random variable with normal distribution (having zero mean and variance equal to σ_a^2). An ARIMA model with *seasonality* is defined as follows:

$$\phi_p(B)\Phi_P(B^s)\nabla^d\nabla^D z_t = \theta_q(B)\Theta_Q(B^s)a_t,$$

where B is the backward shift operator which is defined by $Bz_t = z_{t-1}$ and $\phi_p(B)$, ∇^d, $\theta_q(B)$, $\Phi_P(B^s)$, ∇^D and $\Theta_Q(B^s)$ can be considered as polynomial in the B operator of order p, d, q, P, D, Q respectively.

Thus, a seasonal ARIMA model is synthetically described as ARIMA$(p, d, q)\times (P, D, Q)_s$, where the latter parameters have to be estimated. The reader is referred to Box et al. (2008) for a comprehensive description of ARIMA models.

The most critical disadvantage affecting classical ARIMA models with seasonality is that the effect of exogenous variables on data is overlooked. In the case study under investigation, sales of fresh goods are highly influenced by prices, therefore considering them as exogenous variables may significantly improve the forecasting quality. To this end, we adopt an alternative approach based on TF models.

TF models are based on the assumption that the relation between time series and exogenous variables can be modeled by a transfer function (to be estimated) plus an error vector described by an ARIMA model. More formally,

$$z_t = \frac{\omega(B)B^b}{\delta(B)}x_t + n_t,$$

where the transfer function is defined by v zeros, r poles and a delay b, with $\omega(B)$ and $\delta(B)$ polynomials in B of degree v and r, respectively, and the vector of errors n_t is described by an ARIMA model. Unlike the previous model, in TF models additional parameters have to be estimated, namely parameters v, r and b. Therefore, a TF model will be univocally defined by a tuple $(p, d, q) \times (P, D, Q)_s \times (v, r, b)$, whose values will be identified as discussed in what follows.

12.3.1 A Three-Step Tuning Algorithm

In this Section we provide the outline of the algorithm for tuning the two forecasting models introduced above. The reader is referred to Dellino et al. (2015) for a detailed description. According to Box et al. (2008), given a data set, the best forecasting model can be identified following a three-step framework: (i) model identification, (ii) model estimation and (iii) diagnostic check.

This is a very general framework in which every step can be implemented in many different ways, see for instance Höglund and Östermark (1991). We propose to implement the model identification through an exhaustive approach in which every parameter ranges in predefined intervals. Given a tuple $(p, d, q) \times (P, D, Q)_s$ for ARIMA models and a tuple $(p, d, q) \times (P, D, Q)_s \times (v, r, b)$ for TF, the numerical value of parameters is estimated according to the maximum likelihood principle.

The data set comes from a supply chain comprised of 30 retailers (small and medium sized stores) located in Apulia region (Italy) and a set of 300 fresh and perishable packaged products. The available data is made of three year sales for each store from 2011 to 2013. Data set is formed—for each item/store pair—by gathering Point-of-Sales (POS) information from the store's cash register systems (daily sales and prices) and inventory records (daily stock on hand and deliveries) from the retailer's information systems.

For the sake of implementation of the diagnostic check, we divide—as usual—the given data set into two different sets: a training set and a test set (Box et al. 2008). The training set represents the set of observations used to estimate the forecasting model and its parameters. Once the model has been tuned, we derive forecasts on the test set. Then, we compare forecasted sales with real sales over this set, that provides the forecasting horizon. Thus, the diagnostic check is implemented by means of two kinds of performance indicators, in-sample and out-of-sample, that are used to determine the best model.

In the proposed case study, our models have a forecast horizon of seven days and adopt a training set of three months. The size of the test set is chosen as large as the forecast horizon.

The in-sample indicators are computed on the training set and are mostly used as lack of fit measures, based on the information entropy and parsimony of models. Specifically, we used the Ljiung-Box test and the Hannan-Quinn Information Criterion (HQC) (Box et al. 2008; Burnham et al. 2002). The out-of-sample indicators are well known statistical indicators for quality and accuracy of forecasting and they are computed on the test set. Let us define f_t as the forecast at time t and let us consider a test set of length n. We compute the following indicators:

Root mean squared error (RMSE)	$\sqrt{\frac{1}{n}\sum_{t=1}^{n}(z_t - f_t)^2}$
Mean absolute error (MAE)	$\frac{1}{n}\sum_{t=1}^{n}\lvert z_t - f_t\rvert$
Maximum absolute error (MaxAE)	$\max_{t=1,\ldots,n}\lvert z_t - f_t\rvert$
Mean absolute percentage error (MAPE)	$100 \cdot \frac{1}{n}\sum_{t=1}^{n}\left\lvert\frac{(z_t - f_t)}{z_t}\right\rvert$
Maximum absolute percentage error (MaxAPE)	$\max_{t=1,\ldots,n}\left\lvert\frac{z_t - f_t}{z_t}\right\rvert$
Coefficient of determination R^2	$1 - \frac{\sum_{t=1}^{n}(z_t - f_t)^2}{\sum_{t=1}^{n}(z_t - \mu)^2}$

where μ is the average value of z_t over the test set. There are many works in the literature concerning with statistical indicators. The reader is referred to Armstrong (2001), Makridakis and Hibon (2000), Makridakis et al. (2008) for a complete overview. Among non-dominated models with respect to in-sample indicators, we select the model with minimum MAE concluding the diagnostic check and, hence, the model selection. Alternative (possibly multi-objective) criteria could be chosen; however, our choice for the MAE was suggested by experts' opinion, as retailers are mainly interested in minimizing the absolute deviation from actual sales.

12.4 The Order Planning Module

Once sales forecasts become available through one of the forecasting models described in the previous Section, they can be considered as a proxy of expected demand and are provided as input to an algorithm whose objective is to identify an optimal order planning policy according to multiple (and often conflicting) objectives, namely minimizing stock-outs and waste, as well as maximizing the quality of service perceived by customers, in terms of product freshness, while keeping residual stock levels under control. More specifically, these indicators are computed as follows:

(A) *waste* provides the overall number of items that must be discarded along the forecasting horizon due to outdating;
(B) *freshness* is computed by tracking the age of the product when sold to the customer, and then averaging over the sales in the forecasting horizon;
(C) *stock-outs* express the cumulative unmet demand at the end of the forecasting horizon;
(D) *residual stock* corresponds to the items remaining in stock at the end of the forecasting horizon.

We formulate a multi-objective optimization problem based on the aforementioned KPIs, in order to derive a plan covering the whole forecasting horizon (typically, a week). The proposed formulation accounts for the following issues:

- Lot size constraint; i.e., orders are allowed only in multiples of a minimum order quantity.
- Fixed delivery date; i.e., orders can be delivered only in given days. The number and frequency of weekly deliveries are established by the supplier and taken as input.
- Lead time, which determines when an order has to be placed in order to meet the delivery requirements.

The problem is computationally tackled through a solver, based on an evolutionary algorithm implemented in Matlab, which adopts a variant of NSGA-II

(Deb et al. 2001). This meta-heuristic approach, characterized by its flexibility, has already been successfully employed in different contexts (Dellino et al. 2007; Naso et al. 2006). As a result of the optimization process, we obtain a set of Pareto optimal solutions (Ehrgott et al. 2005), which are non-dominated with respect to the four identified objectives. Among these solutions, management will choose the final (i.e., to be implemented) order plan either on the basis of additional information or specifying a rule to apply.

Thus, the following step of the order planning module is the selection of a single order plan among all the non-dominated solutions computed by the meta-heuristic approach. To this aim, the Alternative Selection Module is included in the DSS. This module takes two inputs: the Pareto front of optimal solutions and a set of criteria defined by the user. Alternative criteria may be specified in order to take management's preferences into account. Among the most widely used selection rules we cite the following two: (i) we introduce an aggregated objective function such as a weighted sum of the four KPIs, then the non-dominated order policy minimizing the aggregated objective function is selected; (ii) we sort KPIs by relevance and we iteratively select the best order plan according to the first KPI, then to the second KPI and so on. It is clear that almost any kind of criterion may be introduced in the DSS. Moreover, the management's priority may change over time or it may depend on the specific kind of item and store, thus the DSS should be as much flexible as possible in order to satisfy different user's requirements. The best alternative selection module is introduced to this specific aim: without altering the overall structure and behaviour of the DSS, the user may define the most suitable selection criteria fitting his/her needs. The optimal order proposal is based on the forecasted demand provided by the forecasting module of the DSS.

As demand might deviate from its forecasted value, it may be helpful for the management to evaluate the impact of demand uncertainty on the computed order proposal, in terms of possible KPIs variation. Therefore, we perform an ex-post analysis in which, given the optimal order plan, the DSS simulates demand variability and evaluates KPIs deviation. In particular, we simulate system performances for different realizations (say, N) of daily demand. To this aim, we estimate the distribution of daily demand based on predictions provided by each forecasting model. In particular, we assume daily demand to be normally distributed (Jacobs et al. 2014; Nahmias et al. 2011) with mean equal to the forecasted value and coefficient of variation equal to the MAPE associated to the forecast. Then, we sample N observations from the estimated distribution and compute the KPIs associated to the optimal order proposal when demand equals each sampled realization. Comparing the value of the KPIs for the base scenario to the N values of the same KPIs for the alternative scenarios, we are able to estimate how sensitive the order plan selected by our DSS is to demand variability. In particular, a small variability in the KPIs would be preferable to denote a stable order plan with a limited impact of demand variability on system's performances (Dellino et al. 2010, 2012, 2015).

12.5 The Delivery Optimization Module

In this Section we detail a mathematical formulation to address the delivery problem. The forecast of the expected sales for each store, along with the order plans, are an input for the problem. Thus, the only issue is how orders have to be satisfied and, in particular, how many trucks are necessary to accomplish the delivery service. Our main operative assumption is that there exists a consolidated list of missions for serving stores. A mission is a sequence of stores that have to be served, i.e., every mission contains information about the stores that have to be visited during the delivery trip starting from and coming back to the depot. In the worst case, the set of missions is comprised of all possible subsets of stores that have to be served. Hence, without loss of generality, we assume that the latter list is known by the decision maker and it represents an input for the mathematical model. The rationale of the model is that the decision maker has to select the best missions among all the available ones in order to fulfill all the orders with the minimum number of trucks.

Two operative constraints are included into the model: the truck capacity and a maximum work shift length (in hours) for each driver. The latter is a realistic constraint and depicts a classical trade off between the maximum number of trucks and the maximum amount of working hours for each driver.

We assume that a set L of missions, a set H of trucks ($|H| = |L|$), a set K of items and a set I of stores are given. Let us introduce the following input data:

- m_{li} Parameter equal to 1 if mission l includes store i and 0 otherwise
- o_{ik} Quantity of item k ordered by store i
- cap_h Capacity of truck h
- τ Unitary handling cost for unloading goods from a truck
- d_l Travel time associated to mission l
- WS Work shift (in hours) of trucks' drivers

For ease of formulation we define a matrix mission–store $M = L \times I$, such that $M = (m_{li})$.

Furthermore, we define the following variables:

- y_h Equals 1 if truck h is used and 0 otherwise
- α_l Equals 1 if mission l is used and 0 otherwise
- q_{ikl} Quantity of item k delivered to store i through mission l
- c_l Load shipped through mission l
- T_l Time to complete mission l, including travel and unloading time

The delivery planning problem is formulated as follows:

$$\min_{y} \sum_{h\in H} y_h \tag{1}$$

$$\sum_{l\in L|m_{li}=1} q_{ikl} = o_{ik} \quad \forall i \in I, k \in K \tag{2}$$

$$\sum_{i\in I|m_{li}=1} \sum_{k\in K} q_{ikl} = c_l \quad \forall l \in L \tag{3}$$

$$c_l \leq cap_h \quad \forall h \in H, l \in L \tag{4}$$

$$\alpha_l \leq \mathcal{M} \cdot \sum_{i\in I|m_{li}=1} \sum_{k\in K} q_{ikl} \quad \forall l \in L \tag{5}$$

$$\mathcal{M} \cdot \alpha_l \geq \sum_{i\in I|m_{li}=1} \sum_{k\in K} q_{ikl} \quad \forall l \in L \tag{6}$$

$$\sum_{i\in I|m_{li}=1} \sum_{k\in K} q_{ikl} \cdot \tau + d_l \cdot \alpha_l = T_l \quad \forall l \in L \tag{7}$$

$$\sum_{l\in L} T_l \leq WS \cdot \sum_{h\in H} y_h \tag{8}$$

$$\sum_{h\in H} y_h \leq \sum_{l\in L} \alpha_l \tag{9}$$

$$y_h \in \{0,1\}, \alpha_l \in \{0,1\}, c_l \geq 0, T_l \geq 0 \quad \forall l \in L \tag{10}$$

$$q_{ikl} \geq 0 \quad \forall i \in I, k \in K, l \in L \tag{11}$$

The objective function (1) is the minimization of the number of trucks necessary to accomplish the service. Constraint (2) imposes that the quantity of item k shipped to store i by means of all missions l is equal to the ordered quantity, while constraint (3) defines the meaning of the auxiliary variable c_l, i.e., the total amount of load shipped by a truck that carries out mission l. The latter can not exceed the truck capacity, as stated by constraint (4). Constraints (5) and (6) define the logical relation between variables α_l and q_{ikl}: for every mission l if no item is delivered to any store then mission l is not selected and α_l equals 0, otherwise if at least one q_{ikl} is positive variable α_l is forced to be 1. This relation is expressed by means of a big-$\mathcal{M}$, i.e., a sufficiently large constant that, along with the binary variable α_l, allows to formulate the following logical relation: $\alpha_l = 1 \quad \forall\, l \Leftrightarrow \exists i, k \text{ s.t.} q_{ikl} = 1$. Constraint (7) defines the auxiliary variable T_l as the total duration of every mission l, including travel time ($d_l \cdot \alpha_l$) and unloading time ($\sum_{i\in I|m_{li}=1} \sum_{k\in K} q_{ikl} \cdot \tau$). Note that, if a mission l is not selected, both q_{ikl} and α_l are 0 and the duration $T_l = 0$.

Constraint (8) imposes that the average duration of each selected mission does not exceed the work shift of truck drivers: the constraint can be restated as $\frac{\sum_{l \in L} T_l}{\sum_{h \in H} y_h} \leq WS$ in which the left-hand side is the ratio between the total missions' duration and the number of truck/drivers. Finally, constraint (9) forces the total number of trucks to be less than or equal to the total number of selected missions. This constraint is necessary because it may happen that, in case of infeasible solutions due to a too strict work shift constraint, the model may compute a solution with a number of trucks that is larger than the number of missions only to increase the right-hand side of constraint (8) and make it satisfied. Imposing constraint (9) we assure that, in the optimal solution provided by the formulation, for each truck there is always at least one mission to carry out. Constraints (10) and (11) define the variables of the mathematical formulation.

A possible future extension of the model may include the mission duration into the objective function and a multi-objective mathematical formulation may be considered to explicitly take into account the trade off between size of truck fleet and work shift.

12.6 Computational Results

The proposed approach is composed of subsequent stages and the accuracy of the results obtained in the early phases plays a crucial role. In this Section we report on the application of the proposed DSS to an illustrative case study based on a set of real sales data. We focus on the role of the DSS for sales forecasting and order planning in fresh food supply chain management in order to avoid or limit the food waste also with respect to other relevant KPIs. The data collection on the delivery planning problem is still ongoing and we plan to test also the module devoted to solve that issue as soon as all required information will become available. However, this does not affect the results of the conducted analysis.

The sales data set comes from a supply chain, located in the Apulia region (Italy), and comprised of 30 retailers (i.e., small and medium size stores) and a set of 300 fresh and perishable products. The available data is made of three year sales for each store from 2011 to 2013 and refers to fresh packaged products belonging—in different percentages—to the following eight food categories: yogurt and cheese (50%), milk/cream (16%), pasta (10%), desserts (8%), cold cuts (5%), specialties (2%), ready meals (1%), others (8%).

For ease of presentation, we selected three representative products with different features. We chose products showing the highest sales volume belonging to the two biggest categories (namely, yogurt and cheese and milk/cream), as they appear to be the most representative, covering about 2/3 of the whole data set:

- milk: 1 L of milk, selected for being a common fresh item, with a very short shelf life;
- mozzarella cheese: 250 g of mozzarella cheese, having a medium shelf life and supplied as single item.

Moreover we select also a product having high sales volumes and a medium-long shelf life:

- smoked salmon: 200 g of smoked salmon.

The selected products have different shelf life and lot sizes, and are required to be supplied in specific delivery days, in particular:

- milk: shelf life = 4, lot size = 12, delivery days = {Mon, Wed, Thu, Sat}
- mozzarella cheese: shelf life = 18, lot size = 1, delivery days = {Mon, Thu}
- smoked salmon: shelf life = 51, lot size = 10, delivery days = {Tue, Sat}.

The modules composing the DSS are implemented in the Matlab environment. All the proposed forecasting models are tested and results on the selected illustrative cases are reported.

After the multi-objective optimization algorithm computes the Pareto front of solutions, it is necessary to define a criterion for the best alternative selection module of the DSS. In our experimental analyses, we test the weighted sum of KPIs as the selection criterion. Note that different combinations of weights can be used to reflect the corresponding management's policies. A first campaign of experiments refers to the case in which the decision maker judges the waste reduction as prioritary with respect to the other KPIs. The results are summarized in Tables 12.1, 12.2 and 12.3.

Table 12.1 Results for milk

	ARIMA	TF
(v,r,b)		(1,1,0)
(p,d,q)	(1,0,0)	(0,0,0)
$(P,D,Q)_S$	$(1,1,1)_7$	$(1,1,1)_7$
RMSE	2.8	3.7
MAE	2.5	3.1
MaxAE	4.6	7.0
MAPE(%)	14.2	19.5
MaxAPE(%)	30.6	58.2
R^2	92.4	86.4
time	233	224

	ARIMA	TF
Thu	60	48
Sat	36	36
Mon	36	48
Wed	12	12
st_out	0	0
waste	0	0
fresh	2.5	2.2
resid	14	14

Table 12.2 Results for mozzarella cheese

	ARIMA	TF
(v,r,b)		(2,1,0)
(p,d,q)	(1,1,2)	(2,0,0)
$(P,D,Q)_S$	$(1,1,1)_7$	$(1,1,0)_7$
RMSE	4.7	3.1
MAE	3.9	2.6
MaxAE	7.8	5.7
MAPE(%)	44.5	31.9
MaxAPE(%)	66.8	54.9
R^2	8.8	59.5
time	245	231

	ARIMA	TF
Mon	32	24
Thu	34	27
st_out	0	1
waste	0	0
fresh	2.3	2.5
resid	0	0

Table 12.3 Results for smoked salmon

	ARIMA	TF
(v,r,b)		(1,1,0)
(p,d,q)	(1,0,0)	(0,0,0)
$(P,D,Q)_S$	$(1,1,1)_7$	$(1,1,1)_7$
RMSE	6.9	5.2
MAE	5.0	4.1
MaxAE	13.7	10.2
MAPE(%)	15.5	13.1
MaxAPE(%)	35.2	26.7
R^2	80.2	88.9
time	219	241

	ARIMA	TF
Tue	110	110
Sat	80	80
st_out	0	3
waste	0	0
fresh	2.3	2.2
resid	4	0

For each product, on the left we compare ARIMA and TF in terms of the above described out-of-sample indicators, along with the computational time required by the three-step tuning algorithm; on the right we report the quantities ordered in the delivery days according to the order planning algorithm and the corresponding KPIs, namely *waste* for waste, *fresh* for freshness, *st_out* for stock-outs, and *resid* for residual stock. Our results clearly show that the considered models provide accurate forecasts in a reasonable computational time. Even if there is no outperforming method between ARIMA and TF, the latter appears as the best compromise between forecasting quality and computational time. Moreover, TF is the most flexible model since it allows to potentially take many different exogenous variables into account. Indeed, many elements, such as promotions or festivities, may have an impact on sales and the possibility to consider their contribution in the forecasting process represents an important feature. Note that the order plans obtained by using sales forecast as input are not remarkably different across models, except for mozzarella cheese. Indeed, in this case the lot size is one piece, thus the orders are exactly equal to sales forecast and it follows that the difference among the two models is more relevant. Finally, we observe that the behaviour of the order plans in terms of waste as well as the other three KPIs is very effective. The only critical result seems to be the stock-outs for smoked salmon, but it can be observed that 31 and 21 pieces correspond to slightly more than 15 and 10% of weekly sales, respectively.

We observe that three out of four KPIs tend to limit the ordered quantity, that is freshness, residual stock and waste; thus, a specific training activity may focus on properly weighting the four KPIs in order take into account this effect and represent the decision maker criteria.

Next, we analyse the variability of the four selected KPIs resulting from demand uncertainty. To this aim, we sample $N = 100$ different realizations of the daily demand over the planning horizon, using the estimated probability distribution. This results in as many scenarios, including the base scenario. Then, we simulate system's performance in terms of its identified KPIs when different demands occur, as expressed by the N alternative scenarios, while the order proposal remains fixed at the optimal level associated to the base scenario. We note that this analysis is usually very fast (less than 1 s, on average), so its contribution to the overall computational cost is negligible.

For the sake of shortness, we discuss in details only the results of the analysis conducted for milk (complete results for other products are reported at the end of the Section) summarising them through box plots, showing variability of the KPIs due to demand uncertainty along the forecasting horizon. According to the previous notation, we recall that (A) corresponds to overall waste at the end of the planning horizon; (B) expresses the average freshness (in days starting from the delivery date) when the product is sold; (C) measures the overall stock-out as number of lost sales during the planning horizon; (D) accounts for the residual stock at the end of the planning horizon. Figure 12.3 depicts the box plots associated to the KPIs for milk, which can be delivered every odd day of the week. Blue dots superimposed to the box plots correspond to the KPI values achieved when demand equals forecasted sales (assumed as the base scenario). Possible outliers have been identified and represented by red crosses. We notice that, according to the decision maker wishes, in this case zero waste (A) is obtained by all forecasting models across the N scenarios, as well as no stock-outs occur (C), while freshness (B) remains around 1.5 days: this may be a satisfactory result, since milk has a shelf life of 4 days, and it will be always sold within the next half a day following the delivery day.

For the same case study and the same product, when the decision maker does not assign a high priority to the waste reduction the relative weights among KPIs can conduct to a different tradeoff solution. This solution is also affected by the combination of the effects of demand predictions, product characteristics and optimization process. Figure 12.4 reports on the results associated to the following order plans for the two forecasting approaches, respectively:

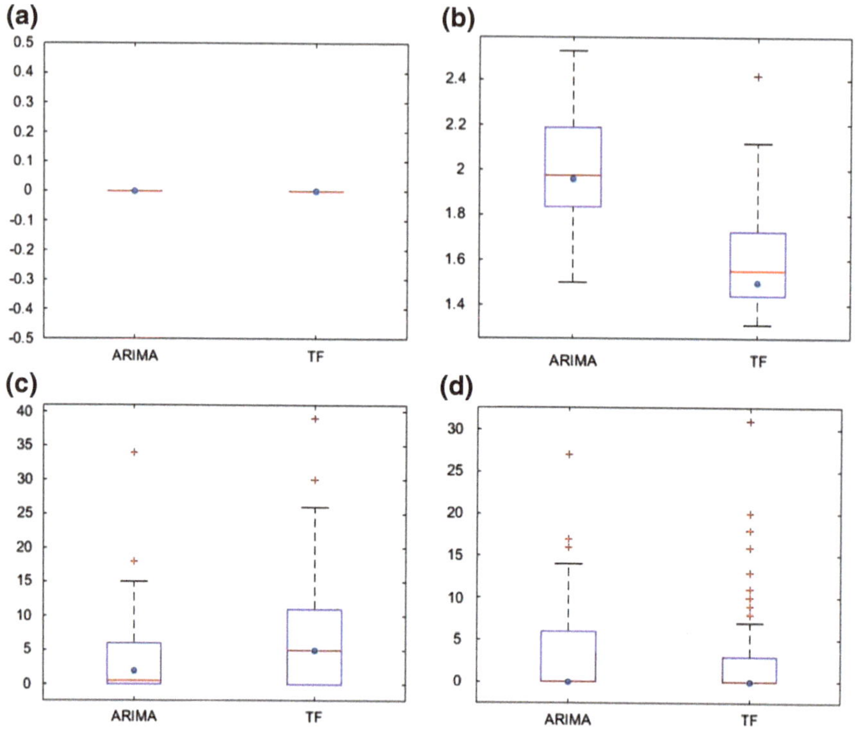

Fig. 12.3 Sensitivity analysis for the case with priority on waste reduction. Product: *milk*. KPI: **a** waste, **b** freshness, **c** stock-outs, and **d** residual stock

- Model: ARIMA, Product: milk,
  delivery days (delivered lots) = {Mon (7), Wed(2), Thu(2), Sat(2)}
- Model: TF, Product: milk,
  delivery days (delivered lots) = {Mon (3), Wed(3), Thu(4), Sat(2)}

In this case the TF approach is still able to provide a desired solution avoiding waste, while the ARIMA approach identifies a different tradeoff among the KPIs, accepting an amount of waste while showing lower result stock-outs; TF, in turn, ensures better freshness.

The uncertainties affecting the demand forecasts are described as error distributions. On the basis of a demand forecast, the order planner receives, for each day the distribution of the forecasting values and the prediction value (i.e., the expected value of the distribution). In order to contrast the risk of high stock-out rates, on the basis of these daily forecasting distribution, decision makers often use an overestimated demand value in order to guarantee a given probability $\alpha\%$ of demand satisfaction. This approach can be considered a risk-averse behavior of the decision maker when the service level in terms of stock-out rates is considered relevant. Of course this behavior will contrast the other considered KPIs. As an illustrative

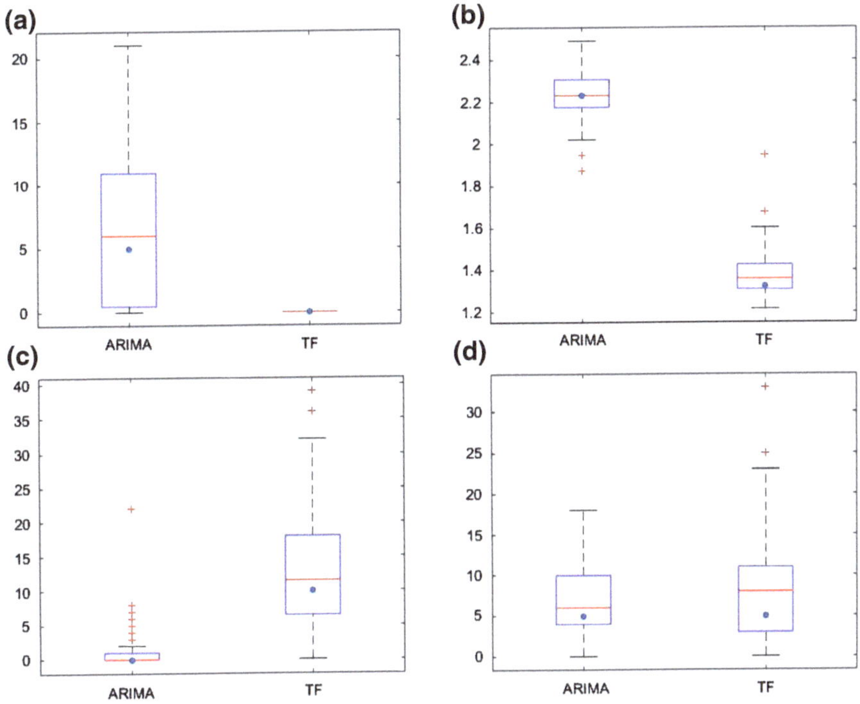

Fig. 12.4 Sensitivity analysis for the case without priority on waste reduction. Product: *milk*. KPI: **a** waste, **b** freshness, **c** stock-outs, and **d** residual stock

example, we report again the case for milk in which the value $\alpha = 90$ is considered, keeping the same setting adopted in the previous analysis concerning the KPIs weights. The results of this analysis are reported in Fig. 12.5 associated to the following order plans for ARIMA and TF respectively:

- Model: ARIMA, Product: milk,
  delivery days (delivered lots) = {Mon (8), Wed (3), Thu (2), Sat (2)}
- Model: TF, Product: milk,
  delivery days (delivered lots) = {Mon (4), Wed (5), Thu (4), Sat (5)}

As expected, the two solutions show bigger orders to cope with a higher demand level. Also in this case, the approach based on TF seems to offer a better performance, except for the residual stock. In particular, using the prediction based on TF it is possible to avoid waste. However, a more accurate analysis on the behavior of the different possible settings of the DSS would require a long-run investigation to adequately consider also the inter-temporal effects (e.g., concerning the role played by residual stocks).

For the sake of completeness, in Tables 12.4 and 12.5 we report the results of the simulation analysis for the other two selected products: mozzarella cheese

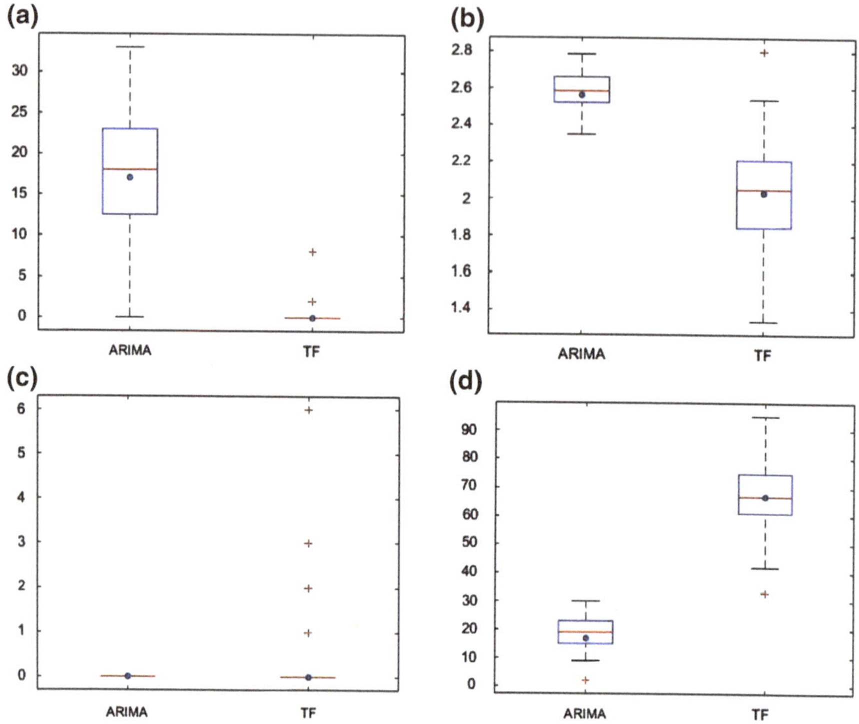

Fig. 12.5 Sensitivity analysis for the case with demand overestimation. Product: *milk*. KPI: **a** waste, **b** freshness, **c** stock-outs, and **d** residual stock

and smoked salmon. The first two columns of the tables report the forecasting model and the considered case (i.e., priority on waste (W), no priority on waste (NW), demand overestimation (DO)), respectively. Then the orders for each allowed delivery day are shown. The fourth column indicates the KPI and the following columns report its value for the base scenario, its minimum value followed by 1st, 2nd, and 3rd quartile values. The maximum value is reported in the last column.

Both ARIMA and TF models provide no waste for all the three cases (W, NW and DO); in fact, this result is expected when assuming to have low inventory levels at the beginning of the forecasting horizon, because both mozzarella cheese and smoked salmon have a shelf-life longer than one week. A long-run simulation analysis will be needed to further investigate the performance of the two models in terms of waste reduction for the different cases proposed. The DO case has worse freshness (B) and residual stock (D) than the other two cases, with TF being slightly better than ARIMA, especially on D. Again, the long-run simulation might provide further insights in this respect.

Table 12.4 Simulation analysis

Model	Case	Orders (Mon/Thu)	KPI	Base	Min	1stQ	2ndQ	3rdQ	Max
ARIMA	W	32/34	A	0	0	0	0	0	0
			B	2.3	1.7	2.0	2.3	2.7	3.4
			C	0	0	0	1	7.5	18
			D	0	0	0	1	9	22
ARIMA	NW	27/30	A	0	0	0	0	0	0
			B	2	1.6	1.8	2.1	2.4	3.2
			C	9	0	2	9	16.5	34
			D	0	0	0	0	2	5
ARIMA	DO	51/48	A	0	0	0	0	0	0
			B	3.2	2.2	2.8	3.2	3.5	4.3
			C	0	0	0	0	0	0
			D	33	8	26	34	41.5	62
TF	W	24/27	A	0	0	0	0	0	0
			B	2.5	2.2	2.4	2.5	2.7	3.1
			C	1	0	0	1.5	6	14
			D	0	0	0	2	6.5	16
TF	NW	20/26	A	0	0	0	0	0	0
			B	2.4	2.1	2.3	2.4	2.5	2.7
			C	6	0	2	6	10	19
			D	0	0	0	0	3	7
TF	DO	33/35	A	0	0	0	0	0	0
			B	3	2.3	2.8	3	3.2	3.7
			C	0	0	0	0	0	0
			D	16	0	12	17	23	31

12.7 Conclusive Discussion

In this Chapter we propose a DSS for sales forecasting and order planning in fresh food supply chain management. The DSS combines a pre-processing module to identify seasonality and noise emerging from historical sales, a forecasting module to derive sales forecasts, supported by a module for automatic tuning the forecasting models, and a multi-objective optimization module equipped with a best alternative selection module, to derive the best order proposal based on a set of KPIs. Two different forecasting models were considered and tested on a set of sample products. Our results clearly show the benefits of deriving an optimal order proposal based on sales forecasting, explicitly accounting for demand variability. The proposed analyses highlight the capability of the DSS to absorb possible variations in the demand, thus limiting their impact on the order planning phase. Another advantage offered by this DSS relies on its flexibility, as it is designed to be user-interactive and to run alternative approaches in terms of forecasting and model settings, depending on the characteristics of the data set, the products and the stores.

Table 12.5 Simulation analysis

Model	Case	Orders (Tue/Sat)	KPI	Base	Min	1stQ	2ndQ	3rdQ	Max
ARIMA	W	11/8	A	0	0	0	0	0	0
			B	2.3	1.8	2.1	2.3	2.4	2.6
			C	0	0	0	0	6	15
			D	4	0	0	5	13	32
ARIMA	NW	9/9	A	0	0	0	0	0	0
			B	2.0	1.7	1.9	1.98	2.1	2.2
			C	14	0	9	16	21	38
			D	8	0	4	9	14	29
ARIMA	DO	13/10	A	0	0	0	0	0	0
			B	2.7	2.2	2.6	2.7	2.8	3.1
			C	0	0	0	0	0	0
			D	44	12	35	44.5	53	72
TF	W	11/8	A	0	0	0	0	0	0
			B	2.2	2.0	2.2	2.2	2.3	2.5
			C	3	0	0	3	8.5	21
			D	0	0	0	1	8	19
TF	NW	12/8	A	0	0	0	0	0	0
			B	2.4	2.1	2.3	2.5	2.5	2.8
			C	0	0	0	0	0	0
			D	7	0	3	8	15.5	34
TF	DO	13/10	A	0	0	0	0	0	0
			B	2.6	2.4	2.5	2.7	2.7	3.0
			C	0	0	0	0	0	0
			D	37	19	32.5	38	45.5	64

The overall computational cost for the DSS, resulting from running its three main modules, remains compatible with real (off-line) applications. When focussing on the variability of system's performance caused by demand uncertainty, the order proposal associated to TF forecasts seems to be more robust. Besides, TF is the most flexible model and allows to potentially take many different exogenous variables into account. Indeed, many elements, such as promotions or festivities, as well as intermittent demand, may have an impact on sales and considering their contribution in the forecasting process may represent a future research direction. Moreover, the TF model definitely dominates the actual management forecasting system. This result further supports the usefulness of adopting rigorous forecasting methods, rather than relying only on experience-based rules, which might leave specific hidden trends unnoticed. Additional benefits are related to order policy implications, as more accurate forecasts enable to reduce potential stock-outs and, even more important for fresh products, outdating and wastage.

Future research may develop along different directions. It may improve forecast accuracy by means of non scale-dependent indicators and considering the impact of other exogenous variables. Further research efforts may also focus on developing more sophisticated tools for parameter tuning of models. Additional computational experiments can test the DSS approach, including the delivery optimization module, in a long-run simulation analysis (e.g., over one year) to evaluate the overall performance of the proposed DSS.

A larger problem that occurs at the distribution stage of a fresh food supply chain is that of rejected shipments and unsold items. They can turn into waste if alternative buyers cannot be found in time. These problems may occur usually due to a short shelf-life by the time the products are delivered to the store. Sometimes these items are brought to food banks or other operators able to distribute them (e.g., if they have the capacity and the time to take and dispatch them). Thus, another research direction should be devoted to adapt and possibly extend the proposed DSS in order to support the cooperation between retailers, customers, food banks, and other operators. This cooperation includes (i) sharing of information (or even predictions) about the residual shelf-life of products in a network of stores; (ii) the possibility to determine an optimal plan for picking and collecting operations; and (iii) support the delivery and dispatching operations to potential final customers.

Acknowledgements This work was partially supported by the E-CEDI project, funded by Apulia Region under the POR FESR 2007–2013 grant, Asse I, Linea 1.2–Azione 1.2.4.

References

Ahumada O, Villalobos JR (2009) Application of planning models in the agri-food supply chain: a review. Eur J Oper Res 195:1–20

Aiello G, Enea M, Muriana C (2014) Economic benefits from food recovery at the retail stage: an application to Italian food chains. Waste Manage 34:1306–1316

Alvarez J, Johnson R (2011) Doug Rauch: solving the American food paradox. Harvard Bus Rev 9:12–22

Armstrong JS (2001) Principles of forecasting: a handbook for researchers and practitioners. Springer

Babai M, Ali M, Boylan J, Syntetos A (2013) Forecasting and inventory performance in a two-stage supply chain with ARIMA(0, 1, 1) demand: theory and empirical analysis. Int J Prod Econ 143:463–471

Beske P, Land A, Seuring S (2014) Sustainable supply chain management practices and dynamic capabilities in the food industry: a critical analysis of the literature. Int J Prod Econ 152:131–143

Borade A, Sweeney E (2015) Decision support system for vendor managed inventory supply chain: a case study. Int J Prod Res 53:4789–4818

Box GEP, Jenkins GM, Reinsel GC (2008) Time series analysis: forecasting and control. Wiley

Brandenburg M, Govindan K, Sarkis J, Seuring S (2014) Quantitative models for sustainable supply chain management: developments and directions. Eur J Oper Res 233:299–312

K. P. Burnham, D. R. Anderson, *Model Selection and Multimodel Inference: A Practical Information-Theoretic Approach*, Springer, 2002

Darlington R, Rahimifard S (2006) A responsive demand management framework for the minimization of waste in convenience food manufacture. Int J Comput Integr Manuf 19:751–761

Darlington R, Staikos T, Rahimifard S (2009) Analytical methods for waste minimisation in the convenience food industry. Waste Manage 29:1274–1281

Deb K (2001) Multi-objective optimization using evolutionary algorithms. John Wiley & Sons

Dellino G, Kleijnen JPC, Meloni C (2010) Robust optimization in simulation: Taguchi and response surface methodology. Int J Prod Econ 125(1):52–59

Dellino G, Kleijnen JPC, Meloni C (2012) Robust optimization in simulation: Taguchi and Krige combined. INFORMS J Comput 23(4):471–484

Dellino G, Laudadio T, Mari R, Mastronardi N, Meloni C (2015) Sales forecasting models in the fresh food supply chain. In: Proceedings of the international conference on operations research and enterprise systems (ICORES-2015), pp 419–426

Dellino G, Lino P, Meloni C, Rizzo A (2007) Enhanced evolutionary algorithms for multidisciplinary design optimization: a control engineering perspective. Hybrid Evolutionary Algorithms, Volume 75—Series Studies in Computational Intelligence. Springer, pp 39–76

Dellino G, Meloni C (2015). Uncertainty management in simulation-optimization of complex systems. Algorithms and applications. Springer. Operations Research/Computer Science Interfaces Series, vol. 59

Dotoli M, Fanti MP, Meloni C, Zhou MC (2003) A decision support system for the supply chain configuration. Proc IEEE Int Conf Syst, Man and Cybern 3:2667–2672

Dotoli M, Fanti MP, Meloni C, Zhou MC (2005) A multi-level approach for network design of integrated supply chains. Int J Prod Res 43(20):4267–4287

Dotoli M, Fanti MP, Meloni C, Zhou MC (2006) Design and optimization of integrated E-supply chain for agile and environmentally conscious manufacturing. IEEE Trans Syst Man Cybern, Part A 36(1):62–75

Ehrgott M (2005) Multicriteria optimization. Springer-Verlag

Fleischmann B, Meyr H, Wagner M (2002) Advanced planning. In: Stadtler H, Kilger C (eds) Supply chain management and advanced planning, 2nd edn. Springer, Berlin, pp 71–96

Garrone P, Melacini M, Perego A (2014) Opening the black box of food waste reduction. Food Policy 46:129–139

Höglund R, Östermark R (1991) Automatic ARIMA modelling by the Cartesian search algorithm. J Forecast 10(5):465–476

Huang M-G (2013) Economic ordering model for deteriorating items with random demand and deterioration. Int J Prod Res 51:5612–5624

Hyvärinen A, Oja E (2001) Independent component analysis. Wiley

Kaipia R, Dukovska-Popovska I, Loikkanen L (2013) Creating sustainable fresh food supply chains through waste reduction. Int J Phys Distr Logist Manage 43(3):262–276

Jacobs FR, Chase RB (2014) Operations and supply chain management. McGrawHill/Irwin

Ma Y, Wang N, Che A, Huang Y, Xu J (2013) The bullwhip effect on product orders and inventory: a perspective of demand forecasting techniques. Int J Prod Res 51:281–302

Makridakis S, Hibon M (2000) The M3-Competition: results, conclusions and implications. Int J Forecast 16(4):451–476

Makridakis S, Wheelwright SC, Hyndman RJ (2008) Forecasting methods and applications. Wiley India Pvt, Limited

Meyr H (2002) Forecast methods. In: Stadtler H, Kilger C (eds) Supply chain management and advanced planning, 2nd edn. Springer, Berlin, pp 379–390

Muriana C (2015) Effectiveness of the food recovery at the retailing stage under shelf life uncertainty: an application to Italian food chains. Waste Manag 41:159–168

S. Nahmias, *Perishable Inventory Systems*, Springer, 2011

Naso D, Turchiano B, Meloni C (2006) Single and multi-objective evolutionary algorithms for the coordination of serial manufacturing operations. J Intell Manuf 17(2):251–270

Rijpkema WA, Rossi R, Van Der Vorst JGAJ (2014) Effective sourcing strategies for perishable product supply chains. Int J Phys Distr Logist Manage 44(6):494–510

Soysal M, Bloemhof-Ruwaard JM, Meuwissen MPM, Van Der Vorst JGAJ (2012) A review on quantitative models for sustainable food logistics management. Int J Food Syst Dyn 3 (2):136–155

Wang X, Li D, O'Brien C (2009) Optimisation of traceability and operations planning: an integrated model for perishable food production. Int J Prod Res 47(11):2865–2886

Van Der Vorst JGAJ, Beulens AJM, De Wit W, Van Beek P (1998) Supply chain management in food chains: improving performance by reducing uncertainty. Int Trans Oper Res 5(6):487–499

van Donsellar K, van Woensel T, Broekmeulen R, Fransoo J (2006) Inventory control of perishables in supermarkets. Int J Prod Econ 1041:462–472

Chapter 13
Participatory Planning in Organic Solid Waste Management: A Backcasting Approach

Roberta Sisto, Edgardo Sica, Mariarosaria Lombardi and Maurizio Prosperi

Abstract The valorisation of the organic fraction of municipal solid waste (OFMSW) represents a relevant matter for local governments that may result in significant economic and environmental benefits. In particular, defining the most cost-effective and environmentally friendly OFMSW management strategy should be based upon the active involvement of local stakeholders in order to allow policymakers to take into account all possible environmental, social, technological, and financial OFMSW-related problems. In this framework the present chapter aims at outlining a long-term management plan for OFMSW in the case of the south-eastern Italian municipality of Foggia. To this end we have employed an adapted participatory backcasting experiment based upon a double-step procedure. By means of a focus group with experts on OFMSW management issues at the municipal level, we firstly identified the desired end point and the relative expected obstacles and opportunities. These were then discussed during a workshop organised with a group of local stakeholders, who identified and proposed all possible actions to be carried out in the short, medium, and long term to reach the identified end point. Such a participatory approach should contribute to reducing the bounded rationality and the subjectivity affecting decision-making processes as well as to broaden the knowledge base and to achieve a greater transparency in the definition of OFMSW management strategies.

Keywords Waste management · Participatory approach · Long-Term strategy

R. Sisto · E. Sica (✉) · M. Lombardi
Department of Economics and Star*Energy Group, University of Foggia,
Largo Papa Giovanni Paolo II, 71121 Foggia, Italy
e-mail: edgardo.sica@unifg.it

M. Prosperi
Department of the Sciences of Agriculture, Food, and Environment,
University of Foggia, Via Napoli, 25, 71122 Foggia, Italy

P. Morone et al. (eds.), *Food Waste Reduction and Valorisation*,
DOI 10.1007/978-3-319-50088-1_13

13.1 Introduction

In recent years municipal solid waste management systems have received a great deal of attention from public opinion and policymakers due to the serious consequences that improper solid waste management can pose to human health and the environment. Indeed, uncontrolled or inappropriate waste handling can cause many problems in terms of water and soil pollution, as well as in terms of increased levels of greenhouse gas emissions, which contribute to climate change (Smith et al. 2015). In general terms municipal solid waste (MSW) represents the waste generated from households, institutions, and commercial activities (such as offices, schools, hotels, restaurants, hospitals, etc.) and includes food, garden waste, paper, plastic, textile, metal, and glass. Although its composition can change according to a number of factors (e.g., population density, economic well-being, seasonality), the organic fraction of municipal solid waste (OFMSW), resulting from food residues and garden waste, represents the highest proportion. OFMSW can reach up to 70% of the MSW composition, and its uncontrolled decomposition can cause contamination of the natural environment (Albanna 2013). Macias-Corral et al. (2008) report that the decomposition of one metric ton of OFMSW can release up to 110 m^3 of carbon dioxide (CO_2) and up to 140 m^3 of methane (CH_4). By contrast, OFMSW valorisation can result in relevant environmental benefits in terms of reduced greenhouse gas emissions and decreased leachate quantities. Moreover, from a life-cycle perspective, OFMSW can produce valuable compost, renewable energy, and biomaterials, depending on the processing method.

In this context finding an effective strategy for dealing with OFMSW represents a relevant challenge for local governments, which are commonly in charge of providing waste management services to their citizens. Indeed, to achieve an environmentally friendly and cost-effective OFMSW management strategy, able to respond to the needs of local communities', local policymakers have to take into account a number of environmental, social, technological, and financial factors in their decisions concerning collection services, disposal infrastructure, waste valorisation, and recycling programmes. The identification of the most appropriate OFMSW management strategy should, therefore, be based on the involvement of all stakeholders (Patel et al. 2007), preferably through a 'participatory approach'. These 'social experiments' involve bringing stakeholders together so that they can discuss specific issues, become informed about them, and arrive at a strategy for taking action (Webler and Tuler 2002). More specifically, our work is based on the hypothesis that stakeholders are usually keen, though sometimes reluctant, to express their opinions and to discuss them openly. Hence, they need a structured technique that is able to foster their participation, stimulate the interaction, and provide a coherent and effective synthesis of the process, leading to a robust strategy, which could represent consistent support for public decision makers. Therefore, the present chapter aims at defining a long-term management plan of OFMSW in the case of the south-eastern Italian municipality of Foggia by using an adapted participatory backcasting experiment (Sisto et al. 2015).

The structure of the chapter is as follows. Section 13.2 explores the definition, characteristics, and legislative framework of OFMSW. Section 13.3 discusses the participatory backcasting tool. Section 13.4 deals with the case study. Finally, Sect. 13.5 ends with some concluding remarks.

13.2 Definition and Characteristics of the Organic Fraction of Municipal Solid Waste (OFMSW)

In the European Union (EU), the concept of waste has evolved over time from material to be disposed of to a resource to be valorised. In this context some important goals about waste management have been integrated into the EU environmental policy through a very extensive and complex set of laws. In this chapter we refer only to the most relevant and recent regulations, such as the European Commission's *Roadmap to a Resource Efficient Europe* (European Commission 2011), the EU *Waste Framework Directive* (EU 2008), and the former *Landfill Directive* (EU 1999), which has driven in large part the Italian legislation in this sector. All the above regulatory tools promote a range of waste management targets and broader goals until 2020 (European Environmental Agency 2013).

The Italian definition of waste (which largely corresponds to the EU legislation) is included in the *Environmental Act* Legislative Decree no. 152 of 3 April 2006, as a replacement of Legislative Decree no. 22 of 5 February 1997 (the so-called *Ronchi Decree*). The latter (which ratified Directives 91/156/EEC and 91/689/EEC) has for almost a decade represented the basis for the Italian legislation on waste management by defining producers' duties and producing a number of implementation documents that represent the relative operating tools (DG Internal Policies of the Union Policy 2006, Presidente della Repubblica 1997, 2006).

The fourth part (art. 183) of the *Environment Act* defines waste, in compliance with the definition of the *Waste Framework Directive* 2008/98/EC, as follows: 'Any substance or object which the holder disposes of or intends or is required to dispose of'. The above definition is still founded upon the word *dispose*, as already happened in the *Ronchi Decree*. However, compared to the latter, the Legislative Decree of 2006 introduced a novelty in the model of integrated waste management, that is, the *time criterion*. In other words, it establishes the moment when the discipline of waste management must be applied, namely 'until the end of the recovery operations' (art. 181). As for the *Ronchi Decree*, waste is classified according to its origin (as urban or special) and considering its dangerousness (as dangerous or not). Art. no. 178 of Legislative Decree no. 152/2006 states that waste management is a public interest activity, and consequently, it must be managed in a rational and sustainable way. This is to ensure high-level protection of the environment as well as human health by means of efficiently using material and ensuring that the consumption of renewable and non-renewable resources (as well as its impact) does not exceed the 'carrying capacity' of the environment

(Presidente della Repubblica 2006; Reichel et al. 2013). Accordingly, the concept of integrated waste management includes all activities aimed at managing the whole supply chain of municipal waste, from production to final disposal or return within the consumption cycle through recycling. Waste, therefore, must be recovered or disposed of without causing any harm to human beings or the environment. Specifically, it should not pose any risk to water, air, soil, fauna, or flora; cause problems through noise or odours; or damage the landscape or places of special interest, protected in accordance with the current legislation (Bovino 2014). In order to comply with the above roles, waste management should occur according to the following hierarchy: reduce, reuse, recycle, recover energy, and dispose (Fig. 13.1).

Some years later, by ratifying *Waste Framework Directive* 2008/98/EC, Legislative Decree no. 205/2010 amended and integrated some parts of the former decree, establishing the priority through which any waste typology should be managed. In this framework specific attention was given to OFMSW (Ciceri 2012), which is defined in Art. 183 of Legislative Decree 205/2010 as 'biodegradable waste from gardens and parks, food and kitchen waste from households, restaurants, catering and retail premises and similar waste from food industry, differentially collected' (Presidente della Repubblica 2010). Such definition deals with OFMSW by separate collection, providing for upstream separation by the user. The effect is a relevant reduction of landfill disposal and, as a result, a significant improvement of the quality of the environment.

In Italy, OFMSW has constantly increased over recent years; according to the most recent data, it represented the main commodity fraction collected separately (about 43% of the total amount of urban solid waste) in 2014 (ISPRA 2015). Its degradation can cause a significant environmental impact due to odour emissions, methane release into the atmosphere, leachate into the soil, and consequent increases in relative restoration costs. For this reason it is crucial to avoid any

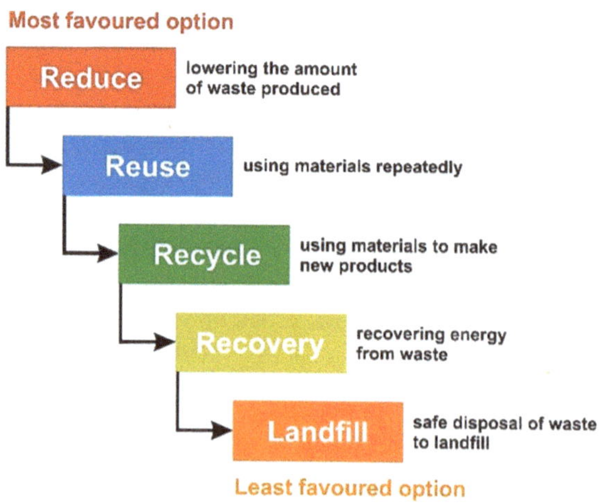

Fig. 13.1 Waste management hierarchy (*Source* SPI 2016)

possible contamination with other product fractions, directing the organic fraction flows to dedicated disposal systems (ISPRA 2015). Indeed, it is worth noting that Directive 99/31/EC (which was ratified in Italy by means of Legislative Decree no. 36/2003) established that by 2016 the biodegradable waste for disposal in landfills must be only 35% of the total biodegradable urban waste produced in 1995 (EU 1999; Presidente della Repubblica 2003; Placentino et al. 2013).

The set of legislative and technical management factors, together with the disposal requirements for OFMSW, has brought economic operators to identify appropriate technologies and facilities in order to treat/dispose of it. Indeed, OFMSW is the most polluting part of all urban waste yet, at the same time, the most valuable fraction since, when properly valorised, it may be used to produce green energy and organic matter for soil, thus improving its fertility. In this context, due to its high humidity, the technologies used for this purpose are aerobic and anaerobic digestion, which are based on biological processes (Atrigna et al. 2010). Such processes last, respectively, 90 and 40 days for obtaining the final product. The aerobic digestion plant leads to the production of compost (soil amendment), while the anaerobic one produces biogas (biofuels for heat and electricity generation and/or for the automotive sector) and digestate (soil amendment) (Vismara et al. 2010). Both technologies allow meeting the targets established by the *Ronchi Decree* in the framework of waste-integrated management in order to prevent waste production and promote the recovery of materials and energy.

13.3 A Tool for Involving Local Stakeholders in Decision-Making About Long-Term Issues

As highlighted in the introduction, this study is based on the literature on participatory approaches, proving the effectiveness of participatory tools in managing long-term, complex socio-technical issues (such as environmental ones) across various world settings (Giordano et al. 2005; Antunes et al. 2006; Lopolito et al. 2011; Sisto et al. 2015).

In addition, moving from a single decision maker to a multiple decision maker setting increases the complexity of the analysis. The decision maker requires a high-quality strategy definition to understand the problem and its complexities. To this aim participatory approaches can help to include multiple perspectives in the decision-making process.

Given the complexity of bioeconomic issues, the development of a bio-based industry is a long-term project. This characteristic makes the bio-based industry and the bioeconomy at large, surrounded by major uncertainties, both economic and social in nature.

In general, participatory tools refer to the involvement in planning and decision making of those involved in, affected by, knowledgeable of, or having relevant expertise in or experience of the issue at stake. Furthermore, the analysis takes into account the conflicts amongst different interest groups that have diverse objectives, criteria, expectations, and so on. This increases the legitimacy of decisions taken, which can save time in the long run due to lower resistance amongst stakeholders (Thrupp et al. 1994). A helpful tool in decision-making is the development of scenarios.

In particular, scenarios can be used to analyse a large number of uncertain future environmental and socioeconomic challenges (Priess and Hauck 2014). In addition, as highlighted by Hagemann et al. (2016), they can support the establishment of policy frameworks and the decision making of policymakers who want to take into account a long-term perspective. There are several types of scenarios. Börjeson et al. (2006), for instance, distinguish three scenario categories: *predictive*, *explorative*, and *normative*.

13.3.1 Participatory Backcasting

Backcasting falls under normative scenarios. It aims to describe desired goals or futures and to analyse how they could be achieved. Since the publication of a seminal article on backcasting by John B. Robinson in 1982, backcasting studies have evolved in significant ways. Attention has especially focussed on areas of environmental and resources policy. Indeed, the whole question of sustainability has been addressed in terms of backcasting (e.g., Dreborg 1996).

Although backcasting was not intended to be a bottom-up participatory method, it has been adapted and is increasingly often used as a participatory method, which makes it possible to include local community and stakeholders' knowledge in the process (Carlsson-Kanyama et al. 2008; Kok et al. 2011; Svenfelt et al. 2011). In a participatory backcasting exercise, participants typically describe their desired end conditions and then work backwards towards milestones and policy actions that are needed to achieve that future (Salter et al. 2010).

There are two main characteristics that most backcasting methods have in common. The first is their normative nature, and the second is their 'working backwards from a particular desired future end point' (Robinson 2003, p. 842). This often translates into methods that at least include a first step, during which desirable images of the future are developed, and a second step, during which these images are analysed by working backwards (Höjer and Mattsson 2000).

The result is typically a number of actions fulfilling possible futures (scenarios) that present a solution to a societal problem, with a discussion of which changes would be needed in order to reach these future images. In other words, the aim of a backcasting exercise is to encourage searches for new paths along which development can take place (Höjer and Mattson 2000).

13.4 The Case Study

13.4.1 Description

Looking at national data from ISPRA (2015), the total amount of OFMSW in Italy recovered in composting and anaerobic digestion plants in 2014 amounted to approximately 4.9 million tons. More specifically, 4.4 million tons were delivered to composting facilities, while 454,000 tons were treated in anaerobic digestion plants. The organic fraction of the recycling was 83% of the total waste delivered to composting plants and 52% of that directed to anaerobic digestion. In the same year the per capita value of valorised OFMSW at the national level was equal to 80 kg/inhabitant, recording very different levels in the three major geographic areas: 124 kg/inhabitant in the north, 59 kg/inhabitant in the centre, and 34 kg/inhabitant in the south.

However, such a picture does not provide a faithful representation of the OFMSW collection since the reduced number of plants in the central and southern regions implies that large amounts of waste move to the northern Italian areas. At the moment there are 279 composting plants (179 in the north, 44 in the centre, and 56 in the south), while the anaerobic digestion plants for biogas production amount to 29 (26 in the north and 3 in the south). There are also 20 plants that combine the anaerobic and aerobic processes, mostly located in the northern part of the country. These systems are increasingly spreading, and in 2014, they treated a total of almost 928,000 tons of waste.

In this framework our investigation focuses specifically on the case of Foggia, an Italian municipality of approximately 150,000 inhabitants in the south-eastern region of Apulia (Fig. 13.2).

This municipality represents a very interesting case study because the management and utilisation of OFMSW is a desirable policy target. On the one hand, this purpose could represent a potential way to revitalise the local economy; on the other, it is a chance to cope with the energy objectives of the European Commission, which aims at a substantial reduction in overall dependence on petroleum feedstocks in the next decade. Despite this, until 2014, the dominant collection system in the municipality was an undifferentiated waste collection system; only at the end of 2015 did the local government begin to experiment with a separate collection system, starting in some peripheral residential areas. In such pilot neighbourhoods the old waste containers were replaced by smaller and differently coloured bins (black for general waste, brown for organic waste, green for glass, yellow for plastic and aluminium, and white for paper). Bins were provided with a lock whose key was delivered to any household. Presently the local government is going to expand the above separate collection system to the whole city, although the experiment has not been particularly successful thus far, and the installation and the use of the new bins is being opposed by some citizens.

Fig. 13.2 Location of the case study (*Source* Our elaboration 2016)

13.4.2 Application of the Participatory Backcasting Methodology

The study was conducted in the winter of 2016 by adopting an adapted participatory backcasting approach in a country (Italy) with little tradition of these types of participatory methods. Two main aspects were considered. First, as the duration of a workshop is a critical variable affecting the participation rate, the workshop length was limited to half a day. The structure of the proposed participatory approach aimed at increasing the participation and engagement by parties that otherwise might be badly represented or have no role in long-term strategy definition. Second, in order to maximise stakeholders' involvement, the methodology combined the workshop with questionnaires. This modified backcasting approach guaranteed that the same stakeholders would participate in the strategy definition (Sisto et al. 2015).

13.4.2.1 Stakeholders' Engagement

Involving a representative sample of the whole population was not the objective. Rather, the aim was to involve people committed to the management of OFMSW at the municipal level. In line with the literature (van Asselt and Rijkens-Klomp 2002; Quist and Vergragt 2006; Kok et al. 2011), we engaged one representative person from each of the five groups of local stakeholders (Table 13.1):

Table 13.1 Local OFMSW stakeholders

Stakeholders	Role
Municipal bodies	Integrate EU and national legislation into Foggia municipality laws
	Provide effective campaign models
Local public sector—local authorities, community wardens, agro-energy working groups, researchers, environmentalists	Coordinate joint work within and across local authorities for consistent approaches, share best practice, and maximise value for money
	Provide a holistic environmental approach, not just waste benefits Influence local communities, local government, and business sector
Residents and communities, school, university	Act as communication channels and engage with other residents to change their behaviour
Manufacturers and retailers	Address food waste issues in manufacturing and retail
	Prioritise socially and environmentally responsible investments
	Assess and follow environmental best practices
Disposal/treatment contractors (AMIU Puglia)	Provide local, cost-effective, and environmentally sound treatment facilities
	Provide accurate data and regular performance updates

Source Adapted from Lamb and Fountain (2010)

- Policymakers
- The public sector (e.g., local authorities, community wardens, schools, hospitals)
- Residents and communities
- Manufacturers and retailers
- Disposal/treatment contractors.

13.4.2.2 The Preliminary Focus Group with Experts

In order to uncover the final end point, obstacles, and opportunities affecting the management of food waste and its utilisation in the municipality of Foggia a focus group with experts on municipal OFMSW management was organised two weeks before the workshop with the stakeholders. The workshop aimed at identifying the desired end point of the strategy and the expected obstacles and opportunities. This procedure can be considered as a means not only to shorten the participatory workshop but also to engage a significant number of stakeholders.

The workshop was organised in a neutral environment, with the involvement of five experts, coordinated by one facilitator. The role of the facilitator was to balance the dialogue amongst the participants, avoiding excess leadership by just a few members and helping the group to reach a good degree of consensus about the key concepts they discussed during the meeting.

First, a brief introduction to the research topic and the relevance of a participatory approach to building a long-term strategy was given. Following this the first focus question was: 'Regarding the area of Foggia, what kind of OFMSW management do you imagine for the future of this territory in the next twenty years, that is, in 2035?'. This open question enabled us to collect information about the experts' expectations and needs regarding the future of the area in which they work and live.

Consequently, the participants were asked to vote for one of three alternative future end points for the year 2035:

- Compost production
- Biogas and digestate production
- Production of bioproducts (i.e., products of organic origin with high added value).

Admittedly, these three options are not necessarily the only ways of defining the future of OFMSW management for the investigated area. They were based on knowledge that the authors had gained in previous research experiences (e.g., in the STAR* AgroEnergy EU Project) and meetings with experts on food waste management.

Answers were ranked to determine the end point that would be used in the workshop. The selected end point was *production of biogas and digestate.*

The same procedure was adopted to identify the most relevant opportunities and obstacles. We submitted a list of generally relevant influence factors derived from a bioeconomy literature analysis (Costello and Finnell 1998; Roos et al. 1999; Rosch and Kaltschmitt 1999; IEA 2003; McCormick and Kaberger 2007; Snakin et al. 2010; van Vliet and Kok 2015; Sisto et al. 2015), fostering (creating opportunities for) and obstructing (creating obstacles in the way of) the development of the bioeconomy, starting from OFMSW management. Participants were asked to rank, according to relevance, seven opportunities and seven obstacles. In addition, they could add obstacles and opportunities that they thought were missing from the list.

The most voted obstacles were as follow:

- Regulation barriers
- Excessive bureaucracy
- Lack of political clearness
- Credit access
- Social acceptance of industrial OFMSW transformation plants
- Poor institutional support
- Weak legislative coordination at different institutional levels.

The selected opportunities were as follow:

- Abundance of OFMSW
- EU public funding

- Technical-scientific support from research institutes and universities
- Growth of environmental concerns with attention to green solutions
- Use of digestate according to a circular economic scheme
- Demand for thermal energy for domestic and industrial use
- Priority of EU policies towards biofuels.

The results of this round were elaborated to define the structure of the backcasting workshop.

At the end of the focus group, participants were asked to fill in a short questionnaire about their perception of the agreeableness of the meeting.

13.4.2.3 The Backcasting Workshop

Some weeks later, at the beginning of February 2016, we organised a workshop with the OFMSW stakeholders of the municipality of Foggia. The seven participants sat around a table with a facilitator who guided the discussion and a collaborator who took notes about the atmosphere and interactions amongst the participants (Fig. 13.3).

First, participants were given an explanation of the backcasting approach. Then, they were asked to imagine travelling ahead in time to the year 2035 and to visualise the situation of Foggia municipality's waste management, where life is much less resource demanding and more sustainable than now. This was a way to introduce the most adequate atmosphere to present the expert focus group's results of the desired end point and the most relevant obstacles and opportunities. The participants introduced themselves and described their main concerns with respect to OFMSW management in the Foggia municipality. Then, they plotted the obstacles and opportunities selected in the previous focus group on a timeline (present 2016–future 2035).

Subsequently, the participants were provided three Post-its, on which they wrote down three possible actions aimed at mitigating/removing the obstacles or taking

Fig. 13.3 The stakeholders' workshop

advantage of the opportunities. This was an individual task that they completed in 10 min.

Finally, the participants briefly discussed each action and put them on the timeline drawn on a chart, starting from the present (2016) and continuing up to 2035. Several actions were redundant, and some were slightly modified, according to the comments and suggestions emerging during the discussion. In all cases a very high degree of consensus was obtained, and every action was approved by all participants.

These actions were put on the timeline (Fig. 13.4). If the participants thought that an action could deal with more than one opportunity or obstacle, they drew lines between them to show relationships. At the end of this step, a volunteer presented the group's results.

Finally, at the end of the workshop, an evaluation questionnaire was delivered to all participants in order to receive feedback on the process. On this questionnaire, which all the stakeholders completed anonymously, they were asked to express their opinion of the results of the workshop and the adopted methodology. This was done to measure the degree of consensus on the final choice, which could affect how well stakeholders support the final decision in the future and may reflect how well members believe their opinions are taken into account by their leaders and policymakers (Miller and Monge 1986).

In total the duration of the workshop was about two hours, and all participants expressed a positive opinion about the exchange of knowledge they had with the others. The results of the workshops and the details of the feedback are presented and discussed in the next section.

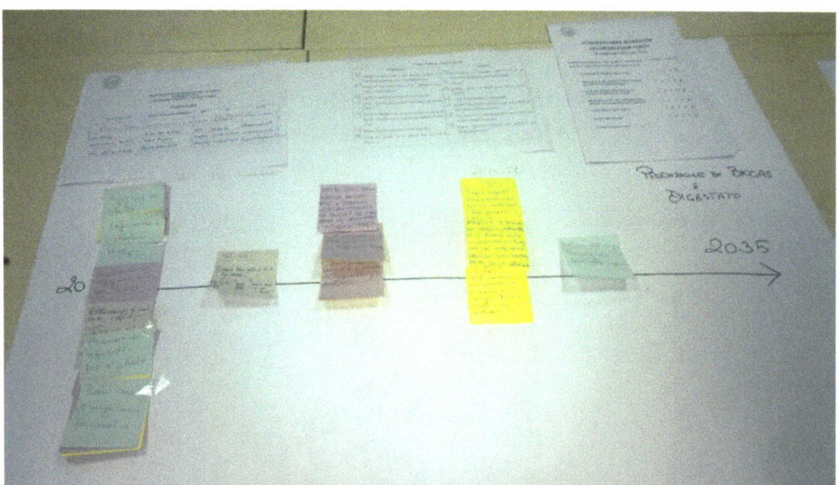

Fig. 13.4 Actions identified by the stakeholders during the backcasting workshop

13.5 Results

The proposed actions, approved by the participants during the discussion, are mostly concentrated at the beginning of the timeline as a consequence of the participants' awareness that several constraints are obstructing the developmental path and that opportunities could especially be captured in the mid- and long terms.

As follows, the main actions that emerged during the discussion are briefly described.

- Actions to be done in the short term:
  - (S-1) The differentiated waste collection should be improved, preferably through a door-to-door collection, whereby it would be possible to identify the people violating the correct disposal rules.
  - (S-2) The use of digestate deriving from anaerobic digestion processing of food waste should be encouraged for agronomic purposes. Targeted demonstrations and training actions should address farmers in their transition from the use of chemical fertilisers to digestate. In fact, at the moment, farmers are still not aware of the fertilising properties of organic matter, especially for large-field application (e.g., wheat, open-air tomato crops).
  - (S-3) The municipal waste agency should undertake a relevant investment to substitute obsolete waste bins with more efficient ones, taking into account the different types of users (e.g., household or restaurant/cafeteria) and different types of waste.
  - (S-4) Education campaigns should be targeted at local citizens, aimed at raising the awareness of the relevance of energy savings and a circular economy.
  - (S-5) Fines to citizens responsible for incorrect waste separation should be enforced. This is expected to promote the awareness of citizens regarding the relevance of waste separation, as well as possibly improve the quality of collected waste.
  - (S-6) Public–private partnerships should be promoted in order to take advantage of financial support for the promotion of innovative technologies for food waste collection and treatment.
- Actions to be done in the midterm:
  - (M-1) Information and communication campaigns targeted at local communities affected by the creation of food waste conversion plants. In particular, local communities should be made aware that the waste they produce can be converted into an economic resource only if: (a) the waste collection is properly disposed of (i.e., the waste is not contaminated by high polluting substances) and (b) most of the products and by-products obtained through the anaerobic conversion process are used by local users (e.g., electric and heating energy, digestate). In practice, visiting tours of farmers' associations

and local administrators to well-established plants could be organised to provide a real representation of best practices.
  - (M-2) Creation of integrated platforms for the collection and treatment of food waste, to be mixed with other types of not-dangerous organic wastes, in order to create a stable feedstock suitable for anaerobic digestion.
  - (M-3) Reinforcement of the role of the public university, playing a neutral role in the assessment of the public benefits and costs deriving from the activation of the full bioenergy value chain, from the collection of food waste through the conversion process until the full use of electric and thermal energy and the distribution of digestate for agronomic purposes. The university should promote scientific dissemination, as well as the creation of small-scale pilot plants, and should stimulate the public debate to emphasise the advantages of green technologies applied to food waste valorisation initiatives.
  - (M-4) Substantial financial support provided by the regional government for the creation of treatment plants, as well as to support private firms for the valorisation of different types of wastes (e.g., glass, plastics, paper). The virtuous circle can be completed only if the waste is correctly differentiated and all types of waste are valorised.

- Actions to be done in thc long term:
  - (L-1) Encouragement of the substitution of current sources of energy running manufacturing firms with (renewable) thermal or electric energy produced by food waste conversion plants.
  - (L-2) Definition of a long-term policy agenda aimed at both planning an integrated and comprehensive waste management strategy and promoting a sustainable development path.

At the end of the meeting, the participants were asked to express their opinion regarding the achievements of the focus group. As is shown in Table 13.2, participants expressed a very positive opinion, meaning that they were able to express themselves in a free and democratic environment.

13.6 Conclusions

Finding an effective strategy for dealing with OFMSW represents a relevant matter for local governments since OFMSW valorisation may result in significant environmental benefits in terms of reduced greenhouse gas emissions and decreased leachate quantities. However, the achievement of an environmentally friendly and cost-effective OFMSW management strategy should be based on the active involvement of all stakeholders, which would allow local policymakers to take into account the different environmental, social, technological, and financial OFMSW-related problems.

Table 13.2 Participants' opinions (n = 7)

	Counts of 'good' (score = 4)	Counts of 'excellent' (score = 5)	Mean score
Was the objective of the meeting clear?	3	4	4.57
Was the methodology adequate with respect to the strategy definition?	4	3	4.43
Do you think that the results of the meeting will provide some useful suggestions to policymakers?	4	3	4.43
Do you think that the meeting was a good opportunity to develop new relationships or to reinforce existing ones?	5	2	4.29
Was the meeting agreeable?	1	6	4.86
What is your opinion about the duration of the meeting?	4	3	4.43

Note The participants' opinions were evaluated through a Likert scale, defined as follows: 'insufficient' (1), 'sufficient' (2), 'moderately fair' (3), 'good' (4), and 'excellent' (5)

In this context the present chapter has aimed at defining a long-term management plan of OFMSW in the case of the south-eastern Italian municipality of Foggia by using an adapted participatory backcasting experiment based upon a double-step procedure.

The results of the proposed methodology are very encouraging. The participants were highly enthusiastic about the workshop. Moreover, the timeline could help policymakers to plan actions over time. This is a very important and relevant outcome, especially in areas where policymaking is negatively affected by poor governance or lack of institutional network coordination. In addition, other advantages of the proposed participatory approach could be summarised in:

- The reduction of bounded rationality and subjectivity affecting the decision-making process
- The enlargement of the knowledge base
- Greater transparency of the whole process.

However, it is important to highlight that because the results of a focus group with experts would condition the subsequent workshop, this is a very sensitive phase for both the respondents and the discussed topics because they could affect the following workshop and the quality of its results. Therefore, it is important to keep in mind a strong caveat: The quality of a decision is strongly dependent on the quality of the process that leads to it.

References

Albanna M (2013) Anaerobic digestion of the organic fraction of municipal solid waste. In: Malik A, Grohmann E, Alves M (eds) Management of microbial resources in the environment. Springer, Berlin

Antunes P, Santos R, Videira N (2006) Participatory decision-making for sustainable development—the use of mediated modelling techniques. Land Use Policy 23:44–52

Atrigna M, Canditelli M, Faustini N, Pescheta G (2010) La gestione della frazione umida/biodegradabile dei rifiuti urbani. Available on website at: http://old.enea.it/produzione_scientifica/pdf_EAI/2010/5/GestioneUmidoRifiuti.pdf. Retrieved on: 03 Dec 2015

Börjeson L, Höjer M, Dreborg KH, Ekvall T, Finnveden G (2006) Scenario types and techniques: towards a user's guide. Futures 38:723–739

Bovino C (2014) Rifiuti e Imballaggi. Manuale Ambiente 2014 a cura di: AA.VV, Milanofiori Assago (MI), IPSOA, 393–558

Carlsson-Kanyama A, Dreborg KH, Moll HC, Padovan D (2008) Participative backcasting: a tool for involving stakeholders in local sustainability planning. Futures 40:34–46

Ciceri G (RSE) (2012) The waste resource in Italy—IEA task 36. Available on our website at: http://www.ieabioenergytask36.org/vbulletin/attachment.php?attachmentid=302&d=1362675836. Retrieved on: 15 Dec 2015

Costello R, Finnell J (1998) Institutional opportunities and constraints to biomass development. Biomass Bioenergy 15:201–204

DG Internal Policies of the Union Policy—Department: Economic and Scientific Policy (2006) Status of implementation of EU Environmental Laws in Italy, IP/A/ENVI/IC/2006-183. Available on website at: http://www.europarl.europa.eu/RegData/etudes/etudes/join/2006/375865/IPOL-ENVI_ET(2006)375865_EN.pdf. Retrieved on: 15 Dec 2015

Dreborg KH (1996) Essence of backcasting. Futures 28:813–828

European Commission (2011) Communication from the commission to the European parliament, the council, the European economic and social committee and the committee of the regions, Roadmap to a Resource Efficient Europe, Brussels, 20 Sept 2011, COM(2011) 571 final

European Environmental Agency (EEA) (2013) Managing municipal solid waste—a review of achievements in 32 European countries. EEA report, no. 2/2013, Copenhagen, 2013. ISSN 1725-9177. doi:10.2800/71424

European Union (1999) Council Directive 1999/31/EC of 26 April 1999 on the landfill of waste, OJ L 182, 16 July 1999, pp 1–19

European Union (2008) Directive 2008/98/EC of the European Parliament and of the Council of 19 November 2008 on waste and repealing certain Directives, OJ L 312, 22 Nov 2008, pp 3–30

Giordano R, Passarella G, Uricchio VF, Vurro M (2005) Fuzzy cognitive maps for issue identification in a water resources conflict resolution system. Phys Chem Earth 30:463–469

Hagemann N, Gawel E, Purkus A, Pannicke N, Hauck J (2016) Possible futures towards a wood-based bioeconomy: a scenario analysis for Germany. Sustainability 8:98. doi:10.3390/su8010098

Höjer M, Mattson LG (2000) Determinism and backcasting in future studies. Futures 32(7):613–634

International Energy Agency (2003) Energy to 2050. Scenarios for a sustainable future. OECD, Paris

ISPRA (2015) Rapporto Rifiuti Urbani. Available at: http://www.isprambiente.gov.it/files/pubblicazioni/rapporti/rifiuti-urbani-2015/RapportoRifiutiUrbani_Ed.2015%20n.230_Vers.Integrale.pdf. Retrieved on 10 Dec 2015

Kok K, van Vliet M, Barlund I, Dubel A, Sendzimir J (2011) Combining participative backcasting and exploratory scenario development: experiences from the SCENES project. Technol Forecast Soc Chang 78:835–851

Lamb G, Fountain L (2010) An investigation into food waste management. Waste in action. http://www.actiondechets.fr/upload/medias/group_b_report_compressed.pdf. Retrieved on 10 Feb 2016

Lopolito A, Nardone G, Prosperi M, Sisto R, Stasi A (2011) Modeling the bio refinery industry in rural areas: a participatory approach for policy options comparison. Ecol Econ 72:18–27

Macias-Corral M, Samani Z, Hanson A, Smith G, Funk P, Yu H, Longworth J (2008) Anaerobic digestion of municipal solid waste and agricultural waste and the effect of co-digestion with dairy cow manure. Bioresour Technol 99(17):8288–8293. doi:10.1016/j.biortech.2008.03.057. Epub 14 May 2008

McCormick K, Kaberger T (2007) Key barriers for bioenergy in Europe: economic conditions, know-how and institutional capacity, and supply chain coordination. Biomass Bioenergy 31:443–452

Miller KI, Monge PR (1986) Participation, satisfaction, and productivity: a meta-analytic review. Acad Manag J 29(4):727–753

Patel M, Kok K, Rothman DS (2007) Participatory scenario construction in land use analysis: an insight into the experiences created by stakeholder involvement in the Northern Mediterranean. Land Use Policy 24:546–561

Placentino CM, Bruno E, Busetti F, Lacarbonara F, Gramegna D, Blonda M, Assennato G (ARPA PUGLIA) (2013) Management of urban biodegradable waste in Puglia: features and crtitical issues in Italian context. In: e-Proceedings of international conference on biochars, composts, and digestates. Production, characterization, regulation, marketing, uses and environmental impact (BCD 2013), 17–20 Oct 2013. Old Central Post Office Building, Bari, Italy

Presidente della Repubblica (1997) Decreto Legislativo 5 febbraio 1997, n. 22, 'Attuazione delle direttive 91/156/CEE sui rifiuti, 91/689/CEE sui rifiuti pericolosi e 94/62/CE sugli imballaggi e sui rifiuti di imballaggio', Gazzetta Ufficiale n. 38 del 15 febbraio 1997

Presidente della Repubblica (2003) Decreto Legislativo 13 gennaio 2003, n. 36, Attuazione della Direttiva 1999/31/CE relativa alle discariche di rifiuti, Gazzetta Ufficiale n.59 del 12-3-2003 - Suppl. Ordinario n. 40

Presidente della Repubblica (2006) Decreto Legislativo 3 aprile 2006, n. 152, 'Norme in Materia Ambientale, Gazzetta Ufficiale n. 88 del 14 aprile 2006

Presidente della Repubblica (2010) Decreto Legislativo 3 dicembre 2010, n. 205, 'Disposizioni di attuazione della Direttiva 2008/98/CE del Parlamento europeo e del Consiglio del 19 novembre 2008 relativa ai rifiuti e che abroga alcune direttive', Gazzetta Ufficiale n. 288 del 10 dicembre 2010

Priess JA, Hauck J (2014) Integrative scenario development. Ecol Soc 19(1):12. doi:10.5751/ES-06168-190112

Quist J, Vergragt P (2006) Past and future of backcasting: the shift to stakeholder participation and a proposal for a methodological framework. Futures 38:1027–1045

Reichel A, Ferraris M, Paleari S (2013) Municipal waste management in Italy. EEA (European Environmental Agency). Available on website at: https://www.eea.europa.eu/publications/managing-municipal-solid-waste/italy-municipal-waste-management. Retrieved on: 15 Dec 2015

Robinson J (2003) Future subjunctive: backcasting as social learning. Futures 35:839–856

Roos A, Graham RL, Hektor B, Rakos C (1999) Critical factors to bioenergy implementation. Biomass Bioenergy 17:113–126

Rosch C, Kaltschmitt M (1999) Energy from biomass: do non-technical barriers prevent an increased use? Biomass Bioenergy 16:347–356

Salter J, Robinson J, Wiek A (2010) Participatory methods of integrated assessment—a review. Adv Rev 1:697–717 (Sept/Oct 2010)

Sisto R, van Vliet M, Prosperi M (2015) Puzzling stakeholder views for long-term planning in the bio-economy: a back-casting application. Futures. doi:10.1016/j.futures.2015.04.002

Smith RL, Sengupta D, Takkellapatia S, Lee CC (2015) An industrial ecology approach to municipal solid waste management: II. Case studies for recovering energy from the organic fraction of MSW. Resour Conserv Recycl 104(2015):317–326

Snakin JP, Muilu T, Pesola T (2010) Bioenergy decision-making of farms in Northern Finland: combining the bottom-up and top-down perspectives. Energy Policy. doi:10.1016/j.enpol.2010.06.002

SPI (Sustainable Process Improvement) http://www.sustainableprocessimprovement.co.uk/about-spi/. Retrieved on 18 Feb 2016

Svenfelt A, Engstrom R, Svane O (2011) Decreasing energy use in buildings by 50% by 2050. A backcasting study using stakeholder groups. Technol Forecast Soc Chang 78:785–796

Thrupp LA, Cabarle B, Zazueta A (1994) Participatory methods in planning & political processes: linking the grassroots & policies for sustainable development. Agric Hum Values 11(2–3):77–84

van Asselt MBA, Rijkens-Klomp N (2002) A look in the mirror: reflection on participation in integrated assessment from a methodological perspective. Glob Environ Change 12:167–184

van Vliet M, Kok K (2015) Combining backcasting and exploratory scenarios to develop robust water strategies in face of uncertain futures. Mitig Adapt Strat Glob Change 20:43–74

Vismara R, Malpei F, Centemero M (2010) Biogas da rifiuti solidi urbani. Palermo, Dario Flaccovio Srl

Webler T, Tuler S (2002) Unlocking the puzzle of public participation. Bull Sci Technol Soc (22):179. doi:10.1177/02767602022003002. url:http://bst.sagepub.com/content/22/3/179

Chapter 14 The Role of Social Networks in the Diffusion of Bio-Waste Products: The Case of Mulching Films Derived from Organic Waste in Province of Foggia

Angela Barbuto, Antonio Lopolito, Myriam Anna Scaringelli and Giacomo Giannoccaro

Abstract In this work we consider the biodegradable mulching film containing soluble bio-based substances (SBOs), as a new Sustainable Agricultural Practice (SAP) potentially useful in both broadening the spectrum of Organic Fraction of Municipal Solid Wastes (OFMSW) management, and improving agricultural sustainability. Of course, the exploitation of such advantages depends on the actual adoption of the novelty from a critical mass of users. Among the various factors influencing this process, we stress the importance of interpersonal channels involving a face-to-face exchange. This implies the fact that people adopt an innovation when sufficient information has reached them, and shows the relevance of the role of social networks in the diffusion of innovations. Specifically, the network position of an actor affects the power and influence he can exert on its immediate neighbors as well as on the collective behavior of the members. This influence can be viewed as a strategic resource for innovation diffusion purpose in a marketing or policy context. The success as injection points, namely, the actors where the novelty is first inoculated, is typically measured as the proportion of actors who adopts the innovation at the end of the process. Following this line of reasoning, the aim of this work is to identify the network characteristics associated with effective injection points. In order to capture the network characteristics of the actors we used typical Social Network Analysis (SNA) measures. From an operative perspective, our purpose is to find the SNA measures associated with high adoption rates. However, being the innovation process new in nature, there are not

A. Barbuto · A. Lopolito (✉) · M.A. Scaringelli
Department of Agricultural, Food and Environmental Sciences,
University of Foggia, Via Napoli, 25, 71122 Foggia, Italy
e-mail: antonio.lopolito@unifg.it

G. Giannoccaro
Department of Agricultural and Environmental Science, University of Bari
"Aldo Moro", Via G. Amendola 165/a, 70126 Bari, Italy

P. Morone et al. (eds.), *Food Waste Reduction and Valorisation*,
DOI 10.1007/978-3-319-50088-1_14

available experimental data to conduct this kind of analysis. Therefore, we chose to simulate the diffusion process among agents by means of an Agent Based Model (ABM) depicting a population of farmers. The model was calibrated on real world data gathered from the case of a network of specialist vegetables producers in the Province of Foggia. Both SNA measures and rate of adoption are simulated data. The results achieved represent the basis for the breaking down of a tailored SAP diffusion strategy within an environmental and sustainability oriented development policy in a rural context, like the one studied. In particular, this study offers valuable hints on the kind of spreaders that should be enrolled, indicating, at the same time, the path for further research. This includes a more in depth analysis on various structure of networks (e.g. very dense and very sparse, very randomized and very regular, with high and low medium degree) and the investigation on the effects of the number of exposures of the agents to the promotional strategy.

Keywords Organic Fraction of Municipal Solid Wastes • Diffusion of Innovation • Sustainable Agricultural Practice • Social Network Analysis • Agent Based Model

14.1 Introduction

In the last years the volume of bio-waste has increased considerable, hence the issue of bio-waste valorization has become a priority for governments, environmental and social organizations, businesses and academics (Morone et al. 2015). Recent studies have demonstrated that the organic fraction of municipal waste can be used to produce biodegradable materials. Research carried out at the University of Torino over the last 7 years for instance, has demonstrated that urban and agriculture wastes are source of soluble bio-based substances (SBOs) that can be used for different applications among which the best performances were shown by reagents for the production of biodegradable plastics, that are particularly appropriate for the manufacture of mulching films used to cover agricultural soils (Montoneri et al. 2011).

Hence, in this work we consider the biodegradable mulching film containing SBOs, as a new Sustainable Agricultural Practice (SAP) that has a dual function:

- to broaden the spectrum of Organic Fraction of Municipal Solid Wastes (OFMSW) management;
- to improve agricultural sustainability.

Concerning to the first function, the OFMSW is humid, therefore the decomposition in the landfill generates leaching and fermentation phenomena and production of bad odors. In particular, the leachate from landfills where biodegradable waste is conferred, if not properly managed, can be a source of heavy contamination of groundwater and water bodies. In addition, the methane produced within the body of the landfill, released into the atmosphere, has a climate-changing effect because, as the carbon dioxide absorbs infrared radiation emitted from the 'hot' surface of the Planet (Assmuth et al. 1992). SBOs are additional materials that can be obtained from the OFMSW treatment through aerobic and anaerobic digestion, two processes that are necessary to reduce the environmental impact of OFMSW disposal and to turn waste into useful products for a sustainable agriculture. In particular, SBOs are obtained from the compost and digestate idrolysis.

Regarding to the second function, the adoption of SAPs may represent an important opportunity to increase the environmental sustainability of agricultural sector (Reimer et al. 2012). In particular mulching technique has several advantages, as well as the higher soil temperature, the weed pressure reduction, the moisture conservation, the more efficient use of soil nutrients, the reduction of certain insect pests, the increase of crop yields (Kasirajan and Ngouajio 2012). Moreover the use of films derived from organic substances can reduce the considerable waste disposal problem after their use. In fact most mulching films are currently produced from petroleum-based plastics, usually polyethylene, and a lot of farmers consider the illegal practice of the films' on-site burning since the high transportation cost and landfill tipping fees (McCraw and Motes 1991; Kasirajan and Ngouajio 2012). Instead biodegradable mulching films at the end of their life, can be integrated directly into the soil reducing the large production of waste in agriculture (Scaringelli et al. 2016).

The first and second function of mulching films containing SBOs are particularly important in an area like the Province of Foggia (Apulia Region, Italy), the geographic field of application for this study. In fact, it is one of the most extended agricultural area in Italy, with 495,111.10 hectares of utilized agricultural area (UAA) (3.9% of national UAA) and it is a land largely intended to crops (99.9% of the total companies) (ISTAT 2010), and at the same time is one of the five areas in Italy, where there is the highest rate of pollution resulting from the incineration of agricultural waste (ISPRA 2013), with a production of 10,254 Mg of carbon monoxide (CO), 473 Mg of nitrogen oxides (NOX) and 488 Mg of methane (CH4) (ISPRA 2013). Therefore, this is an area where it is necessary to improve the agricultural sustainability.

Regarding the organic fraction of municipal solid waste (MSW) collecting in the last years, in particular between 2011 and 2014, there were, an increase of about 500 thousand tons (+9.7%) of separate collection of organic (wet + green), like in the other Apulia province. This means that in this Province too, like in the other of Italy, we face more and more the problem of how to manage this organic fraction derived from MSW, therefore, the production of mulching films containing SBOs derived from the hydrolysis of the digestate and the compost obtained from OFMSW treatment, represent an additional opportunity for OFMSW management.

Concerning the potential adoption of this new SAP, according to Tey et al. 2014 there are many studies that have attempted to understand which factors influence it.

Analyzing these studies Tey and the other authors identified different factors from which depends SAP adoption in agriculture and divided them into six dimensions: socio-economic factors; agro-ecological factors; informational factors; psychological factors; institutional factors. The socio-economic dimension includes factors like, gender, farmer's age, educational levels, and some farm-specific characteristics (farm size, farming experience, access to financement etc.). The second dimension concerns variables like the practice of organic farming and geographical location. Very important are informational and psychological factors too, which made up the fourth and fifth dimensions, like usefulness of information, intention to adopt, habits and perceived attributes (Rogers 2003). The last dimension regards institutional factors, as well as organizational membership, participation in institutional arrangements, participation in certification programs that ultimately refers to the farmer's presence in different social networks.

In this context, the aim of this chapter is to underline which role the social networks may play in the diffusion of agricultural innovation, in particular focusing on farmers' attitude towards the adoption of mulching films derived from organic waste. Here, we stress the fact that the network position of an actor affects the power and influence he can exert on its immediate neighbors as well as on the collective behavior of the members. This influence can be viewed as a strategic resource for innovation diffusion purpose in a marketing or policy context. Bearing this in mind, we want to identify the network characteristics associated with effective injection points, namely, the actors where the novelty is first inoculated. The success for an injection point can be measured as the proportion of actors who adopt the innovation at each times of the process (usually half and final time). The network characteristics of an actor are typically captured by centrality and position measures elaborated by means of Social Network Analysis (SNA) tools. Thus, our purpose is to find the centrality and position measures associated with high adoption rates. However, being the innovation process new in nature, there are not available experimental data to conduct this kind of analysis. Therefore, we chose to simulate the diffusion of the SBOs mulching films using an Agent Based Model (ABM) based on real world data. By means of the model we obtained simulated SNA measures and rate of adoption. These data allow to test if the best spreaders are also those who have central position within the network.

The chapter is structured as follows. In Sect. 14.2 is depicted the theoretical framework of this work. In particular, in Sect. 14.2.1 a brief literature about the influence of the social network on the innovations diffusion is discussed. Sect. 14.2.2 is dedicated to a brief overview of works that use ABM both to include the role of social networks in diffusion of a novelty and to study the effect of different promotional strategies that take into account of this important role. The Sect. 14.3 describes the methodological framework, especially the model and the social network indicators, the data collection method and the case study used for the identification of the model parameters. In Sect. 14.4 the results are shown. Some concluding remarks are provided in Sect. 14.5.

14.2 Theoretical Framework

14.2.1 Diffusion Networks

There are several scholars that highlights the importance of interpersonal networks in the innovation diffusion process,[1] among which Rogers points out the importance of interpersonal channels in persuading an individual to adopt an innovation. Another means of information diffusion are Mass media channels as well as radio, television and newspapers that according to Rogers are often the most rapid and efficient means to create awareness-knowledge, in other words to inform the potential adopters about the existence of a novelty. However, the interpersonal channels are salient in speeding up the adoption of the process. Therefore, mass media can be seen as creators of primarily knowledge, while interpersonal networks are the best means to persuade individuals to adopt or refuse an innovation (Rogers 2003). In literature, there are different works exploring the effects of interpersonal networks on diffusion dynamics. These models, called "who-to-whom" studies, (Coleman et al. 1966; Rogers and Kinkaid 1981) are based on the fundamental assumption that consumers adopt an innovation when they collected sufficient information about the novelty from their peers. This mechanism causes the S-shaped diffusion curve. As described by Rogers (2003:272): "The s-shaped adopter distribution rises slowly at first when there are few adopters in each time period. It then accelerates to a maximum until half of the individuals in the system have adopted. It then increases at a gradually slower rate as the few remaining individuals finally adopt". A general conclusion of who-to-whom studies is that the most important determinant of who talks to whom in diffusion networks is the heterophily/homophily degree intended as the space and social distance between the potential adopters.

In particular, Rogers, based on De Tarde (1903), Lazarsfeld and Merton (1954) works, claims that '…homophily is the degree to which pairs of individuals who interact are similar in certain attributes, such as beliefs, education, social status, and the like.' (Rogers 2003, p. 305). Heterophily is the opposite of homophily, in fact, Rogers (2003, p. 306) defines it as '…the degree to which pairs of individuals who interact are different in certain attributes' and states that when two individuals have common meanings, beliefs, and a mutual language, their communications have more chances to be effective. Even if heterophilous communication channels are not

[1]According to Rogers (2003:169): "innovation decision process consists of five stages; Knowledge occurs when an individual (or other decision making unit) is exposed to the innovation's existence and gains some understanding of how it functions; Persuasion occurs when an individual (or other decision making unit) forms a favorable or unfavorable attitude toward the innovation; Decision occurs when an individual (or other decision-making unit) engages in activities that lead to a choice to adopt or reject the innovation; Implementation occurs when an individual (or other decision making unit) puts an innovation into use; Confirmation occurs when an individual (or other decision making unit) seeks reinforcement of an innovation-decision already made, but he or she may reverse this previous decision if exposed to conflicting messages about the innovation".

efficient as homophilous one, according to Granovetter's (1973) theory of "the-strength-of-weak-ties, they are very important in carrying information about innovations. An extension of whom-to whom studies is represented by the so called threshold models (Granovetter 1978; Markus 1987) in which the decision of adopting the innovation of the agents belonging to a network of potential adopters, depends on the proportion of their neighbors that have already adopted. According to Rogers (2003), thresholds models have been developed on the basis of the social learning theory. Moreover Rogers say that the great exponent of this theory is Professor Albert Bandura (1977) and he define this theory as: "a social-psychological theory according which an individual learns from another by means of observational modeling" (Rogers 2003:342). Besides giving a definition of this theory he assert that it can be directly applied to diffusion networks. From Rogers definition of social learning theory we can conclude that its fundamental idea is that the potential adopter decides whether or not to buy the new product, not only on the basis of his own preferences but also on the decisions of his neighbors in the social network. Hence the adoption behavior of one member influences the adoption decision of another member (Rogers 2003). In diffusion networks an important role is played by opinion leaders. According to Rogers (2003, p. 388) '… opinion leadership is the degree to which an individual is able informally to influence other individuals' attitudes or overt behavior in a desired way with relative frequency' Therefore, opinion leaders are those individuals in the network that are better able to influence others' opinions about innovations. Moreover, the effects of their roles and position in the network have been deeply analysed (Weimann et al. 2007). Centrality (Berelson and Steiner 1964; Czepiel 1974; Valente 1996) and other relational features of opinion leaders, such as innovativeness and their interpersonal influence, may significantly affect their effectiveness as spreaders. Finally, according to Deutsch and Gerrard (1955) it is possible to discriminate two main types of interpersonal influence: informational and normative influence. In particular, the former can be seen as the acceptance of the information from others as evidence of reality. For example, consumers can be directly influenced by opinion leader through advice and verbal directions about their use of the new product. The latter is the tendency to conform to the expectations of others. Grewal et al. (2000) distinguish a different importance of each type of influence on the basis of the product and the situation. In particular, they highlight the importance of informational influence for privately consumed goods, and those of both type of influence for publicly consumed goods. Recent works, like Deroïan (2002), show that opinion leaders can accelerate the innovation diffusion process and are potentially interesting in supporting marketing and political strategies.

From the previous discussion it can be concluded that diffusion is a dynamic process where social influence has a crucial role (Kiesling et al. 2012). This process has been modeled through traditional aggregate models based on differential equation formulation (Bass 1969; Chatterjee and Eliashberg 1990) and in recent years by Agent Based Models (ABMs). Aggregate models such as the Bass model (Bass 1969) can address only some theoretical issues because they do not explicitly consider consumers' heterogeneity and the complex dynamics of social processes that are at the basis of diffusion processes. At the contrary, ABMs are able to represent complex emergent phenomena (Lopolito et al. 2013), such as the diffusion of a novelty in a socio-economic system.

In particular, according to Kiesling et al. (2012): "ABMs, differently from traditional differential equation models of innovation diffusion offer researchers the opportunity to explicitly model the interactions that exert social influence, and thereby allow them to take the structure of social interactions into account". This is important, because, as remarked by Katz (1961, p. 16):

> "It is as unthinkable to study diffusion without some knowledge of the social structures in which potential adopters are located as it is to study blood circulation without adequate knowledge of the veins and arteries".

In particular, due to the fact that in the ABMs of diffusion, the atomic element is the single agent and not the social system as a whole, they are particular indicated to explicitly represent consumers' heterogeneity, their social interactions, and their decision making processes. Moreover, ABMs can include several marketing variables providing the opportunity to decision makers to test different strategies for diffusion of innovation in what-if experiments.

A good literature review of the ABMs for diffusion of innovation is presented in Kiesling et al. (2012). Among the different works analyzed by these authors we focus on those using ABM both to explore the structural effects of social network topology and the effectiveness of different promotional strategies in terms of acceleration of the diffusion process. The following section briefly reviews these selected works.

14.2.2 Agent Based Model for Diffusion of Innovations

There are several studies that, comparing different relation network topologies between potential adopters, demonstrate the importance of Word of Mouth (WoM) process in diffusion of innovation, as Goldenberg et al. (2001) that highlight how the diffusion process is dominated by WoM rather than by advertising. In particular they compare the probability of an agent to be informed on the existence of an innovation by weak-tie WoM, strong-tie WoM and the exposure to different marketing efforts. They demonstrate that that, after the early stage of the diffusion process, the effect of promotional strategies (e.g., advertising) quickly decreases, and strong and weak ties become the main important propellers of adoption. These results support Rogers theory according to which advertising is more effective in the initial knowledge stage of the innovation diffusion process, than in the following persuasion and adoption stages where WoM becomes the main driver of adoption.

In a following work Goldenberg with Moldovan (2004) extended the above mentioned model introducing three different type of consumers which spread different kind of information about the innovation: uninformed (that not propagate information), adopter (that propagate positive information), or resistor (that propagate negative information). Moreover, they divided the population into three groups:

- opinion leaders, who may be the innovation adopters;
- resistance leaders, who may only reject the innovation;
- regular consumers that may be influenced by positive and negative WoM.

In this model adoption decision is influenced with a certain probability, by positive WoM or advertising, while rejection depends only on negative WoM. Varying the proportion of opinion and resistance leaders in the market as well as the probabilities of being influenced by advertising and positive/negative WoM the authors concluded that resistance leaders will reduce sales significantly, on the basis of their relative number and the power of their social influence. Regarding the effectiveness of advertising, they find that it has a small and nonlinear effect on market size in markets where both opinion and resistance leaders play a role. Finally, the authors demonstrate that if opinion leaders are activated in advance on unfocused advertising messages, they can mitigate the destructive effect of resistance leaders and significantly enhance the market size.

The importance of a focused advertising strategy is also highlighted by Alkemade and Castaldi (2005). They investigate whether firms can use information about the network structure of the potential adopter and their personal characteristics to project a successful directed-advertising strategy. The authors concentrate their attention on fashionable products modeling both "exposure" and "over-exposure" thresholds The former represents the minimum proportion of adopters in the neighborhood necessary to the adoption decision of the single agent and the latter is the maximum proportion of neighbors that adopt the product, beyond which the adoption decision is inhibited because the proportion of adopters in the social network is too large, therefore the innovation is no more "fashionable". Authors assume that firms are not fully aware of the communication structure among the potential adopters, therefore to identify efficient strategies, they use a genetic algorithm targeting individual consumers to model the strategy search and the learning behavior of the firms. In their work,testing different scenarios with varying assumptions on the consumption decision, Alkemande and Castaldi show that in a sparse network, cascades occur even when consumers' exposure threshold is high while in a denser network ,the probability that cascades take place and the critical exposure threshold decrease. In addition, the authors find not significative differences between the small-world and regular networks critical exposure thresholds. Finally, in their work they compare diffusion rates of a dynamic advertising strategy and a random advertising results, demonstrating the overperformance of the former.

Goldenberg et al. (2007) investigate on advertising strategies in the context of positive and negative WoM, comparing linear and concave advertising strategies. According with previous works the authors model adoption as a probabilistic transition between three states, adoption, rejection, and none, on the basis of the.

probabilities of being influenced by positive WoM, advertising, and/or negative WoM. The network underlying the simulations is a dynamic small-world-type composed by permanent strong ties and randomly changing weak ties. In the model the size of strong ties and weak ties, the percentage of disappointed consumers, and the probability of being influenced by advertising and positive/negative WoM via strong/weak ties, vary, to create the experimental conditions. Findings indicate that the presence of dissatisfied consumers and of weak ties, can cause considerable damage to long-term profits, since this consumers create an invisible diffusion of product rejection.

Another important study of 2007 on the effect of different promotional strategy is Delre et al. (2007), which investigate this issue in terms of "final market penetration" and "time to takeoff". They test these strategies with respect to brown goods (i.e., electronics) and white goods markets (i.e., household products). Results can be summarized as follows: (a) if there is no promotional support and/or a wrong timing, the product diffusion may fail; (b) the best targeting strategy is to contact distant, small and cohesive groups of potential adopters; and (c) the best timing of a promotion varies on the basis of the different durable categories (white goods, like for example kitchens and laundry machines; brown goods, like TVs and CD players).

An interesting promotional strategy, studied by Valente and Davis (1999), points on the important role of highly connected individuals in a network of potential adopters (opinion leaders) and uses them as a marketing or policy instrument to speed up the diffusion process. In particular the authors find that a diffusion campaign initiated by opinion leaders is more efficient in terms of diffusion acceleration rather than one carried out by random or marginal agents. Like the previous work Delre et al. (2010) investigate the effectiveness of a promotional strategy based on the opinion leader engagement. Results indicates that the highly interconnected agents have the important function to inform their neighbors about the new products, but they do not have an important role in influencing the decision process of consumers. They also find that in markets in which such highly connected actors do not exist, diffusion hardly happens. For this kind of markets, it is more appropriated a direct-to-consumer advertising to stimulate the diffusion of the novelty in the network.

Like Valente and Davis (1999) and Delre et al. (2010) we analyze, through an agent based simulation, the effect of a promotional strategy on the diffusion of an innovation, in particular a new SAP, in a network of potential adopters. This promotional strategy consists in the inoculation of the information about the innovation to a set of injection-points (first-informed agents in a social network) (Banerjee et al. 2013). We use a diffusion model based on a threshold function. Differently from the previous works that select the best spreaders in the network on the basis of the degree centrality measure (Delre 2010) or on the basis of the nomination from the others network agents (Valente and Davis 1999) we consider others network measures, as well as betweenness, closeness and clustering coefficient (Wasserman and Faust 1994). In particular, we seek to find the most appropriate network measure to choice the injection points. Moreover, like Goldenberg et al. (2001) we use a static network. In order to calibrate the model we use real relational data obtained by an interview to a local expert, that will be described in the next section.

14.3 Methodology

14.3.1 Theoretical Model of Diffusion

In order to study the diffusion dynamics of SBOs mulching films in this work we used a simple theoretical model of diffusion. It depicts a network of farmers linked by professionals and social relationships. The basic assumptions are that the single agent (the farmer) is persuaded to adopt the new technology based on

(i) its awareness of the novelty,
(ii) its inertia toward the novelty (that is its resistance to innovate),
(iii) the attitude of its neighbors (that is the influence of others).

Evidently, the agent's attitude (sub ii) depends on its *internal characteristics*, while both the chances to become aware of the novelty (sub i) and the influence received (sub iii) depends on its *environment*. The internal characteristics of the agent, relevant for this model, are the level of education, the preference toward the new technology, and an innovation threshold that, compared with the preference, determine agent's decision on the adoption. The environment features are catch by its number of neighbors, and its level of homophily with each neighbor. Each agent interacts with its environment by means of a series of links connecting it with its neighbors. Each link is bidirectional and allows it to receive/send information and influence from/to its neighbors.

Figure 14.1 depicts the model dynamics. In the figure, a node represents the single farmer, that is the decision-making unit, and the edges represent the bidirectional relations linking the farmer with his neighbors. The process of novelty diffusion can be unpacked in four consecutive phases.

1. An initial set of agents is endowed with the novelty. These agents are conceived as the injection points where the novelty is inoculated. The injection points are not only informed of the new technology but are persuaded to use it, as their preference of the novelty become higher.
2. The injection points pass the information about the novelty and about their preferences toward it to their neighbors who became aware and form their own preferences in turn.
3. The agents informed decide whether to adopt the novelty comparing the level of their preferences with their innovation threshold.
4. In each subsequent period, the agents form or reconsider their preferences on the basis of the information received from its neighbors.

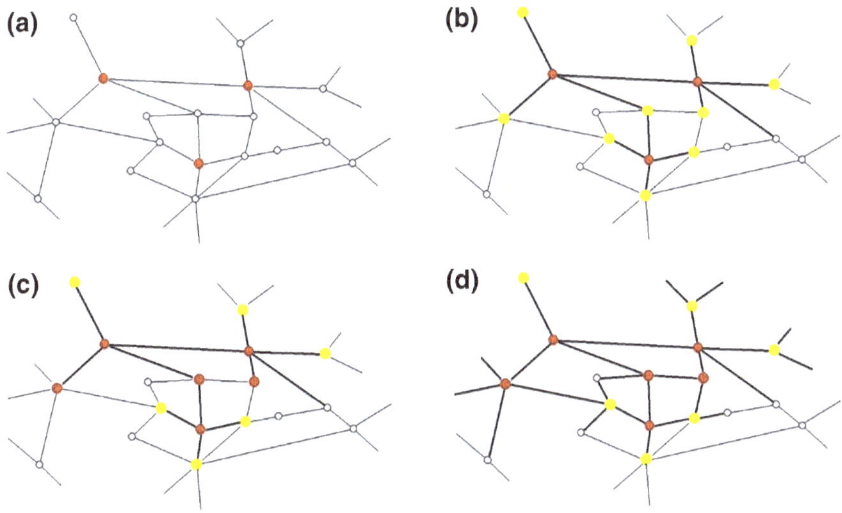

Fig. 14.1 The diffusion model. **a** The injection points are endowed with the novelty (*red circles*). **b** The injection points pass the information to their neighbors (*yellow circles*) who became aware and form their own preferences. **c** The agents informed decide whether to adopt the novelty. **d** The agents informed pass the information to their neighbors in turn (*Source* our elaboration)

The process described in points 3 and 4 repeats until T periods of information passing. In forming/reconsidering their preferences, the each agent consider:

- the preferences of its neighbors toward the novelty (Martins et al. 2009);
- the level of homophily with these neighbors (Mc Pherson 2001; Rogers 2003);
- its level of education that is its capability to interpreter the new information.

In each reiteration, the attitude of the agent depends on its attitude in the previous period, plus the average of the attitude of its neighbors weighted with the homophily level. A correction factor for the agent i (C_i) is used to take into consideration the level of knowledge/education of the agent. It multiplies the average of the attitudes of the neighbors and is calculated as follows (Eq. 14.1):

$$C_i = \frac{E_i}{maxE} \tag{14.1}$$

Where E_i represent the years of education of the agent i; and $maxE$ is the number of years of a complete course of education (that have been fixed at 21) including PhD formation. After forming their references as described, the agents compare their preferences with the persuasion threshold and decide to adopt if the level of their preferences equal or overcome the threshold.

14.3.2 Chart Flow of Methodology

The analysis carried out consists of three main steps. At the first, the innovation market and agents who potentially can adopt the SBOs mulching film were identified. The case study refers to the Province of Foggia (Apulia Region, Italy), that is one of the most extended agricultural area in Italy. A survey was designed to collect information about relational variables affecting farmers' willingness to adopt such new mulching materials. In particular, information related to the existence of cooperative association, business relationship, friendship and kinship, were gathered (for details see Sect. 14.3.3).

Secondly, based on the survey data, the networking structure was built. The parameters of link attributes (i.e. homophily) and agent attributes (i.e. threshold, injection and education) were calculated. To build the network, we considered a link between two firms if they knew each other for professional (i.e. they worked in the same cooperative) or social reasons (i.e. they were friends or acquaintance).

Thirdly we used the NetLogo 5.2 platform (Wilensky 1999) for implementing our model. Provided that the model assumption is that the information about the novelty goes through the agents' connections, the network connections was used to calibrate the simulation model. The basic chart flow of the model, its structure and interaction rules were elaborated *ex-novo* specifically for the scope of this research by the authors. To import the network investigated, we included in the model a routine adapted by the 'Network Import Example' authored by Uri Wilensky and available at the modeling commons platform.[2]

The main output of the simulation is the number of adopters at half and final time. In order to estimate the model time span, we equated the number of model ticks needed to reach a steady state in the level of adopters (that is the tick where the level of adopters does not increase further) to five years time-lapse, that is a reasonable time frame needed for an innovation to reach the maturity level of diffusion (where the late majority adopt the novelty). For the overwhelming majority of the injection points, the steady state in the level of adopters is reached around 45 model periods, therefore in our model a tick is equivalent to 40 days, and a year is split in nine ticks. Since, in the real world, farmers can decide to adopt mulching films once per year, the agents of the model update their adoption decision each nine ticks.

Since the simulation process implies various random variables and routines, we simulated each injection point repeatedly (batches of 20 runs) in order to produce reliable data. In order to guarantee high robustness of results, we took averages of adopters within the batch. Specifically, for each injection point we calculated the mean of adopter at $T = 22$ time steps, representing the half time span, and at $T = 45$ that corresponds to the final model time.

[2]The Network Import Example is downloadable at: http://modelingcommons.org/browse/one_model/2214#model_tabs_browse_procedures.

14.3.3 The Case Study

Because of crop-specific adoptability of mulching technique, the targeted-farmers could be potentially interested in the use of a SAP represented by a SBOs mulching film. The sampling was organized grouping the farmers into four areas according to geographical farm location (Fig. 14.2).

The municipality belonging to each area and the number of farmers interviewed are reported in Table 14.1.

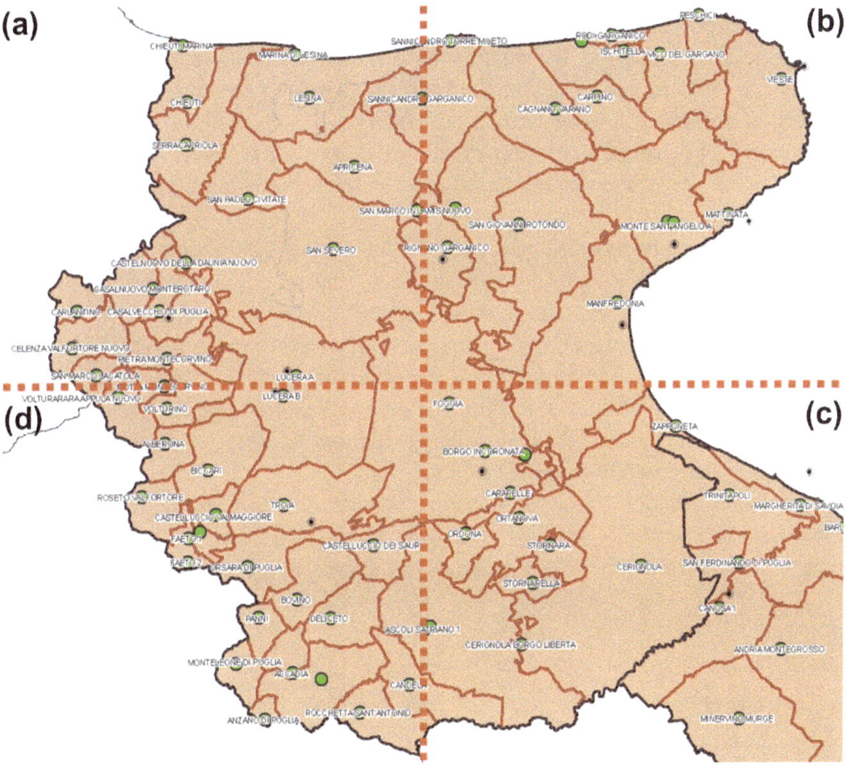

Fig. 14.2 Localization map of observations

Table 14.1 Sample key features

Variable	Mean	Standard deviation	Min	Max
Age (years)	45.74	11.6	24	72
Farm size (ha)	76.96	203.71	4	1805
Employees	13.94	16.76	1	112
Education (years)	10.84	3.53	5	18
Distances between farmers (km)	59.77	40.92	0	198.42

A survey was carried out in each area by means of face-to-face interviews during the month of May 2014. For each farm data on relational and demographic characteristics were gathered. To detect the networks relationships, we also interviewed two technicians working in the context investigated. The observers were asked to:

- identify those farmers belonging to the same cooperatives, in order to detect professional relationships;
- identify who knows who, to reveal social relationships.

Although sample size is small, 2% of censed (ISTAT 2010) farms specialized in vegetables production in the area were gathered. Of course, the aim here is not to produce statistical results with inference aims, but to provide real world data in order to calibrate the model. The suitability of the case study lies in the fact that it provides a cross section of the relational set of farms producing vegetable crops in the area. In order to gather the relevant information, we used a participatory social network approach involving two direct observers. This method is based on the involvement of actors directly implicated in the network investigated, by means of workshops or deep interviews to co-produce a representation of that network (Edwards et al. 2010).

From the sample key features used in the model calibration were obtained. Table 14.2 shows the statistics on farmer's age, farm size, number of employees, farmers' educational levels and average distance among each one.

By means of the questionnaire, farmer's attitudes towards the SBOs novelty were also gathered. Table 14.3 reports the observed attitudes. Farmers are divided into six level of persuasion, according to how far they are from adopting the SBOs mulching film technique (1 most adverse–6 most favorable). Each level has a persuasion score ranging from −1 to 1.

On this information base, we grounded the identification of the key parameters of our model. The parameter identification is reported in Table 14.3. The basic link attribute is the homophily. It represents the degree to which pairs of individuals

Table 14.2 The persuasion level of the farmers

Persuasion level	Observation	Frequency %	Persuasion Score
Adverse to mulching films	13	16.25	−1
Willing to adopt mulching technique (conventional) but adverse to adopt SBOs films	2	2.5	−0.67
Already adopting conventional films but adverse to adopt SBOs films	12	15	−0.33
Non adopting mulching technique but willing to adopt SBOs films	12	15	0.33
Already adopting conventional and willing to adopt SBOs films	16	20	0.67
Already adopting bio-films and willing to adopt SBOs films	25	31.25	1

Table 14.3 Parameter identification

Parameter	Symbol	Description	Value
Global			
Number of agents	N	the number of agents interacting in the model	80
Links attributes			
Homophily	h	It represents the level of homophily of the link's ends	Various [0,1]
Agents attributes			
Threshold	θ	the innovation threshold	Discrete
Injection	I	It is a logic value: true if the agent is an injection point	Boolean
Education	edu	It represents the year of education of the agent	Discrete

Source our elaboration

who interact are similar in certain attributes, such as beliefs, education, social status, and alike. When two individuals share common meanings, beliefs and a mutual language communication between them is more likely to be effective (Rogers 2003).

In our model, we set this parameter on the basis of four socioeconomic attributes (see Table 14.2), instead of a single one. This provides a more robust approach in creating homophilous relationships across the network than choosing any one characteristic as the basis for all homophilous ties (Blau 1984, Mc Pherson 2001, Centola et al. 2007). To each link we assigned an homophily score given by the mean of four homophily index, that are

1. Agent's age (ratio between the minimum and the maximum age of its ends);
2. Farm size (ratio between the minimum and the maximum size of its ends);
3. Number of employees (ratio between the minimum and the maximum number of employees of its ends);
4. Location (given by the maximum distance minus the distance of the ends divided for the maximum distance).

Finally, we used the persuasion score (see Table 14.3) to set the innovation threshold of the agents.

14.4 Results

Figure 14.3 depicts the network of farmers reproduced by the model. The key network features of this web are reported in the Table 14.4. As shown in the table, we deal with a network formed of a unique component (3), not fragmented (5) characterized by a high density (2), where nodes have 16 relations with others in mean (1).

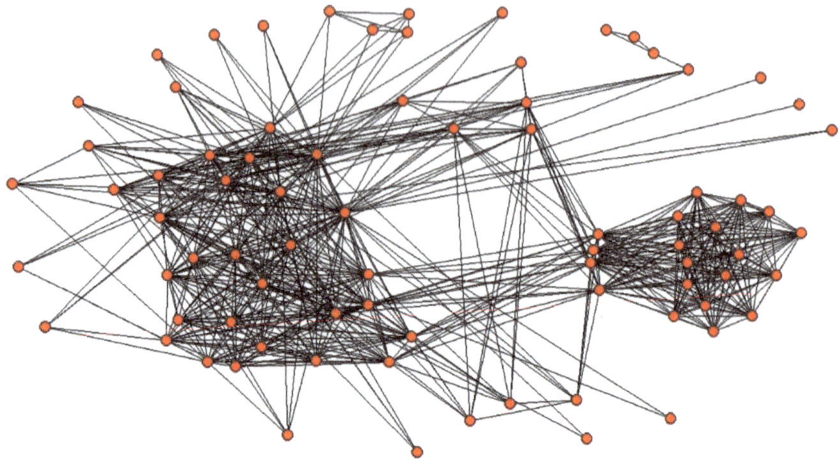

Fig. 14.3 The firms' network (*Source* our elaboration)

Table 14.4 Network cohesion

Network measures	
1 Average Degree	16.2
2 Density	0.205
3 Components	1
4 Connectedness	1
5 Fragmentation	0
6 Average Distance	2.415
7 Diameter	6
8 Overall Clustering Coefficient	0.78

The average distance (6) of two random chosen nodes is low, while the maximum distance revealed (7) is six. On the whole, the network is very clustered (8).

This network context represents the information basis to properly interpreter the model findings. Using this social tissue, the novelty reaches different diffusion outcome in relation to the various injection points used. The injection points vary in terms of position, power and role they play within the network of farmers. Since the objective of this work is to find the actors able to act as effective spreaders of a SAP within a network based on their centrality and position, we calculated several SNA measures of the farmers. The measures adopted are (Wasserman and Faust 1994):

1. the Degree Centrality, that is defined as the number of links of the single node. The degree can be conceived as the immediate potential of a node for influencing the information flowing through the network;
2. the Betweenness, that measures the number of times a node acts as a bridge along the shortest path between two other nodes. It can be an indicator of the influence of an agent on the communication between other agents. The

Table 14.5 SNA measures and adoption rates

Network measure	Mean	St. dev	Min	Max
1 Degree Centrality	16.20	10.10	1.00	45.00
2 Betweenness	55.87	98.27	0.00	554.08
3 Closeness	43.22	8.56	22.01	59.85
4 Local clustering coefficient	0.78	0.24	0.25	1.00

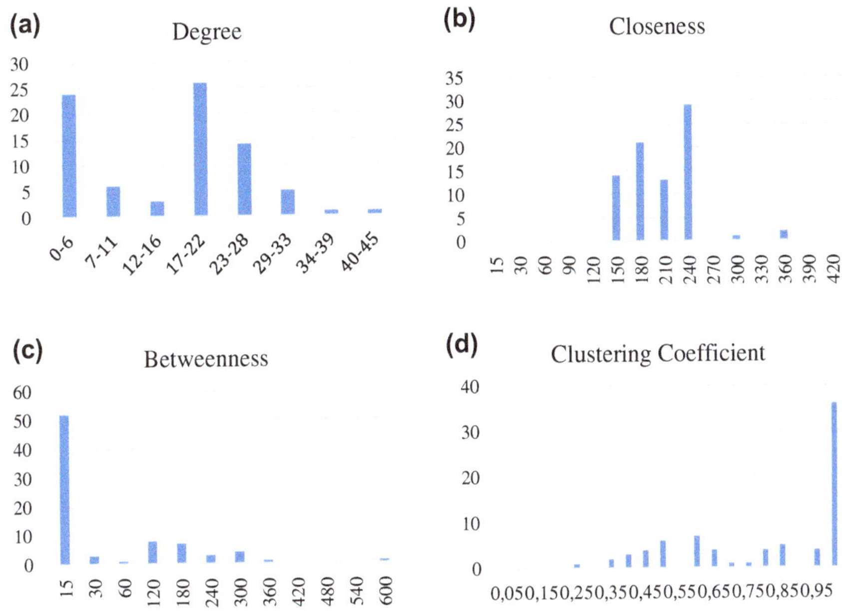

Fig. 14.4 Frequency distribution of the network measures (*Source* our elaboration)

Betweenness of node *i* is calculated as the proportion of the shortest paths of others passing through *i*;

3. the Closeness, that is the reciprocal of the farness of a node. The farness is the sum of the distances of a node from all other nodes. Thus, the closer a node is the lower its total distance from all other nodes;
4. the local Clustering Coefficient, that measures how close is the neighbors of a node to being a completely connected.

Table 14.5 reports the descriptive analysis of these measures. Panels A-D in Fig. 14.4 show the frequency distribution of each index.

The Table 14.5 shows that, in mean, the node has 16 ties with neighbors (1), intercepts 56 shortest path length among others (2), is quite close with others (3), and has very clustered neighborhood (4). The frequency distribution of these measures (Fig. 14.4), confirms that the most part of nodes has a high degree (panel A). On the contrary, the high average value of the betweenness is due to few actors

Table 14.6 Number and rate of adopters at half and final time

Number of adopters	Mean	St. dev	Min	Max
T = 22(%)	10.78 (12)	0.70 (1)	6.00 (6)	11.10 (13)
T = 45(%)	19.40 (23)	2.34 (3)	9.30 (10)	20.90 (25)

Source our elaboration

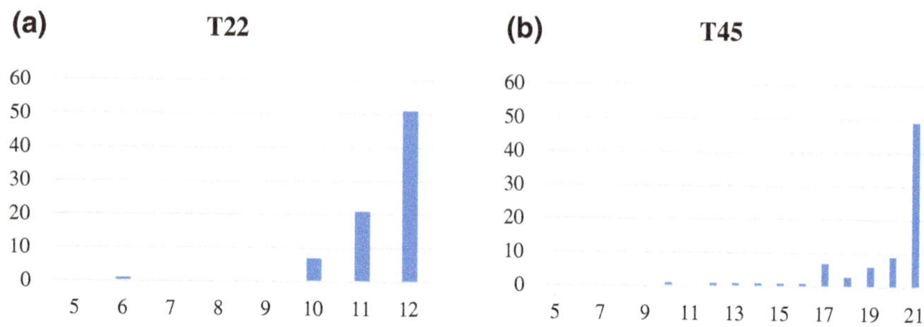

Fig. 14.5 Number of adopters at half and final time (*Source* our elaboration)

with high values, but the most part falls in the first class, thus the norm is a value of 15 or less (panel B). The measures of closeness are concentrated around medium and high values (panel C), while the clustering coefficient is the maximum for 36 actors, with some other actors showing various levels (panel D).

Table 14.6 reports the number of adopters and the adoption rates. The average number of adopters at T = 22 is around 11 and at T = 45 is 19, corresponding respectively to 12% and 23%. With reference to the population, it seems that the level of dispersion is very low (1% and 3% respectively), meaning that the most part of the injection points are flattened around the mean. In the best case ¼ of the total population is reached, that is the best injection point can cover at the most 25% of the network.

Figure 14.5 shows the frequency distribution of the number of adopters at half and final time. In both cases the most part of injection points falls in the highest class. At t = 22 (panel A) we have only one spreader with a very low performance (six adopters), while the rest of the injection points range from 10 to 12 adopters. The results related to the final time (t = 45, panel B) show a greater variation. 14 spreaders reach 17 or less adopters while 19 spreaders fall between 18 and 20 adopters, and 49 injection points reach the level of 21 or more.

These findings suggest, for this specific context studied, that the most part of the work is done by a very dense and clustered network and only little room for improvement is left to the individual characteristics of the injection points. Figure 14.6 shows the effect of the network position of the injections points on the levels of adoption. Panels a-d report the scatter plots of the network measures (x) and the related number of half (blue points) and final adopters (red points) (y).

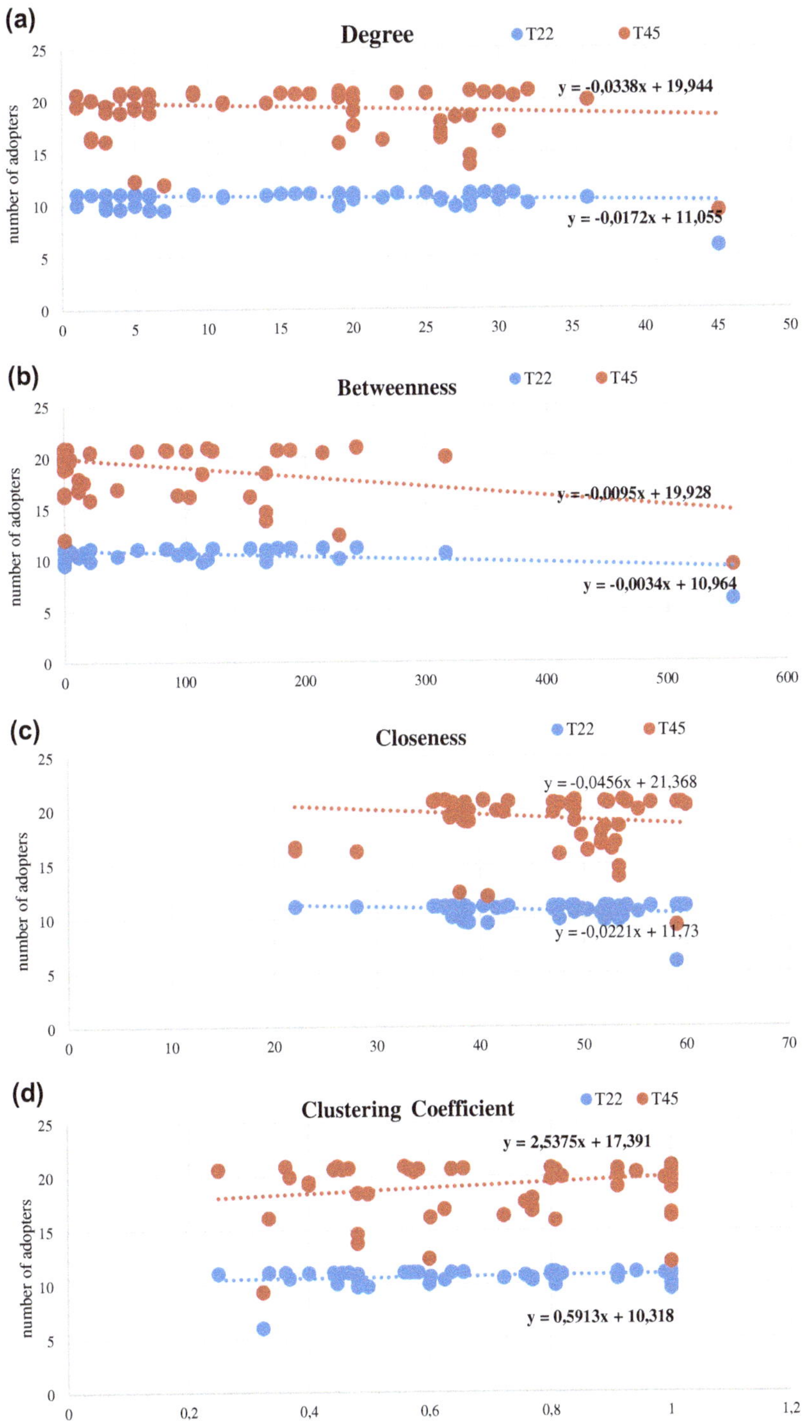

Fig. 14.6 Network measures and level of adoption (*Source* our elaboration)

A first observation is that a high intercept in all diagrams confirms that a very dense network favor the innovation diffusion regardless the points of its inoculation. An interesting point is that the network measures of degree centrality (panel A), betweenness (panel B) and closeness (panel C), that are indices of high information power, surprisingly, are negatively correlated with the adoption rates. This is particularly counterintuitive in the case of Betweenness, that measure the bridge role of the actors. Probably, this is due, in particular to some outliers that have high centrality but a lower performance.

On the contrary, a positive relation is found in the case of the clustering coefficient (panel D). This indicates that, at least in this kind of network, for diffusion purposes, is not primarily important how connected is the injection point and how high is its immediate influence, but how close it is to the its immediate neighborhood proven that the clustering coefficient indicates how much close are the neighbors of the spreader. This sheds light on a second factor behind the capacity of the spreader to influence others, that is the need for a social reinforcement of the innovation stimulus. Indeed, these result highlights that, for an effective diffusion strategy, the influence is important but more important is the effect of reinforcement coming from multiple sources of information that the structure of the network can favor among its members. This mechanism is assured by a clustered neighborhood, which warranties a certain redundancy of the information. In fact, since the innovation adoption is a costly activity requiring reorganization, there is a sort of the inertia to change represented by the innovation costs, and captured in the model by the innovation threshold. In order to overcome this barrier, the single actor needs repeated stimulus to the innovation.

14.5 Concluding Remarks

The environmental sustainability is a central theme of Rural Development Policy 2014–2020. The adoption of SAPs such as the SBOs mulching films may represent an important opportunity to increase the environmental sustainability of the agricultural sector. In fact, as well as the environmental benefits related to the practice of mulching (such as soil conservation in agriculture, savings in the use of water resources, etc.) the adoption of SBOs-containing materials produces two additional positive effects in terms of environmental sustainability. It contributes to the problem of disposing of municipal waste, using the organic part of the waste for the production of new materials (in term of bio waste valorization) and reduces the phenomenon of pollution (emissions of greenhouse) linked to the use of plastic mulching materials, such as the burn-in field of the films after their use (a widespread phenomenon in the province of Foggia).

The objective of this work was to underline which role the social networks may play in the diffusion of agricultural innovation, stressing the fact that the network position of an actor affects the power and influence he can exert on its immediate neighbors as well as on the collective behavior of the members. In particular, bearing in mind this purpose, we analyzed the relationship between different centrality and position measures and the adoption rates in a network of potential adopters. We have done this simulating through an ABM the effect of various strategy of innovation diffusion in a set of potential adopters represented by a real network of specialist vegetables producers in province of Foggia. This network covers the 2% of the entire population of farms specialized on vegetables production in the area. Of course it does not represent the universe of the specialist vegetables producers of this province but gives us an insight into those that happens in this area.

The findings help in outlying the network measures that better catch the features of an effective spreader. In particular results show that the most work for the innovation diffusion is made by the network density. However, we demonstrated that there is a considerable scope for injection points' rule. In particular, we observed that the clustering level is the best predictor of who are the most effective injection points, because clusterization guarantees a minimum level of redundancy, providing some social enforcement.

This information represents the basis for the breaking down of a tailored sustainable agricultural policy. In particular, this study offers some hints on the kind of spreaders that should be enrolled indicating also the paths for further research. The results showed that no injection points have been able to overcome the threshold of a 25% rate of adoption, and this highlights the need to form packs of spreader that influence different areas of the network at the same moment in order to reach higher diffusion performances. This kind of analysis, and the class of model here used are useful in the formation of these groups of spreaders. It is also possible that other centrality measures, here not considered, can be valid predictors of high diffusion performance. It is useful to explore other indexes in further works. A caveat is that the results presented here are based on a little network with specific characteristics as a very high density and clusterization. Hence, a more in depth analysis is required in order to focus on various kinds of network with diverse characteristics. Thus, future works can consider networks with high and low density, very randomized and very regular, with high and low medium degree. Another interesting aspect that should be analyzed in future works is the effect of the number of exposures of each agent to the promotional strategy. These further explorations might give valuable indications to innovators both in the domain of marketing and of policy making.

References

Alkemade F, Castaldi C (2005) Strategies for the diffusion of innovations on social networks. Comput Econ 25(1–2):3–23

Assmuth T, Kalevi K (1992) Concentrations and toxicological significance of trace organic compounds in municipal solid waste landfill gas. Chemosphere 24(9):1207–1216

Banerjee AV, Duflo E, Glennerster R, Kinnan C (2013) The miracle of microfinance? Evidence from a randomized evaluation. Am Econ J: Appl Econ 7(1):22–53

Bass FM (1969) A new product growth model for consumer durables. Manage Sci 15(5):215–227
Deroïan F (2002) Formation of social networks and diffusion of innovations. Res Policy 31 (5):835–846
Berelson B, Steiner GA (1964) Human behavior: an inventory of scientific findings. Harcourt, Brace & World
Blau PM, Beeker C, Fitzpatrick KM (1984) Intersecting social affiliations and intermarriage. Soc Forces 62(3):585–606
Centola D, Gonzalez-Avella JC, Eguiluz VM, San Miguel M (2007) Homophily, cultural drift, and the co-evolution of cultural groups. J Conflict Resolut 51(6):905–929
Chatterjee RA, Eliashberg J (1990) The innovation diffusion process in a heterogeneous population: a micromodeling approach. Manage Sci 36(9):1057–1079
Coleman JS, Katz E, Menzel H (1966) Medical innovation: a diffusion study, Bobbs-Merrill Co, Indianapolis
Copp JH, Sill ML, Brown EJ (1958) The function of information-sources in the farm practice adoption process. Rural Sociology 23(2):146–157
Czepiel JA (1974) Word-of-mouth processes in the diffusion of a major technological innovation. J Mark Res pp 172–180
Delre SA, Jager W, Bijmolt TH, Janssen MA (2007) Targeting and timing promotional activities: an agent-based model for the takeoff of new products. J Bus Res 60(8):826–835
Delre SA, Jager W, Bijmolt TH, Janssen MA (2010) Will it spread or not? The effects of social influences and network topology on innovation diffusion. J Prod Innov Manage 27(2):267–282
Deroïan F (2002) Formation of social networks and diffusion of innovations. Res Policy 31 (5):835–846
De Tarde, G. (1903). The laws of imitation. H. Holt
Deutsch M, Gerard HB (1955) A study of normative and informational social influences upon individual judgment. J Abnorm Soc Psychol 51(3):629
Edwards GP, Zeng B, Saalfeld WK, Vaarzon-Morel P (2010) Evaluation of the impacts of feral camels. Rangeland J 32(1):43–54
Freeman LC, Freeman SC, Michaelson AG (1988) On human social intelligence. J Soc Biol Struct 11(4):415–425
Goldenberg J, Libai B, Muller E (2001) Talk of the network: a complex systems look at the underlying process of word-of-mouth. Mark lett 12(3):211–223
Goldenberg Jacob, Libai Barak, Moldovan Sarit, Muller Eitan (2007) The NPV of Bad News. Int J Res Mark 24(3):186–200
Granovetter MS (1973) The strength of weak ties. Am J Sociol pp 1360–1380
Granovetter M (1978) Threshold Models of Collective Behavior. Am J Sociol 83:1420–1443
Grewal R, Mehta R, Kardes FR (2000) The role of the social-identity function of attitudes in consumer innovativeness and opinion leadership. J Econ Psychol 21(3):233–252
ISPRA (Istituto Superiore per la Protezione e la Ricerca Ambientale) (2013) Italian greenhouse gas inventory 1990–2011. National inventory report, available at http://www.isprambiente.gov.it/files/pubblicazioni/rapporti/Rapporto_177_2013
ISTAT (Italian National Institute of Statistics) (2010) 6th General census of agriculture
Kasirajan S, Ngouajio M (2012) Polyethylene and biodegradable mulches for agricultural applications: a review. Agron Sustain Dev 32(2):501–529
Katz E, Lazarsfeld PF (1966) Personal Influence The part played by people in the flow of mass communications. Transaction Publishers, USA
Katz E (1961) The social itinerary of technical change: two studies on the diffusion of innovation. Human Organ 20(2):70–82
Kiesling E, Günther M, Stummer C, Wakolbinger LM (2012) Agent-based simulation of innovation diffusion: a review. CEJOR 20(2):183–230
Lazarsfeld PF (1944) The controversy over detailed interviews—an offer for negotiation. Public Opin Q 8(1):38–60
Lazarsfeld PF, Merton RK (1954) Friendship as a social process: A substantive and methodological analysis. Freedom Control Mod Soc 18(1):18–66

Lazarsfeld PF, Menzel H (1963) Mass media and personal influence. In: The science of human communication. Wilbur Schramm, pp 94–115

Lopolito A, Morone P, Taylor R (2013) Emerging innovation niches: an agent based model. Res Policy 42(6):1225–1238

Markus ML (1987) Toward a 'critical mass' theory of interactive media universal access, interdependence and diffusion. Commun Res 14(5):491–511

Martins ACR, Pereira CdB, Vicente R (2009) An opinion dynamics model for the diffusion of innovations. Phys A Stat Mech Appl 388(15–16):3225–3232

McCraw D, Motes JE (1991) Use of plastic mulch and row covers in vegetable production. Coop Extension Ser Oklahoma State University. OSU Extension Facts F-6034

McPherson M, Smith-Lovin L, Cook JM (2001) Birds of a feather: homophily in social networks. Annual Rev Sociol pp 415–444

Moldovan S, Goldenberg J (2004) Cellular automata modeling of resistance to innovations: Effects and solutions. Technol Forecast Soc Chang 71(5):425–442

Montoneri E, Boffa V, Savarino P, Perrone D, Ghezzo M, Montoneri C, Mendichi R (2011) Acid soluble bio-organic substances isolated from urban bio-waste. Chemical composition and properties of products. Waste Manag 31(1):10–17

Morone P, Tartiu VE, Falcone P (2015) Assessing the potential of biowaste for bioplastics production through social network analysis. J Clean Prod 90:43–54

Reimer AP, Weinkauf DK, Prokopy LS (2012) The influence of perceptions of practice characteristics: an examination of agricultural best management practice adoption in two Indiana watersheds. J Rural Stud 28:118–128

Rogers EM, Kincaid DL (1981) Communication networks: toward a new paradigm for research. Free Press, New York

Rogers EM (2003) Diffusion of innovations. Free Press, New York

Scaringelli MA, Giannoccaro G, Prosperi M, Lopolito A (2016) Adoption of biodegradable mulching films in agriculture: is there a negative prejudice towards matherials derived from organic wastes? Ital J Agron 11:90–97

Tey YS, Li E, Bruwer J, Abdullah AM, Brindal M, Radam A, Darham S (2014) The relative importance of factors influencing the adoption of sustainable agricultural practices: a factor approach for Malaysian vegetable farmers. Sustain Sci 9(1):17–29

Valente TW (1996) Social network thresholds in the diffusion of innovations. Soc Networks 18 (1):69–89

Valente TW, Davis RL (1999) Accelerating the diffusion of innovations using opinion leaders. Annals Am Acad Polit Soc Sci 566(1):55–67

Wasserman S Faust K (1994) Social network analysis: methods and applications. Cambridge University Press

Weimann G, Tustin DH, Van Vuuren D, Joubert JPR (2007) Looking for opinion leaders: Traditional vs. modern measures in traditional societies. Int J Public Opini Res 19(2):173–190

Wilensky U (1999) {NetLogo}

Chapter 15
Grassroots Innovations and the Transition Towards Sustainability: Tackling the Food Waste Challenge

Valentina Elena Tartiu and Piergiuseppe Morone

Abstract The need for innovative approaches to tackle food waste problem is widely recognized, given its tight links with agriculture, food security, trade, energy, deforestation, and climate change challenges. As a matter of fact, an emerging branch of literature is drawing attention to the value of food waste, reporting both technological aspects of food waste valorisation (by means of case studies and/or pilot-scale laboratory experiences), and how such innovative pathways may contribute to the transition towards sustainable production and consumption systems and a more sustainable waste regime. However, little research efforts have been invested so far in relation to the development and diffusion of innovative approaches addressing the food waste problem and the role of grassroots innovations. Thus, our chapter aims at contributing to this strand of literature, by addressing two main issues:

- how do grassroots movements act and how effective are they in catalysing innovation in the food waste field?
- what are the specific roles that grassroots innovations may play in the transition towards sustainable production and consumption systems and a more sustainable waste regime?

Our investigation draws on the analysis of several case studies of grassroots innovations from European countries, and builds on the multi-level perspective (MLP) approach.

The specific findings of our study could support decision makers in developing tailored strategies to minimize the amount of food wasted along the supply chain and to unlock the enormous potential of food waste that is being landfilled, and also to instil some further investigations related to this strand of food waste literature.

V.E. Tartiu (✉)
TIK, Centre for Technology, Innovation and Culture,
University of Oslo, 0851 Oslo, Norway
e-mail: v.e.tartiu@tik.uio.no

P. Morone
Department of Law and Economics, Unitelma-Sapienza,
University of Rome, 00161 Rome, Italy

P. Morone et al. (eds.), *Food Waste Reduction and Valorisation*,
DOI 10.1007/978-3-319-50088-1_15

Keywords Food waste · Grassroots innovations · Transition towards sustainability · MLP

15.1 Introduction

Food waste is an environmental, socio-economic, and ethical problem that connects with the most critical societal challenges: food security, poverty, energy, deforestation, and climate change. The need to find a sustainable and economically viable way of managing food waste is a main driver behind many recent waste valorisation practices around the world (Ki Lin et al. 2013). Furthermore, there is a general consensus that food waste constitutes a largely underexploited reservoir from which a variety of valuable resources—from chemicals to energy—can be derived (Clark and Luque 2013). As a matter of fact, an emerging branch of literature is drawing attention to the value of food waste, reporting both technological aspects of food waste valorisation (by means of case studies and/or pilot-scale laboratory experiences), and how such innovative pathways may contribute to the transition towards sustainable production and consumption systems and a more sustainable waste regime.

The exact causes of food waste are very much dependent on the specific conditions and local context in a given country (Gustavsson et al. 2011: 1); for instance, 40% of food losses in developing countries occur at the farmer-producer stage of the supply chain, mainly due to an inefficient harvesting, an inadequate local transportation and poor infrastructures, while in developed countries, over 40% of food is wasted at the consumers' level, due to 'a culture which places little value on food, making it "easier" to throw it away and buy more from the overstocked supermarkets' (IIED 2013).

In view of this, grassroots innovations emerge as a suitable option to address the food waste problem, as they may provide answers which are different from mainstream innovations (Monaghan 2009), focusing on the local context, the associated interests and the values of the communities involved (Seyfang and Smith 2007; De Keersmaecker et al. 2012), and seeking 'innovation processes that are socially inclusive towards local communities in terms of the knowledge, processes and outcomes involved' (Smith et al. 2014; cited in Pansera and Sarkar 2016: 2). Furthermore, it is widely acknowledged that grassroots innovations can act as incubators of the social change that is needed to respond to the current societal challenges and the transition towards sustainability (see among others, Feola and Nunes 2014; Smith et al. 2014, 2015; Seyfang and Smith 2007).

However, little research efforts have been devoted so far in relation to the development and diffusion of innovative approaches to address the food waste problem and the role of grassroots innovations. In this respect, we can recall the focus on:

- How recent mobilizations impact the way surplus food is actually managed with respect to sustainable production and consumption (Mourad 2016)
- Potential of food redistribution, in terms of: market opportunities for surplus food (O' Donell et al. 2015), sustainability of retail food recovery (Phillips et al. 2013; Cicatiello et al. 2016), economic and environmental assessment of food rescue operations (Reynolds et al. 2015), the evolution of food donation with respect to waste prevention (Schneider 2013; Priefer et al. 2016)
- The diffusion across space of solidarity purchasing groups (Feola and Butt 2015)
- The mobilisation of values in collaborative consumption (Martin and Upham 2015)
- Grassroots innovations and the sharing economy (see among others Martin et al. 2015; Avelino et al. 2015)
- Civil society roles in transition towards sustainable food (Durrant 2012)
- Dynamics of networks of social economy and civil society actors (Vergragt et al. 2014).

This chapter adds to the existing literature on grassroots innovations and food waste, by addressing two main research questions:

- How do grassroots movements act and how effective are they in catalysing innovation in the food waste system?
- What are the specific roles played by grassroots innovations in the transition towards sustainable production and consumption systems and a more sustainable waste regime?

Our investigation draws on the analysis of several case studies of grassroots innovations from European countries, and builds on the multi-level perspective (MLP) approach. Accordingly, our chapter is structured as follows. In Sect. 15.2, the food waste challenge is depicted using the lenses of the multi-level perspective (MLP) approach. In Sect. 15.3, the multiple forms that grassroots innovation embrace in respect to food waste are introduced, and their effectiveness is assessed. In Sect. 15.4 the many roles that grassroots innovations may play in the transition towards a more sustainable waste regime are discussed. Finally the concluding remarks are highlighted in Sect. 15.5.

15.2 Food Waste and the Transition Towards a Sustainable Waste Regime

It is widely acknowledged that food waste is a dynamic category that needs to be understood in relation to multiple domains (social, economic, environmental), which acts at various levels (local and global contexts) and bears a high degree of complexity in its semantics (i.e. meanings attributed to it) (Evans et al. 2013). In this section, we aim at contributing to this endeavour by using the lenses of the multi-level perspective (MLP) approach and thereby enhancing the understanding about the 'food waste momentum'—that unveiled 'the consequences of a long trajectory of economic expansion, unsustainable resource use and/or "out of control consumerism", and that also holds "the promise of a game-changing", reorientation of our practices, institutions, and policies of resource management' (Evans et al. 2013: 11).

This attempt is divided into two steps. First, we provide a brief overview on the historical phases of food waste dynamics; subsequently, we put the 'food waste momentum' into the multi-level perspective framework.

15.2.1 Food Waste Dynamics—Historical Phases

The food waste dynamics has varied significantly throughout the centuries. According to Evans et al. (2013), we may distinguish between three different historical phases, namely:

(i) *The relative visibility of food waste*—specific for the mid-nineteenth to mid-twentieth century
(ii) *The ubiquitous invisibility*—associated with the post-World War II decades
(iii) *The heightened visibility*—specific for the contemporary period.

A brief summary of these phases is presented in Fig. 15.1.[1] As emphasized in Fig. 15.1, the societal interest on food waste, especially in terms of value of preventing food waste, that characterized the mid-nineteenth and the first half of the twentieth century is replaced by an ubiquitous invisibility of food waste in the decades following the World War II, as a result of the transition towards a new food regime—triggered by a shift in production practices and technologies, farming approaches, food policies[2] and global trading. The all-pervading invisibility is overthrown (starting with mid 2000s) by complex dynamics, such as: the financial crisis, the global food crisis, energy security challenge, corporate land-grabbing phenomenon, deforestation, climate change challenges, etc.—that have brought back the food waste both on the agenda of food policy, and on the social and environmental debates, transforming it into a compelling, critical issue (heightened visibility phase).

The latter dynamics and the 'food waste momentum' are discussed in more details in the next subsection.

[1]For a comprehensive overview, please see Evans et al. (2013).

[2]For instance, food security has been formally turned into a policy matter, that legitimized massive investments into agricultural and food production technologies, and hence, food became abundant and also very cheap.

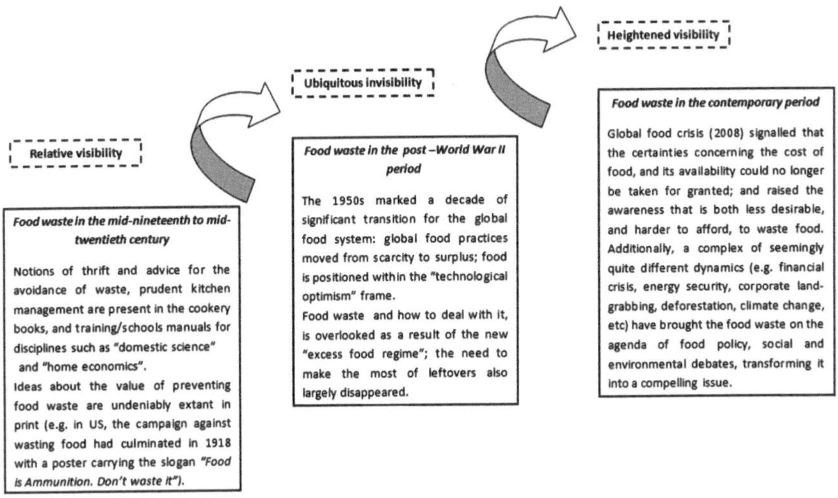

Fig. 15.1 Food waste dynamics—historical phases (after Evans et al. 2013)

15.2.2 Food Waste in the Multi-level Perspective Framework

This subsection aims at laying the ground for *food waste—grassroots innovations* nexus, by enhancing the understanding about the shift towards the heightened visibility phase and furthermore towards sustainable production and consumption systems and a more sustainable waste regime. Thus, we use the lenses of the multi-level perspective approach (MLP), as MLP enables the analysis of complex and non-linear phenomena such as historical and structural changes, including technological transitions, using a multidisciplinary and multidimensional perspective. Therefore, MLP is not just about economics or competing technologies; it involves many other areas of investigation placing the transition process in a well-defined societal space and historical context.

As our investigation is entrenched in a well-established theoretical setting (i.e. sustainability transition; see for instance, Geels and Schot 2007; Markard et al. 2012), before putting the 'food waste momentum' into the multi-level perspective (MLP) framework, we provide next, a very brief overview on MLP.

15.2.2.1 MLP—A Snapshot

The MLP is a heuristic framework, which covers three levels of analysis: the landscape (macro-level), the socio-technical regime (meso-level) and the niche (micro-level). Technological transitions can be explained through the interaction among these three levels as the transition basically entails a shift from an incumbent socio-technical regime to a new one, which is nurtured in the technological niche and prompted at the landscape level. Landscape and niches are derived concepts 'because they are defined in relation to the regime, namely as practices or technologies that deviate substantially from the existing regime, and as external environment that influences interactions between niche(s) and regime' (Geels 2011: 26–27).

In this model, transition occurs whenever a pressure at landscape level destabilizes the regime, thus creating a window of opportunity for pioneering niche-innovations to enter in the mainstream market. In the food waste context the niches innovations could take various forms stretching from the development of new technologies (e.g. smart packaging) to behavioural changes (e.g. new consumption models). This reflects the multivariate nature of the phenomenon under scrutiny, which involves changes occurring at various levels (social, economical, environmental, contextual) all concurring to complete the transition. However, when considering grassroots movements all niches innovations have in common that 'change must spring from below, outside the existing institutions. Social movements should develop alternative structures (cooperatives, communes, ecovillages) and hope that the majority will be influenced by the power of the example' (Geels 2011: 33).

In this framework, niches are the locus of innovation. A niche is like an incubation room in which new and emerging technologies can have a space, which protects them from competition and pressures of the selective process taking place in mainstream markets. Usually rules and procedures in the niches are flexible and not formalized in order to facilitate the emergence of innovation. At the same time, the niches space is highly unstable and characterized by the co-existence of several (and often alternative) niches, which usually lack coordination between them and are in competition among each other. However, not every niche can survive for a long time and only few of them will get to a point where they will really challenge the incumbent sociotechnical regime. A niche should be sufficiently developed and mature in order for this to happen.

As introduced in Sect. 15.2.1, the shift from the ubiquitous invisibility phase to the heightened visibility phase, has been triggered by different and complex dynamics. Following the MLP framework, (depicted above) we can roughly group these complex dynamics, into (see Fig. 15.2[3]):

(i) Events affecting the socio-technical landscape
(ii) Events affecting the socio-technical regime
(iii) Events (movements) influencing the niches—innovations.

In the first category, we can mention, for instance, the *global food crisis* (2008), which signalled that the certainties concerning the cost of food, and its availability could no longer be taken for granted. This, in turn, raised the awareness on waste of food being both less desirable and harder to afford. As a matter of fact, 'since 2008,

[3]Figure 15.2 should be taken as an academic exercise, aimed at illustrating the diversity of events (exerting pressure at different levels), influencing the transition towards a more sustainable food waste regime; a far more systematic content analysis is needed to depict thoroughly, in more details this transition.

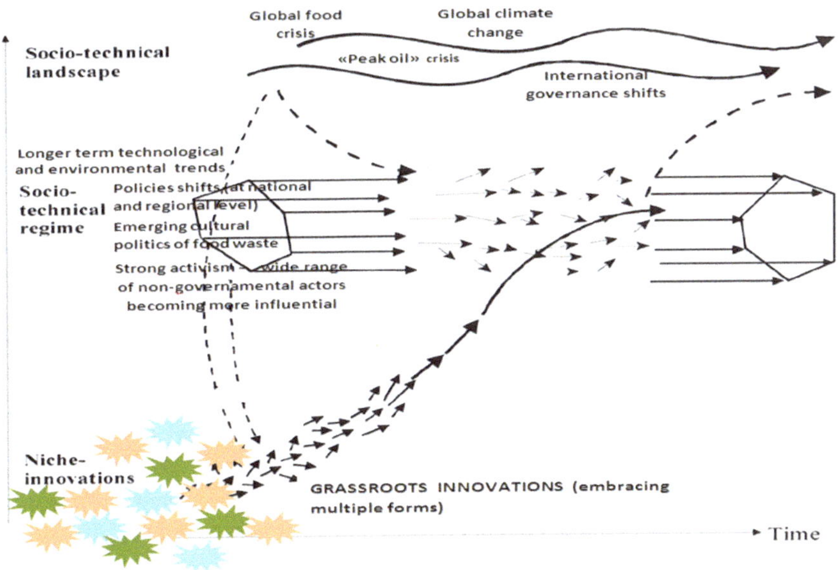

Fig. 15.2 Transition towards a sustainable waste regime. A multi-level perspective (adapted from Geels 2002)

food prices have begun to increase cyclically, and have created an environment where the relative cost of food has become a matter of consumer and public concern' (Evans et al. 2013: 16). Another event affecting the *socio*-technical landscape level is the *global climate change* associated with the level of carbon dioxide (CO_2) released into the Earth's atmosphere. The sharp increase occurred since the beginning of the industrial era,[4] lead (among other things) to severe changes in the world's weather patterns (heat-waves, cyclones, extreme floods, droughts) threatening in this way the food supply chains. However, to date, around a third of all food produced for human consumption is lost or wasted; these food losses and wastes account for about 4.4 gigatonnes of greenhouse gas emissions per year, generating more than four times as much annual greenhouse gas emissions as aviation, and being comparable to total emissions from road transports. As observed by the World Resource Institute, 'to put this in perspective, if food loss and waste were its own country, it would be the world's third-largest emitter—surpassed only by China and the United States' (Hanson et al. 2015).

As for the second group of events, we can recall the various *policies shifts* occurred both at national and regional levels. First and foremost, the EU waste policy set two ambitious targets: reducing biodegradable waste[5] to 35% of 1995 levels by 2016 (or by 2020 for some countries); halving current volumes of food

---

[4]Carbon dioxide levels are now approximately 40% higher than they were at the start of the Industrial Revolution.

[5]Category under which falls also food waste.

waste by 2025. Indeed, these kinds of targets exert a huge pressure on incumbent industries. Along policy shifts also the *strong activism of a wide range of non-governmental actors* should be mentioned. As these actors become more influential, we can recall two significant contributions, namely: the 2009 landmark tour of Tristram Stuart, promoting the book—*Waste: Uncovering the Global Food Scandal*; soon after followed by the first *Feeding the 5000* event in London's Trafalgar Square (in December 2009). With the extensive media outreach and public support, his endeavour had an immediate impact on government and business policy, promoting the activities and messages of partner organisations, including FareShare, This is Rubbish, ActionAid and Save the Children. Since 2009 Tristram and his team (Feedback NGO) played a central role in catalysing the food waste movement around the world, reshaping the world agenda around food waste.[6] A similar action was launched in the USA in 2011, by Jonathan Bloom, who wrote *American Wasteland: Why America Throws Away Nearly Half Its Food (and What We Can Do About it)*. Bloom gathered statistics and highlighted the extent of the food waste problem in a crowd-pleasing and accessible way, triggering the development of several initiatives across US. Since then similar initiatives have been promoted by non-governmental actors at national level worldwide.

Finally, for the third category, we can distinguish between multiple forms of grassroots innovations tackling food waste problem. In the following section we will investigate, by means of comparative case studies, how grassroots innovations evolved in the context of food waste.

15.3 Grassroots Innovations (GRI) Tackling the Food Waste Challenge

Before exploring how grassroots innovations are addressing the food waste challenge, we shall first provide a general overview of this concept as it emerged in the literature.

Hence, in the '70s 'the analysis of grassroots initiatives and social movements has traditionally been centred around political activism and the mobilization of a group of individuals that share personal grievances and deprivation' (McCarthy and Zald 1977 cited in Grabs et al. 2015: 3). Since then the concept has evolved in the academic discourses, and has been associated with various frames, such as: cultural theories (e.g. Bordieu 1984), social capital (e.g. Putnam 1995), open source innovation (e.g. Raymond 1997), collective active frames (e.g. Benford and Snow 2000), new social movement theories (e.g. Touraine 2002), autonomous geographies (e.g. Pickerill and Chatterton 2006), environmental movement (e.g. Curry 2011), social innovation (Howaldt et al. 2013), governance and participation (e.g. Stirling 2011), etc.

---

[6]http://feedbackglobal.org/about-us/.

To date, the most diffused definition in the emerging literature on grassroots innovations for sustainability, belongs to Seyfang and Smith (2007), that used the lenses of New Economics and Socio-Technical Transitions, and defined the concept of 'grassroots innovations' as 'networks of activists and organizations generating novel bottom-up solutions for sustainable development; solutions that respond to the local situation and the interests and values of the communities involved. In contrast to mainstream business greening, grassroots initiatives operate in civil society arenas and involve committed activists experimenting with social innovations as well as using greener technologies' (Seyfang and Smith 2007: 585).

For the purpose of this chapter, we adopt the broader definition of Smith et al. (2014) which includes in grassroots innovations also 'people and organisations coming from outside local communities, such as engineers and designers, but who engage the grassroots in innovation processes, in their ideas from the outset, and put local knowledge and communities in the lead in the framing of a collaborative innovation activity' (Smith et al. 2014: 114).

In order to address our two research questions—(1) how do grassroots movements act and how effective they are in catalysing innovation in the food waste system?; and (2) what are the specific roles played by grassroots innovations in the transition towards a sustainable food waste regime?—we performed a meticulous analysis of several case studies of grassroots innovations addressing the food waste problem in different European countries.

From this analysis emerged the fact that, when it comes to food waste, grassroots innovations embrace multiple forms, which can be roughly grouped into three categories,[7] namely:

(i) Prevention
(ii) Reduction and Reuse
(iii) Valorisation.

As mapping and discussing all the existing grassroots worldwide is beyond the purpose of our study, we briefly depict next, for each category, some of the most renowned ones–as they are empirically well documented.

15.3.1 GRI—Prevention

Grassroots initiatives belonging to this category are aiming at preventing and raising awareness about food waste, by promoting behavioural changes.

[7]In making this distinction, we followed the European Waste hierarchy, and we considered the primary goal of the grassroots. However, many of the grassroots innovations are tackling more than one aspect, and all of them directly or indirectly and in different extents, contribute to awareness raising; therefore this classification should be taken as an academic exercise, aimed at illustrating the diversity of grassroots innovations in respect to food waste.

Feeding the 5000 is one of the *Feedback*'s[8] flagship events aimed at raising awareness on the global food waste problem, and at catalysing the global movement against food waste. At each event, dishes prepared out of food that would have been wasted are served for 5000 people. Furthermore, as part of this event, local communities are encouraged to become involved (either by volunteering or attending in support), while the public is also invited to sign up to the Food Waste Pledge, committing themselves to reduce the amount of food waste they produce and to ask businesses and governments to do the same. The first Feeding the 5000 event has been held in London in 2009, and since then similar events have been organized worldwide (Paris, Sydney, Brussels, Amsterdam, Dublin, etc.).

Disco Soup, is an international grassroots movement that sets off action on food waste, works to 'fill bellies not bins' and to raise awareness about the unconceivable quantity of food that is wasted around the world, by empowering the public to recognize positive solutions to this global challenge. The first Disco Soup event was organized in Germany in 2012 (*Schnippel Disko*) by the Slow Food Youth Network and gathered 300 volunteers that came together to wash, peel, chop and cook fresh but unwanted fruits and vegetables that would otherwise have been discarded, on the music of two DJ's. This event was followed by a larger-scale Disco Soup event, staged in Paris in October 2012. In a span of few months this anti-food waste initiative spread to more than 14 cities in France. Since then, similar events have been held across the globe: Belgium, Colombia, Canada, US, South Korea, Holland, etc.

Wastecooking[9] is a movement aimed at protesting against food waste and serving up a critical stance on consumerism. It was initiated in 2012 (as an art project) by David Groß, an Austrian trained chef, filmmaker and activist, but soon after became a movement, which regularly organizes cook-ins and performances in public spaces, such as: film, music and art festivals and museums. As part of this movement, David Groß organized a Wastecooking tour in 5 European countries (Austria, Belgium, France, Germany and Netherlands) in which only cooked up what others threw out. Along the way, he received support from other activist groups, chefs and scientists. The tour was captured on a film as a five-part television series and an evening-filling documentary film.

Gleaning Network is a grassroots movement initiated in 2012 in UK[10] by Feedback, that became international, involving gleaning activities in countries such as France, Belgium, Greece and Spain. Foremost, this initiative, aims at raising

---

[8]Is an environmental organisation that campaigns to end food waste at every level of the food system. To date governs 5 movements: Feeding the 5000, Gleaning Network, The Pig Idea, Stop Dumping and the FSE Network. For a comprehensive overview, see http://feedbackglobal.org/.

[9]http://www.wastecooking.com/en/#home.

[10]From its start in 2012 to the end of 2015, the Gleaning Network UK gleaned over 188 t of produce (apples, pears, plums, strawberries, cauliflower, cabbages, lettuces, pumpkins and parsnips), equal to over 2 million portions of fruits and vegetables, with over 1000 volunteers across 99 gleaning days. http://feedbackglobal.org/campaigns/gleaning-network/.

awareness about the causes of food waste that occurs at the farm level, and argues for the change of retailers' policies and consumers' perceptions in respect to this type of food waste, by providing practical and implementable solutions to the problems encountered at this stage of the supply chain. Specifically, Gleaning Network coordinates volunteers, farmers and food redistribution charities to save thousands of tonnes of fresh fruit and vegetables that are wasted on farms every year, in order to direct this fresh, nutritious food to people in need.

15.3.2 GRI—Reduction and Reuse

Grassroots innovations belonging to the second category are aiming at reducing the amount of wasted food and at changing the incumbent processes across different stages of the value chain.

Anti-Gaspi (Stop Food Waste) is a grassroots movement initiated by Arash Derambarsh, a local councillor of Courbevoie (France). In December 2014, Derambarsh, joined by volunteers and friends, run a 'field experiment'—an anti-poverty and anti-food waste campaign—in Courbevoie, by recovering unsold supermarket food and distributing it to needy people, including the homeless. This 'field experiment' led to a petition—on Change.org—calling for action against food waste and change of the food retailers policies. The petition sparked the concern of many French citizens, with over 210,000 people signing it and several French celebrities endorsing the cause. Arash's tireless efforts to integrate social action, public and political mood have made its grassroots initiative, swift and effective, contributing substantially to the 'momentum creation' for the adoption of the first anti-food waste law in France. Adopted on first reading at the National Assembly on 9 December 2015 and the first reading in the Senate on February 3, 2016, in each case unanimously, the French law sets a four-step hierarchy to be implemented to limit the loss of food, that prioritizes prevention of food waste, followed by donation or reprocessing of unsold food for human consumption, recovery for animal feed and, finally, use as compost in agriculture or energy recovery, such as biomethane. It requires food retail businesses whose sales area exceeding 400 m^2 to sign a grant agreement with one or more associations of food aid. The law also prohibits any contractual provision that constitutes an obstacle to the gift of food sold under private label, and includes the fight against food waste in the school curriculum and in social and environmental responsibility, feeding into France's plan to reduce food waste by 50% by 2025. All the food retail businesses have to comply with the law by July 2016 in order to avoid penalties, which include both fines and up to two years in jail (Ministère de l'Environnement, de l'énergie et de la Mer 2016).

Given the successful outcome of the grassroots movement in France, the initiative has been taken further, pushing for anti-food laws throughout Europe, and for a European directive concerning the issue. To date, a petition entitled: *Mettons fin au gâchis alimentaire en Europe! #StopFoodWaste*[11] is on Change.org, calling for such legislation to be enacted at the EU level. The petition was launched simultaneously in six European countries: Belgium (by Frédéric Daerden), Greece (by Nikos Aliagas), Italy (by Daniele Messina), Spain (by Manuel Bruscas), Germany (by Claudia Ruthner) and UK (by Tristram Stuart) and overall, has reached so far, over 773,000 signatures in support.

Gueules Cassées[12] (Ugly Mugs) in 2014 a Provencal entrepreneur, Nicolas Chabanne together with a group of fruit and vegetable producers, gave birth to a trademark that promotes 'ugly' fruits and vegetables (damaged, deformed or too small) among consumers to limit food waste.[13] This idea emerged when one of his friends, producer of quality apricots saw them systematically rejected by supermarkets because they were not 'calibrated'—when you see an apricot that provides optimal taste qualities but is thrown because it is too round, not large enough or not of the right colour, it hurts the heart (Chabanne, cited in Fougier 2016). In order to reassert the value of some of these products among consumers, a label which reads 'What's wrong with me? Fruits and Vegetables, less pretty but delicious!' was created to promote these substandard products and to allow them to be sold in the so-called 'classic' distribution chains. The idea led to several professionals contacting them—'We were in fruit and vegetables, but confectioners, butchers and bakers, rang to tell us "we also have our own *gueules cassées*"' (Chabanne, cited in Fougier 2016)–thus, soon after, the *Ugly Mugs* anti-food waste brand was created to raise awareness all around the world on this perfectly good food, and to save it from being senselessly wasted.

The initiative is open to all food producers or craftsmen who want to market products with minor flaws or retailers and shopkeepers who want to get the most from products with short shelf lives. Furthermore, the anti-food waste brand, offers three distinct labels:[14]

- *A label that identifies fruits and vegetables with minor aesthetic flaws*—this allows to value and sell products with small defects in shape, appearance or size
- *A label for products approaching their use-by date*—this −50 or −30% anti-food waste label can be used by any retailer or shopkeeper carrying products with short shelf-lives or products approaching their use-by date
- *A 30% discount anti-food waste label*—for a range of products with small defects in size, shape, or colour (e.g. cheese with slightly irregular edges and cereals, which seemed to have no abnormalities at a glance, but when looked at

---

[11]https://www.change.org/p/mettons-fin-au-g%C3%A2chis-alimentaire-en-europe-stopfoodwaste?source_location=trending_petitions_home_page&algorithm=curated_trending.

[12]*Première marque mondiale antigaspi*–The first anti-waste brand worldwide.

[13]Each year, 17 million tons of perfectly edible produce is not consumed for purely aesthetic reasons (Chabanne 2014).

[14]http://www.lesgueulescassees.org/#!solutions/b0jes.

meticulously do not fulfil the norms). To date, thanks to a sturdy and international mobilization effort, the initiative is being developed globally.[15]

Along with the above noteworthy initiatives, we can recall also the innovative web and mobile applications that encourage and allow consumers to reduce the amount of wasted food. In this sense, we can mention initiatives, such as: *Share your Meal*—a grassroots movement that encourages consumers to share the home cooking with neighbours and/or acquaintances, and to reduce in this way the amount of food thrown away. To date is promoted through online platforms in: Belgium (Thuisafgehaald), Italy (Cucina e condividi), Portugal (Acomida da vizinha), Spain (Comparto plato), Great Britain (Dinner Time), Germany (Teildeinessen), Netherlands (Thuisafgehaald), UK (Share your Meal), Slovakia (Ktominavari); *Bring Food* (Italy)—crowdsourcing web/mobile application that allows consumers, donors, to seamlessly publish offers and easily coordinate collections; *Zéro-Gâchis*[16] (France)—web and mobile platform that allows French consumers to find food products near them at a significant discount (30–70% off) that need to be consumed rapidly (as nearing their sell-by date); *Partage ton frigo (Share your Fridge) App*[17]—allows French consumers to take a picture of what they cannot eat, to name it, and share it on the app's database. After the completion of this step, consumers have just to wait for neighbours to select their leftovers, and arrange for pick-up.

Another type of grassroots initiatives that falls under this category refers to those that connect structures that have surplus food with charities and/or the neediest (FSE Network 2016). For instance: *FoodWe* (Belgium)—allows food professionals to provide, through its online platform,[18] supplies of unsold but still edible food to charitable or civic associations. On this platform, food surplus can be sold at a reduced price; *Taste Before You Waste* (Netherlands)—Collects food that otherwise would be discarded and brings it to different charities; *Foodcloud* (Ireland)—by using the app, or through the website, businesses who have registered with Foodcloud, can upload details of their surplus food and the time period in which the food can be collected; *Neighbourly Food* (Great Britain)–connects professionals (distributors, manufacturers, etc.) having surplus food with charities.

15.3.3 GRI—Valorisation

In this sub-section we focus on those grassroots innovations that have as outcome innovative products derived from food waste valorisation, that generate added value and enable the diverting of waste from landfill. Given the fact that innovative

[15]The US Investment Fund—Global Emerging Markets decided to invest in the development of this concept in US and the Middle East.

[16]https://zero-gachis.com/.

[17]http://www.partagetonfrigo.fr/.

[18]https://www.foodwe.be/.

products may find themselves in different stages of the innovation lifecycle, we report and we discuss next examples that correspond to the various stages: ideation–validation (Agridust), product development—pilot scale (Orange Fiber) and manufacturing (GroCycle and RecoFunghi).

15.3.3.1 Agridust—How Can We 3-D-Print with Food Waste

Main motivation of the grassroots entrepreneur(s)[19] The idea of reusing this type of waste was born because of a present problem in our society: waste, which can be either food or also the raw material used for constructing objects destined to end up in landfill. Inspired by the basic concept of the book *Cradle to cradle*, that is the creation of a new system where there is no longer the concept of waste, Marina Ceccolini, an Industrial Designer, has created Agridust—a biodegradable and atoxic material.[20]

Innovative idea in a nutshell The material consists of 64.5% of waste fruit and vegetables (coffee grounds, peanuts shells, husk tomato, bean pod, orange waste and lemon waste), and the remaining 35.5% of a potato starch-based binder. 'The choice of this binder was not random, as from the beginning I wanted to search a natural binder to create a non-toxic material in all its creation and processing stages' (Chiocchia 2015). Thanks to the latter feature of the material, the processing phase can be distinguished from that of other polymeric materials such as plastics (derived from petroleum processing, which causes serious damage to the ecosystem), as environmental friendly.

For this reason AgriDust is excellent as a substitute of the plastic materials for the production of plant pots and other elements dedicated to the nursery sector, it can also be used to create containers and packaging. Furthermore by controlling its viscosity is suitable as a material (a fine powder) that can be used as 'ink' for the 3D printers, taking advantage of the cold technology (LDM), where the extruder is replaced by a syringe.

Impact 'Considering how many first prints are just tests anyway, and how many prototypes makers—especially novices—often send through the 3D printer before reaching the desired shape and effect, Agridust offers a way to test and enjoy more 3D printing without worrying about the environment—the only concern is that the 3D printed items will not last indefinitely and are considered disposable' (Butler Millsaps 2015).

If taken to a higher level (up-scaling at industrial level) Agridust may facilitate the development of a sustainable 3D Printing industry, by reducing the use of plastics in 3D printing worldwide—'by 2020, experts estimate that we may be using as much as 1.4 million barrels of oil in 3-D printing' (Peters 2015), and thus decreasing the costs required for its treatment or landfilling. Additionally, AgriDust

[19]Industrial Design Student (Marina Ceccolini–Università degli studi di San Marino).

[20]Outcome of Marina Ceccolini's project work (progetto di tesi).

besides giving a second life to the chosen vegetables waste is a biodegradable material, which in turn will never become waste, because it is born with the intent to return the biological nutrients to the nature, revealing itself as advantageous for both human activities and the environment (Ceccolini 2015 cited in Chiocchia 2015).

15.3.3.2 Orange Fiber—Innovative Yarns and Fabrics from Citrus Wastes

Main motives of the grassroots entrepreneur(s)[21] Passion for textiles and attachment to Sicily, their native region, drove them to investigate if they could produce a fabric using the wastes of citrus; and thus provide a possible solution of a problem that in Sicily is a very debated one: disposal of the citrus waste derived from the processing industry (which amounted to over 700 thousand tons per year).

Innovative idea in a nutshell In 2011, from the passion for textiles, and the challenging situation of the Sicilian citrus processing industry, derives the idea of producing innovative yarns and fabrics from citrus waste. From the residues, that is all that remains after the pressing and processing of citrus, is extracted cellulose suitable for the spinning. Using nanotechnology, are produced innovative yarns and fabrics, releasing vitamin A, C and E with beneficial effects on the skin.

From the feasibility study conducted by Politecnico di Milano (Polytechnic University of Milan) develops the patent,[22] (Rubino 2014) which is registered in Italy (2012) and extended internationally. In February 2014 is established Orange Fiber, a startup based in Catania and Trentino,[23] with the goal of creating sustainable fabrics that respond to the innovation need of the fashion brands, by tackling in the same time a challenging issue in Sicily: the disposal of citrus wastes. To do that, the two young entrepreneurs planned to reuse more 700,000 tons of waste that the Italian citrus processing industry produces annually. In September 2014 it is presented the first fabric derived from citrus in the world, consisting of acetate by citrus and silk in two variants: solid satin and lace combined.

In December 2015, thanks to the Smart and Start funding of Invitalia, the first pilot plant for the extraction of cellulose from citrus wastes was opened. Currently, the first lot of fabric has been produced and proposals of top fashion brands to enter the market are undergoing an evaluation process. Furthermore, the two young entrepreneurs plan to build a production plant in Sicily, focusing on creating business partnerships (Orange Fiber, 2015, 2016).

[21]Two young entrepreneurs: Adriana Santanocito−background in fashion design and innovative materials and Enrica Arena−background in communication.

[22]For the first fabrics from citrus wastes in the world, consisting of acetate by citrus and silk.

[23]Two Business Angel, a lawyer and Trentino Development have funded the project.

Impact Orange Fiber is facilitating the development of a sustainable recovery chain of citrus, one of the most problematic waste streams in Sicily, in the Mediterranean region and not only. This will most likely have a positive impact also on other sensitive social aspects of the region, such as unemployment.

15.3.3.3 Growing Mushrooms from Waste Coffee Grounds

To get valuable insights into the up-scaling process of the grassroots innovations (which is still an aspect under-investigated in the literature) and gain a better understanding of the relevance of contextual elements (such as local constraints and web of relations) on innovation diffusion pathways, we decided to perform a comparative analysis between two cases, which are reporting the same type of innovative product, namely: *mushrooms from waste coffee grounds*. Hence, we have selected the Grocycle (England) and the RecoFunghi (Italy).

15.3.3.4 GroCycle (England)

Main motives of the grassroots entrepreneur(s)[24] 'Worldwide more than 1.6 billion cups of coffee are drunk each day and in the UK alone this figure is around 80 million every day. A cup of coffee is at the end of a process where less than 1% of the coffee plant is used. Coffee drinkers only value the beans, and after the brewing process most of the coffee grounds end up in landfill sites. This is a problem that's likely to increase as the UK already has more than 15,000 coffee shops and this number is set to keep on growing.

Taking this waste and turning it into local food is such a simple solution and a huge opportunity. Not only are there sustainability benefits, but it can change people's attitudes and create other opportunities in the process. Although most of the UK's food is consumed in cities, virtually none of it is grown there. Mushrooms are one crop ideally suited to urban agriculture, where both waste and demand are highest. They can be grown in empty spaces and add to urban food security'[25] (GroCycle 2016a).

Innovative idea in a nutshell The idea came out in 2009, as a hobby 'foraging for wild mushrooms'[26] of Adam Sayner (one of founders) and evolved along the years, the entrepreneurs being inspired by the 'scale of opportunity' to study thoroughly the chemical, sensorial, economical and sustainability aspects of their innovative idea. Traditional mushroom cultivation requires an energy intensive process to sterilise the substrate. By using coffee grounds, the two entrepreneurs reuse the energy that has already gone into the brewing process. Each week,

---

[24]Fungi Futures CIC, an innovative social enterprise based in Devon, UK.

[25]GroCycle (2016a), Mushrooms from Coffee Grounds?

[26]http://www.fungi-futures.co.uk/our-story/.

hundreds of kilos of coffee grounds are collected from city cafes and used to grow the Oyster mushrooms (a gourmet sortiment), that subsequently, are delivered to the best restaurants and food outlets in the South West of England.

The GroCycle innovative process delivers mushrooms and fertile compost, diverting tonnes of coffee grounds from landfill. The two entrepreneurs have been growing Oyster mushrooms from waste coffee grounds since 2011, and developed further the innovative idea, by creating the *GroCycle Urban Mushroom Farm*[27]—in order to test new ideas, and designing an user friendly Mushroom Grow Kit–to enable other people to cultivate their own mushrooms at home, by recycling their waste coffee grounds.

Additionally, they provide an online course and people from over 40 countries around the world have been trained so far.

Impact Since the first year of activity, tonnes of coffee grounds[28] have been diverted from landfill. Thanks to GroCycle the collection and recovery of coffee grounds has been shed into a new light, leading to a win-win business model for the region and to an increased environmental awareness.

15.3.3.5 RecoFunghi (Italy)

Main motives of the grassroots entrepreneur(s)[29] 'La sostenibilità, la nostra passione!'—'Sustainability, our passion!'—Concerned about environmental issues, after years of studies, research and testing in the field they managed to create a procedure to recover what, until few years ago, was considered a waste: the coffee grounds, by setting up the first company of this type in Italy (Basilicata Region, South of Italy).

Innovative idea in a nutshell The innovation journey starts in 2010 with the collection of coffee grounds from bars managed by friends, and also from unusual production places such as cellars and attics, followed by the participation with the project *'Recoffee'–mushrooms from coffee grounds*—in the call for tender N.I.D.I. —Nuove Idee di Impresa Innovative (New Ideas of Innovative Enterprises) organized by a special agency of the Chamber of Commerce of Potenza.

In 2011 their business plan was successful and, subsequently the young family of entrepreneurs got a voucher spendable in technical consulting and feasibility studies. This voucher (financial support) allowed them to study thoroughly the chemical and sensorial aspects of the products (the various types of mushrooms) and the economic possibilities for setting up a small scale company, with low environmental impact, that delivers sustainable products. During this phase they discovered that their idea

---

[27]They have converted an unused office building into an urban mushroom farm right in the heart of the city of Exeter (UK) (GroCycle 2016b).

[28]'5 tonnes have been collected in Plymouth and diverted only in the first year' (GroCycle 2016c).

[29]Young family of entrepreneurs with background in Enzymology (Daniele Gioia e Annarita Marchionna–Basilicata Region, South Italy).

of producing mushrooms out of coffee grounds is not that new, but they didn't get discouraged and decided to work towards the development of a manufacturing process that allows the valorisation of the expresso coffee grounds (widespread and typical in Italy) and the production of a variety of highly appreciated mushrooms in the region and all around in South of Italy, namely *cardoncello*.

The results obtained, in all respects have been satisfactory, so they decided to set up a small company, despite the times of crisis. Among all the challenges they had to confront with, the young family of entrepreneurs retains as critical as the one related to the financing of the whole activity. Fortunately, in their case, one credit bank[30] decided to finance almost the entire project. Thus, in January 2013, they could start their activity. Currently, they are producing many varieties of Pleurotus ostreatus and Eryngii (*cardoncello*). The goal, by the end of 2017, is to recover all the coffee grounds produced by the bars in the capital city of the Basilicata region and also to increase the sale of kits (RecoKit) for the production of ultra-fresh mushrooms at home or in the coffee grounds restaurants (RecoFunghi 2016).

Impact RecoFunghi has facilitated the development of a new and sustainable agro-food value chain in the region, by establishing stable partnerships with different commercial actors in the area. Thanks to RecoFunghi the collection and recovery of coffee grounds has been shed into a new light, leading to a win-win business model for the region and to an increased environmental awareness.

15.3.4 Grassroots Channels of Actions and Effectiveness in Catalysing Innovation in the Food Waste System

As depicted above, the range of grassroots innovations in the food waste field is very diverse in terms of goals, scale, space, diffusion pathways, typology of movement brokers and grassroots entrepreneurs. Furthermore, the analysis of several empirical case-studies allowed identifying some distinctive features of the grassroots innovations tackling the food waste problem.

Going back to the first research question of our study (*how GRI act and how effective they are in catalysing innovation*), the empirical evidence collected and presented above allow us to argue that grassroots innovations addressing food waste mostly are:

- Both need-generated and need-oriented
- Tackling mostly overlooked collective needs and concerns
- Aiming at delivering solutions that can be easily customized to the various contexts and settings worldwide—in this sense, most of the reported GRIs were able to articulate their own agenda and vision (e.g. *Anti-Gaspi*, *Gleaning Network*, *Gueules Cassées* etc.).

[30]La Banca di Credito Cooperativo di Laurenzana e Novasiri.

Another important aspect to note is that, in spite of the lack of knowledge and capabilities required for scaling-up the innovation, the grassroots entrepreneurs showed willingness to cooperate with other actors in managing their innovation; aimed at improving local productivity, and overall at strengthening the regional economy over the long term because of their strong attachment to the local community (see for instance the Orange Fiber case).

When considering GRI effectiveness in catalysing innovation activities and policy design, certainly the success of the French grassroots movement (which contributed substantially to the adoption of an anti-food waste law), shed a new light on the grassroots developments, from an overlooked site for innovation, to a visible hub in the realms of food policy and regulation, and emerging cultural politics of food waste, with huge potential in the transition towards a more sustainable waste regime.

From the analysis of the case studies, also emerged the fact that grassroots innovations addressing food waste problem, can act as incubators of the social change needed to respond to the current global challenges—as food waste connects with the most critical societal challenges: food security, poverty, energy, deforestation, and climate change. This remark is in line with the other findings from the grassroots innovations for sustainability literature (see among others, Feola and Nunes 2014; Smith et al. 2014, 2015; Seyfang and Smith 2007).

15.4 Grassroots Innovations Roles in the Transition Towards a Sustainable Waste Regime

In this section we make an effort to reconcile the cases examined above with the MLP framework, attempting to address the second research question on *the specific roles played by grassroots innovations in the transition towards a sustainable food waste regime*.

In all three groups of cases-studies (*Prevention, Reduction and Reuse, and Valorisation*) GRI act primarily at the niche level, creating however also the condition for enhancing landscape pressure aiming at the opening up of windows of opportunities. In fact, in all case-studies investigated, GRI can be seen as emerging and evolving socio-technical niche configurations, as they described at various levels: new arrangements of technologies, new competencies creations, alternative social practices design and development. At the same time, GRI exerts lobbing pressure on policy makers and societal groups at various levels, setting out viable and alternative models of development–e.g. food growing using waste residues on urban micro-sites (GroCycle); new forms of production based on food waste valorisation (Orange Fiber, Agridust); alternative ways of distribution and retail (*Gueules Cassées*, Gleaning Network); mobilising peoples' support for sustainable alternatives through public events, campaigns, online petitions (*Anti-Gaspi*, Wastecooking), etc.—and challenging in this way also the incumbent regime.

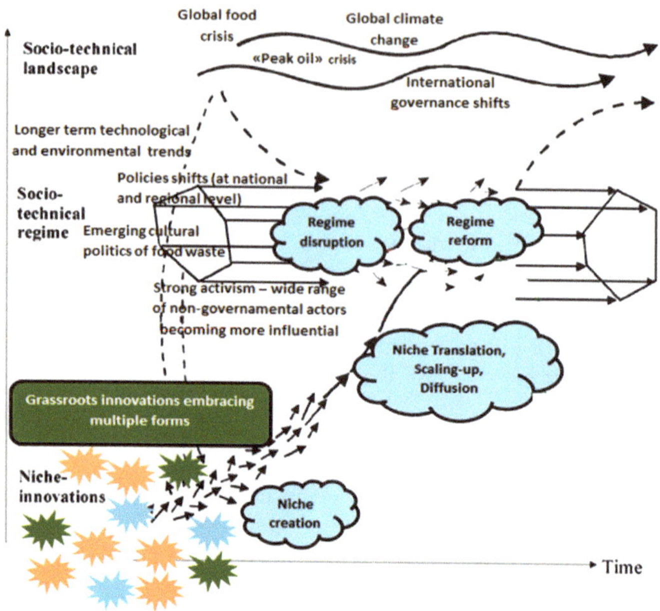

Fig. 15.3 Grassroots innovations roles in the transition towards a sustainable waste regime. A multi-level perspective (adapted from Geels 2002; Durrant 2012)

Given the specific nature of these niches innovations we shall propose a development pattern of GRI niches which differs from standard niches development process (as it was described, for instance, in Lopolito et al. 2013). The proposed path towards niche maturity articulates into four subsequent steps (see Fig. 15.3):

(1) *Niche creation* Grassroots movements can play a role in several core processes of the niche creation—in terms of network formation, learning and *competence building, but also in shielding, nurturing and empowering niche innovations—by providing legitimacy and linking them to broader societal discourses* (Ornetzeder and Rohracher 2013). *We can recall, in this sense, mainly the GRIs reported in* Sect. 15.3.2 *(GRI—Reduction and Reuse) and* Sect. 15.3.3 *(GRI—Valorisation).*

(2) *Niche translation scaling-up and diffusion* By actively contesting unsustainable incumbent arrangements and re-framing debates, GRIs pressure incumbent industries to respond; By lobbying policymakers, staging direct actions and protests, engaging in framing struggles in the media, GRIs mobilize resources and supporters (Durrant 2012). Both GRIs reported in Sect. 15.3.1 (GRI—prevention) and Sect. 15.3.2. (GRI—Reduction and Reuse) have proven to play a significant role in this phase, by advocating specific policy changes, lobbying decision makers, educating and influencing people's behaviour, promoting good practices (e.g. promoting substandard products

—'ugly mugs', and allowing them to be sold in the so-called 'classic' distribution chains).

(3) *Regime disruption* Furthermore, through this mix of strategies, GRIs 'can at times create the initial conditions required for the destabilization of incumbent industrial regimes and their replacement with more sustainable configurations' (Turnheim and Geels 2012; cited in Durrant 2012: 3). When considering GRIs potential contribution to regime disruption phase, certainly the success of the *Anti-Gaspi*—French grassroots movement (which contributed substantially to the adoption of an anti-food waste law), is encouraging, providing a piece of evidence that GRIs can undermine existing unsustainable practices, acting as incubators of the social change needed to respond to the current global societal challenges (food security, poverty, energy, deforestation, and climate change).

(4) *Regime reform* GRIs may encourage the regime actors 'to adopt and embed more sustainable configurations of technologies, practices and organizational arrangements, thus leading to the reform and re-orientation of the incumbent regime' (Durrant 2012: 4). As reported in Sect. 15.3, many of the grassroots innovations are tackling more than one aspect, and all of them directly or indirectly and in different extents, contribute to the regime reform.

15.5 Concluding Remarks

In this chapter we aimed at contributing to an emerging strand of food waste literature, namely, the development and diffusion of innovative approaches to address the food waste problem and the role of grassroots innovations. Specifically, we assessed the role of grassroots movements in catalysing innovation in the food waste system and promoting a transition towards a more sustainable waste regime. We did so by means of an in-depth case studies analysis embedded in the MLP framework. Building on the original definition of grassroots innovations provided by Seyfang and Smith (2007) and extended by Smith et al. (2014), allowed us to define grassroots innovations as networks of activists and organizations generating novel bottom-up solutions for sustainable development, including also people and organisations coming from outside local communities (e.g. engineers and designers) and engaged in the innovation processes from the onset.

From our in-depth analysis of several case-studies of grassroots innovations addressing the food waste problem in different European countries, emerged the fact that grassroots innovations dealing with food waste embrace multiple forms, which we grouped into three categories:

(i) Prevention
(ii) Reduction and Reuse
(iii) Valorisation.

In making this distinction, we followed the European waste hierarchy, and considered the primary goal of grassroots organisations. Indeed, this categorisation should not be considered as a clear-cut as many grassroots innovations are tackling more than one aspect, and all of them directly or indirectly and in different extents, contribute to awareness raising; therefore this classification should be taken as an academic exercise, aimed at illustrating the diversity of grassroots innovations in respect to food waste, and in terms of goals, scale, space, diffusion pathways, typology of movement brokers and grassroots entrepreneurs.

Regardless of these categories, however, all grassroots innovations addressing food waste appeared as both need-generated and need-oriented, tackling mostly overlooked collective needs and concerns, and aimed at delivering viable solutions that could be customized in various contexts and settings worldwide. Their effectiveness in catalysing innovation activities proved to be high, especially when it came to steer policy actions. Certainly, the success of the *Anti-Gaspi*—French grassroots movement (which contributed substantially to the adoption in France of an anti-food waste law), shed a new light on the grassroots developments, from an overlooked site for innovation, to a visible hub in the realms of food policy and regulation, and emerging cultural politics of food waste, with huge potential in the transition towards a more sustainable waste regime.[31]

Our investigation also showed how grassroots innovations can act as incubators of the social change needed to respond to the current global societal challenges, such as: food security, poverty, energy, deforestation, and climate change.

Based on the empirical evidence gathered, we can argue that grassroots innovations addressing food waste, can play a key role in several core processes (niche creation; niche translation, scaling-up and diffusion; regime disruption; regime reform) of the transition towards sustainable production and consumption systems and a more sustainable waste regime.

Furthermore, GRIs configured mostly as niches innovation are able to respond to the societal challenges and at creating the conditions for enhancing landscape pressure. These combined effects could determine the opening up of windows of opportunities which would eventually result in the hoped for sustainable transition.

The findings of our study add to the existing literature on grassroots innovations and food waste, extending the current understanding on how grassroots innovations addressing food waste challenge could support decision makers in developing

[31]Note that the French law is setting an example in Europe and similar bills are being adopted also in other countries. Italy, for example, passed a law in august 2016, which makes it easier for companies and farmers to donate food to charities and is encouraging greater use of 'doggy bags' at restaurants as part of a legislative push to curb the epidemic of food waste. The new anti-food waste Italian law has essentially relaxed regulations that made food donations cumbersome. It has clarified that food may still be donated even if it is past its sell-by date, and allows farmers to transfer produce to charities at no extra cost if it has not been sold. The law also opens the door for companies to donate food that has been mislabelled as long as it does not pose a safety risk. (see: https://www.theguardian.com/world/2016/aug/03/italy-food-waste-law-donate-food—last accessed 22-08-2016).

tailored strategies to minimize the amount of food wasted along the supply chain, and to unlock the enormous potential of food waste that is being landfilled.

Acknowledgements This work was carried out partially under the project "Sustainable path creation for innovative value chains for organic waste products" (SusValueWaste), funded by the Research Programme on Sustainable Innovation in Food and Bio-based Industries (BIONÆR) under the Research Council of Norway.

References

Avelino F et al (2015) Transitions towards new economies? A transformative social innovation perspective, TRANSIT working paper #3. Retrieved online at http://www.transitsocialinnovation.eu/resource-hub/transitions-towards-new-economies-a-transformative-social-innovation-perspective

Benford RD, Snow DA (2000) Framing processes and social movements: an overview and assessment. Ann Rev Sociol 26:611–639

Bordieu P (1984) Distinction: a social critique of the judgment of taste, Harvard University Press. Retrieved online at: https://monoskop.org/images/e/e0/Pierre_Bourdieu_Distinction_A_Social_Critique_of_the_Judgement_of_Taste_1984.pdf

Butler Millsaps B (2015) Italian design student creates 'Agridust,' uses food scraps and compost for 3D printing, 3D print–the voice of 3D printing technologies. Retrieved online at https://3dprint.com/55358/agridust-food-3d-printing/

Chabanne N (2014) Gueules Cassées: promotion of ugly fruit and vegetables by SOLS & FRUITS, Journal RESOLIS. Retrieved online at http://www.resolis.org/fiche-pratique/gueules-cassees-valorisation-des-fruits-et-legumes-moches-par-sols-fruits/419/uk

Chiocchia E (2015) Agridust, il materiale biodegradabile ottenuto dagli scarti alimentary, design playground. Retrieved online at http://www.designplayground.it/2015/03/agridust-il-materiale-biodegradabile-ottenuto-dagli-scarti-alimentari/

Cicatiello C, Franco S, Pancino B, Blasi E (2016) The value of food waste: an exploratory study on retailing. J Retail Consum Serv 30:96–104

Clark JH, Luque R (2013) Valorisation of food residues: waste to wealth using green chemical technologies. Sustain Chem Process 2013. doi:10.1186/2043-7129-1-10

Curry P (2011) Ecological ethics: an introduction, 2nd edn. Polity Press, Cambridge

De Keersmaecker AEK et al (2012) Towards scaling up grassroots innovations in India: a preliminary framework. Retrieved online at http://repository.tudelft.nl/islandora/object/uuid:fc2d3631-9881-4a73-8a42-acf1d185e111?collection=research

Durrant R (2012) Civil society roles in transition: towards sustainable food? Sustainable lifestyles research group (SLRG). Working paper 02-14, ISSN:2050-4446. Retrieved online at http://www.sustainablelifestyles.ac.uk/projects/community/sustainability-transitions

Evans D, Campbell H, Murcott A (2013) A brief pre-history of food waste and the social sciences. Sociol Rev 60(S2):5–26

Feola G, Butt A (2015) The diffusion of grassroots innovations for sustainability in Italy and Great Britain: an exploratory spatial data analysis. Geogr J 1–18

Feola G, Nunes R (2014) Success and failure of grassroots innovations for addressing climate change: the case of the transition movement. Glob Environ Change 24:232–250

Fougier E (2016) Lutte contre le gaspillage alimentaire, tour d'horizon des initiatives. Retrieved online at http://wikiagri.fr/articles/lutte-contre-le-gaspillage-alimentaire-tour-dhorizon-des-initiatives/8438

FSE Network (2016) Food surplus entrepreneurs. Retrieved online at http://fsenetwork.org/entrepreneurs/

Geels FW (2002) Technological transitions as evolutionary reconfiguration processes: a multi-level perspective and a case study. Res Policy 31:1257–1274

Geels FW (2011) The multi-level perspective on sustainability transitions: Responses to seven criticisms. Environ Innov Societal Transitions 1(1):24–40

Geels F, Schot J (2007) Typology of sociotechnical transition pathways. Res Policy 36:399–417

Grabs J et al (2015) Understanding role models for change: a multilevel analysis of success factors of grassroots initiatives for sustainable consumption. J Cleaner Prod:1–14

GroCycle (2016a), Mushrooms from coffee grounds? Retrieved online at https://grocycle.com/food-for-thought/

GroCycle (2016b) GroCycle urban mushroom farm: a model for urban agriculture. Retrieved online at https://grocycle.com/urban-mushroom-farm/

GroCycle (2016c) GroCycle—about us. Retrieved online at https://grocycle.com/about-us/

Gustavsson J et al (2011) Global food losses and food waste: extent, causes and prevention, Food and Agriculture Organization of the United Nations, Interpack 2011, Rome, Italy. Retrieved online at http://www.fao.org/docrep/014/mb060e/mb060e.pdf

Hanson C et al (2015) What's food loss and waste got to do with climate change? A lot, actually, World Resources Institute. Retrieved online at http://www.wri.org/blog/2015/12/whats-food-loss-and-waste-got-do-climate-change-lot-actually

Howaldt J, Kopp R, Schwarz M (2013) Social innovations as drivers of social change—Tarde's disregarded contribution to social innovation theory building, Social Frontiers. Retrieved online at http://www.transitsocialinnovation.eu/content/original/Book%20covers/Local%20PDFs/99%20SF%20Howaldt%20Kopp%20and%20Schwarz%20Social%20innovations%20as%20drivers%20of%20social%20change%202013.pdf

International Institute for Environment and Development (IIED) (2013) Food waste: can grassroots initiatives stop us throwing good food in the bin? Retrieved online at http://www.iied.org/food-waste-can-grassroots-initiatives-stop-us-throwing-good-food-bin

Ki Lin CS et al (2013) Food waste as a valuable resource for the production of chemicals, materials and fuels. Current situation and global perspective. Energy Environ Sci 6:426–464

Lopolito A, Morone P, Taylor R (2013) Emerging innovation niches: an agent based model. Res Policy 42(6–7):1225–1238

Markard J, Raven R, Truffer B (2012) Sustainability transitions: an emerging field of research and its prospects. Res Policy 41:955–967

Martin CJ, Upham P, Budd L (2015) Commercial orientation in grassroots social innovation: insights from the sharing economy. Ecol Econ 118:240–251

Martin CJ, Upham P (2015) Grassroots social innovation and the mobilisation of values in collaborative consumption: a conceptual model. J Cleaner Prod:1–10

McCarthy JD, Zald MN (1977) Resource mobilization and social movements: a partial theory. Am J Sociol 82(6):1212–1241

Ministère de l'Environnement, de l'énergie et de la Mer (2016) La lutte contre le gaspillage alimentaire. Retrieved online at http://www.developpement-durable.gouv.fr/La-lutte-contre-le-gaspillage.html

Monaghan A (2009) Conceptual niche management of grassroots innovation for sustainability: the case of body disposal practices in the UK. Technol Forecast Soc Chang 76:1026–1043

Mourad M (2016) Recycling, recovering and preventing "food waste": competing solutions for food systems sustainability in the United States and France. J Clean Prod 126(2016):461–477

O'Donnell TH et al (2015) New sustainable market opportunities for surplus food: a food system-sensitive methodology (FSSM). Food Nutr Sci 6(10):883–892

Orange Fiber (2015) La magia dell'economia circolare. Retrieved online at http://www.orangefiber.it/la-magia-delleconomia-circolare/

Orange Fiber (2016) Concept. Retrieved online at http://www.orangefiber.it/#

Ornetzeder M, Rohracher H (2013) Of solar collectors, wind power, and car sharing: comparing and understanding successful cases of grassroots innovations. Global Environ Change (23) 5:856–867
Pansera M, Sarkar S (2016) Crafting sustainable development solutions: frugal innovations of grassroots entrepreneurs. Sustainability 8(51):1–25
Peters A (2015) Now you can 3-D-print with food waste, just like you always wanted. Retrieved online at http://www.fastcoexist.com/3044409/now-you-can-3-d-print-with-food-waste-just-like-you-always-wanted#1
Phillips C et al (2013) Understanding the sustainability of retail food recovery. PloS One 8(10), Published online 2013 Oct 10. doi:10.1371/journal.pone.0075530
Pickerill J, Chatterton P (2006) Notes towards autonomous geographies: creation, resistance and self-management as survival tactics. Prog Hum Geogr 30(6):730–746
Priefer C, Joerissen J, Bræutigam KR (2016) Food waste prevention in Europe—a cause-driven approach to identify the most relevant leverage points for action. Resour Conserv Recycl 109:155–165
Putnam RD (1995) 'Bowling alone: America's declining social capital. J Democracy 6(1):65–78
Raymond ES (1997) The Cathedral and the bazaar. Retrieved online at http://www.unterstein.net/su/docs/CathBaz.pdf
RecoFunghi (2016) La sostenibilità, la nostra passione! Retrieved online at http://www.recofunghi.com/l-azienda/una-storia-vera.html
Reynolds CJ, Piantadosi J, Boland J (2015) Rescuing food from the organics waste stream to feed the food insecure: an economic and environmental assessment of Australian food rescue operations using environmentally extended waste input-output analysis. Sustainability 7 (4):4707–4726
Rubino M (2014) Orange fiber, vestirsi con gli scarti delle arance. In Italia si può, quando c'è stoffa. Retrieved online at http://www.repubblica.it/scienze/2014/10/04/news/orange_fiber_vestirsi_con_gli_scarti_delle_arance_in_italia_si_pu_quando_c_stoffa-97155159/?refresh_ce
Seyfang G, Smith A (2007) Grassroots innovations for sustainable development: Towards a new research and policy agenda. Environ Polit 16(4):584–603
Smith A, Fressoli M, Thomas H (2014) Grassroots innovation movements: challenges and contributions. J Clean Prod 63:114–124
Smith A et al (2015) Making the most of community energies: Three perspectives on grassroots innovation. Environ Plann A 48(2):407–432
Schneider F (2013) The evolution of food donation with respect to waste prevention. Waste Manag Special Thematic Issue Urban Min 33(3):755–763
Stirling A (2011) Pluralising progress: from integrative transitions to transformative diversity. J Environ Innov Societal Transit 1(1):82–88
Touraine A (2002) The importance of social movements. Soc Mov Stud 1(1):89–95
Turnheim B, Geels F (2012) Regime destabilisation as the flipside of energy transitions: lessons from the history of the British coal industry (1913–1997). Energy Policy 50:35–49
Vergragt P, Akenji L, Dewick P (2014) Sustainable production, consumption, and livelihoods: global and regional research perspectives. J Clean Prod 63:1–12

FSC
www.fsc.org
MIX
Papier | Fördert gute Waldnutzung
FSC® C083411

Kolibri 360 GmbH
Ferdinand-Jühlke-Straße 7
99095 Erfurt, Deutschland
produktsicherheit@kolibri360.de